T0337842

Petrochemistry

Petrochemistry

Petrochemical Processing, Hydrocarbon Technology, and Green Engineering

Martin Bajus
Faculty of Chemical & Food Technology, Slovak University of Technology, Bratislava, Slovakia

This edition first published 2020

Registered Offices
John Wiley & Sons, Inc., 111 River Street, Hoboken, NJ 07030, USA
John Wiley & Sons Ltd, The Atrium, Southern Gate, Chichester, West Sussex, PO19 8SQ, UK

Editorial Office
The Atrium, Southern Gate, Chichester, West Sussex, PO19 8SQ, UK

For details of our global editorial offices, customer services, and more information about Wiley products visit us at www.wiley.com.

Wiley also publishes its books in a variety of electronic formats and by print-on-demand. Some content that appears in standard print versions of this book may not be available in other formats.

Library of Congress Cataloging-in-Publication Data applied for

Hardback ISBN: 9781119647768

Cover Design: Wiley
Cover Image: © Avigator Fortuner/Shutterstock

Set in 9.5/12.5pt STIXTwoText by SPi Global, Chennai, India

Printed and bound by CPI Group (UK) Ltd, Croydon, CR0 4YY

10 9 8 7 6 5 4 3 2 1

To
my loving wife Mária
and grandchildren,
Rebeka, Kristína,
Jakub, and David

Contents

About the Book

The main purpose of this book is to bring alive concepts forming the basis of chemical technology and to give a solid background for innovative process development. This is done by treatment of actual practice processes, which all present one or more challenges that chemical engineers have to deal with during the development of these particular processes and which are often still challenges. It is not my intention to treat chemical technology in an encyclopedic way. The emphasis is on concepts rather than on facts. Hopefully, this approach will stimulate students in chemical engineering and also those who play a large role in the field, such as graduates and technologists who work in or are interested in chemical technology. The next generation should invent and develop novel operations and processes.

Preface

This textbook deals with petrochemical conversion processes for petroleum and natural gas fractions as produced by refinery operations that are covered in the chain of my previous textbooks concerned with hydrocarbon technology, the last of which is titled *Organic Technology and Petrochemistry*.

Following an introduction that shows the growing importance of petrochemical processes in hydrocarbon technology, this textbook presents the fundamentals of chemical mechanisms as the basis of those processes. These disciplines are thermodynamics, chemical kinetics, reactor calculations, and industrial catalysts.

In this book, I have changed some things. The petrochemical industry is based on four basic pillars, all deriving from petroleum and natural gas. Alternative feedstocks and alternative technologies mainly arise from the abundance of cheap propane, ethane, and methane, as well as from shale gas and stranded gas. In particular, the production of propylene has grown as more and more steam crackers shift from naphtha feed to lighter shale condensates. This is especially true in the United States, where shale gas exploitation has grown exponentially. Although biomass and waste streams are also mentioned here, they are believed to be useful as a complement to rather than a complete replacement for fossil resources in the near future.

Petrochemistry's four pillars are as follows:

- *Pillar A*
  - Production of lower alkenes
  - Production of lower alkenes from other sources
  - Petrochemicals from C_2–C_3 alkenes
- *Pillar B*
  - Production of BTX aromatics
  - Chemicals from BTX aromatics
- *Pillar C*
  - C_1 technology
- *Pillar D*
  - Diversification of petrochemicals such as petrochemicals containing oxygen, halogens, nitrogen, and sulfur
  - Derived processes and products such as polymers, agrochemicals, surfactants, dyes, textile chemicals, and related products

- Fuels, lubricants, and additives on a petrochemical basis
- Application of petrochemicals in other chemical technologies

From the wealth of processes a selection had to be made. I have attempted to do this in logical way. Knowledge of some processes is essential for the understanding of the culture of the chemical engineering discipline. Examples are the major processes in oil refineries and the production of base chemicals from synthesis gas. Chemical engineers have been tremendously successful in bulk chemicals technology (methanol, Fischer–Tropsch synthesis, ammonia). For some other sectors, this was not the case in the past, but today they are becoming more and more important. Major examples are hydrogen technology, biorefineries, recycling technologies, microchannel technologies, and nanotechnology. Therefore, these areas are treated in separate chapters. More recently, emphasis has shifted to energy technology, sustainable technology, and, related to that, process intensification. These subjects are also touched upon.

In all chapters, the processes treated are represented by simplified flow schemes. For clarity, these generally do not include process control systems, and valves and pumps are also omitted in most cases. It is expected that after having read the book, students will be able to think in terms of conceptual process designs.

At the Slovak University of Technology, in the Faculty of Chemical and Food Technology, the text is the basis for some courses and supplementary for others:

- Petrochemistry
- Organic Technology and Petrochemistry
- Petroleum and Hydrocarbon Technology
- Energy Materials and Technologies
- Recycling Technologies
- Natural Gas
- Alternative Fuels
- Catalysis

It is hoped that this text will help give chemical engineers sufficient feeling for chemical technology and chemists for chemical engineering. Needless to say, I would highly appreciate any comments from users.

Martin Bajus
Bratislava, Slovakia

Acknowledgments

Many thanks to my wife Mária for supporting my efforts in bringing together these concepts in the form of a book and also for her direct participation in the generation of graphics.

General Literature

Aguado, J. and Serrano, D. (1999). *Feedstock Recycling of Plastic Wastes.*, RSC Clean Technology Monographs (ed. J.H. Clark). Cambridge: Royal Society of Chemistry.

Albright, L.F., Crynes, B.L., and Corcoran, W.H. (1983). *Pyrolysis: Theory and Industrial Practice*, 266. New York: Academic Press; p. 233.

Arpe, H. (2010). *Industrial Organic Chemicals*, 5e. New York: Wiley.

Arpentinier, P., Cavani, F., and Trifiro, F. (2001). *Chemical, Catalytic & Engineering Aspects*, The Technology of Catalytic Oxidation. Paris: Editions Technip.

Arpentinier, P., Cavani, F., and Trifiro, F. (2001). *The Technology of Catalytic Oxidation, 2. Safety Aspects*. Paris: Editions Technip.

Austin, G.T. (1984). *Shreve's Chemical Process Industries*. New York: McGraw Hill Book Company, Finfth Edition.

Bajus, M. (1987). Thermal conversion of petroleum hydrocarbons into petrochemicals. D.Sc thesis. Slovak University of Technology, Bratislava, Slovak Republic.

Bajus, M. (1989). Sulfur compounds in hydrocarbon pyrolysis. *Sulfur Reports* 9 (1): 25–71.

Bajus, M. (2002). *Organická technológia a petrochémia, Uhľovodíkové technológie*. Bratislava: Vydavateľstvo STU.

Bajus, M. (2019). *Chemické procesy v organickej technológii a petrochémii (Chemical Process in Organic Technology and Petrochemistry)*. e-Book. Bratislava: Vydavateľstvo Spektrum STU.

Bajus, M., Lederer, J., and Bělohlav, Z. (2016). Hydrocarbon technology, trends and outlook in petrochemistry. Plenary lecture at the 4th International Conference on Chemical Technology.

Bajus, M., Malatinský, T., Buláková, Ľ. et al. (2006). *150 rokov plynárenstva na Slovensku*. Bratislava: Vydal Slovenský plynárenský a naftový zväz.

Bisio, A. and Boots, S. (1995). *Encyclopedia of Energy Technology and the Environment*, vol. 1–4. Wiley.

BP. (2019). BP statistical review of world energy (1995–2018). http://bp.com/statisticalreview.

Brown, R.C. (ed.) (2011). *Thermochemical Processing of Biomass, Conversion into Fuels, Chemicals and Power*. Wiley.

Burnham, A.K. (2017). *Global Chemical Kinetics of Fossil Fuels*. Springer.

van Santen, R.A., van Leeuwen, P.W.N.M., Moulijn, J.A., and Averill, B.A. (eds.) (1999). *Catalysis: An Integrated Approach*, second, revised and enlargede, vol. 123. Elsevier.

Cenka, M. (2001). *Obnovitelné zdroje energie*. Praha: FCC Public.

Centi, G., Cavani, F., and Trifiró, F. (2001). *Selective Oxidation by Heterogeneous Catalysis*. New York: Kluwer Academic/Plenum Publishers.
Comyns, A.E. (1999). *Encyclopedic Dictionary of Named Processes in Chemical Technology*, 2e. Boca Raton London: CRC Press.
Corma, A. and Martinez, A. (2002). *Handbook of Porous Solids* (eds. F. Schutz, K.S.W. Sing and J. Wetkamp), 2825. Weinheim: Wiley-VCH.
Coutler, J.C. (2004). *Encyclopedia of Energy*, vol. 1–6. San Diego, California: Elsevier Inc., Academic Press.
Ehrfeld, W., Hessel, V., and Loewe, H. (2000). *Microreactors-New Technology for Modern Chemistry*, 9. Weinheim: Wiley-VCH.
El-Gendy, L.S. (2018). *Biodesulfurization in Petroleum Refining*. Scrivener Publishing: Wiley, 1200 pages.
Germain, J.E. (1969). *Catalytic Conversion of Hydrocarbons*. London and New York: Academic Press.
Harker, J.H. and Backhurst, J.R. (1981). *Fuel and Energy*. London: Academic Press.
Holt, K., Jewell, L., Niemantsverdriet, H. et al. (2017). Designing new catalysts for synthetic fuels. *Faraday Discussions* 197: 353–388.
Hougen, O.A. and Watson, K.M. (1947). *Chemical Process Principles*, vol. 3, 887–886. New York: Wiley.
Hsu, C.S. (2019). *Petroleum Science and Technology*. Springer International Publishing, 487 pages.
Hsu, C. and Robinson, P.R. (eds.) (2017). *Springer Handbook of Petroleum Technology*. Springer International Publishing AG.
Humphrey, J.L. and Keller, G.L. (1997). *Separation Process Technology*. New York: McGraw-Hill.
Hussein, K.A. and Alsahlavi, M. (2013). *Petroleum Economics and Engineering*, 3e. CRC Press.
Jones, D.S.J. and Pujado, P. (eds.) (2006). *Handbook of Chemical Processing*. Springer.
Ke, L., Chunshan, S., and Velu, S. (2010). *Hydrogen and Syngas Production and Purification Technologies, AIChe*. Hoboken: Wiley.
Kelland, M.A. (2014). *Production Chemicals from the Oil and Gas Industry*. CRC Press.
Kizlink, J. (2011). *Technologie chemických látek a jejich použití (Technology of Chemical Compounds)*, 4ee. VUTIUM.
Lee, S. (2012). *Biofuels and Bioenergy*. CRC Press.
Luque, R. (2013). *Producing Fuels and Fine Chemicals from Biomass Using Na*. CRC Press.
Luque, R. and Leung-Yuk Lam, F. (2018). *Substainable Catalysis: Energy-Efficient Reactions and Applications*. Wiley-VHC, Verlag GmbH and CoKGaA.
Malwell, I.E. and Stork, W.H.J. (2001). Introduction to zeolite science and practise. In: *Studies Science and Catalysis*, 2e, vol. 137 (eds. H. van Bekkum, E.M. Flamigen, P.A. Jacobs and J.C. Jansen), 747. Amsterdam: Elsevier.
Meyers, R.A. (2005). *Handbook of Petrochemicals Production Processes*. Inc., New York: The McGraw-Hill Companies.
Minteer, S. (2006). *Alcoholic Fuels*. Boca Raton: CRC, Taylor and Francis.
Dawe, R.A. (ed.) (2000). *Modern Petroleum Technology*, 6e, vol. 1, Upstream. Chichester: Wiley.
Lucas, A.G. (2000). *Modern Petroleum Technology*, 6e, vol. 2, Downstream. Chichester: Wiley.
Moulijn, J.A., Makkee, M., and Van Diepen, A. (2005). *Chemical Process Technology*. Wiley.

Parmaliana, A., Sanfilippo, D., Frusteri, F. et al. (eds.) (1998). *Natural Gas Conversion V*, vol. 119. Elsevier.

Iglesia, E., Spivey, J.J., and Fleisch, T.H. (eds.) (2001). *Natural Gas Conversion VI*, vol. 136. Elsevier.

Noronha, F.B., Schmal, M., and Sousa-Aguiar, E.F. (eds.) (2007). *Natural Gas Conversion VIII*, vol. 167. Elsevier.

Olah, G.A. and Molnar, A. (1995). *Hydrocarbons Chemistry*, 632. New York: Wiley.

Olivé, G.H. and Olivé, S. (1984). *The Chemistry of the Catalyzed Hydrogenation of Carbon Monoxide*. Berlin, Heidelberg: Springer – Verlag.

Pastorek, Z., Kára, J., and Jevič, P. (2004). *Biomasa obnovitelný zdroj energie*. Praha: FCC Public.

Wauquier, J.-P. (ed.) (1995). *Petroleum Refining, 1. Crude Oil, Petroleum Process Flowsheets*. Editions Technip.

Wauquier, J.-P. (ed.) (2000). *Petroleum Refining, 2. Separation Processes*. Editions Technip.

Leprince, P. (ed.) (1995). *Petroleum Refining, 3. Conversion Process*. Editions Technip.

Trambouze, P. (ed.) (2000). *Petroleum Refining, 4. Materials and Equipment*. Editions Technip.

Favennec, J.P. (ed.) (2001). *Petroleum Refining, 5. Refinery Operation and Management*. Editions Technip.

Pines, H. (1981). *The Chemistry of Catalytic Hydrocarbon Reactions*. Academic Press.

Rojey, A., Jaffret, C., Cornot-Gandolphe, S., and Durand, B. (1997). *Natural Gas Production Processing Transport*, 429. Editions Technip.

Rossillo-Calle, F., de Groot, P., and Hemstock, S.L. (eds.) (2007). *The Biomass Assessment Handbook, Bioenergy for a Sustainable Environment*. London: Earthscan.

Sabu, T. (2014). *Biomaterial Applications: Micro to Nanoscales*. CRC Press.

Shah, Y.T. (2014). *Water for Energy and Fuel Production*. CRC Press.

Sheldon, R.A. (1982). *Chemicals from Synthesis Gas, D*. Dordrecht, Holland: Reidel Publishing Company.

Skrzypek, J., Sloczynski, J., and Ledakowicz, S. (1994). *Methanol Synthesis*. Sp. Z o.o, Warszawa: Polish Scientific Publishers PWN.

Sorensen, B. (2005). *Hydrogen and Fuel Cells, Emerging Technologies and Applications*. Amsterdam: Elsevier, Academic Press.

Spath, P.L. and Dayton, D.C. (2003). Preliminary screening – technical and economic assessment of synthesis gas to fuels and chemicals with emphasis on the potential for biomass-derived syngas. National Renewable Energy Laboratory.

Speight, J.G. (2011). *Handbook of Industrial Hydrocarbon Processes*, 1e. Burlington, MA: Elsevier Inc.

Speight, J.G. (2014). *The Chemistry and Technology of Petroleum*, 5e. CRC Press.

Spitz, H.P. (1988). *Petrochemicals, the Rise of an Industry*. Wiley.

Stocchi, E. (1999). *Industrial Chemistry*, vol. 1. New York: Ellis Horwood Limited.

Sylvester, E. (2010). Course: chemical technology (organic), lecture 7, aromatic production. https://www.academia.edu/33302746/Course_Chemical_Technology_Organic_Module_VII_Lecture_7_Aromatic_Production.

Uner, D. (2017). *Advances in Refining Catalysis*, Chemical Industries, 1e. CRC Press, 422 pp.

Varfolomeev, S.D. (2014). *Chemical and Biochemical Technology: Material and Processing*. CRC Press.

Vertes, A.A., Qureshi, N., Blaschek, H.P., and Yukawa, H. (eds.) (2010). *Biomass to Biofuels, Strategies for Global Industries.* Wiley.

Veselý, V., Mostecký, J., Bajus, M. et al. (1989). *Petrochémia (Petrochemistry).* Alfa Bratislava.

Wikipedia. (2019). Petrochemical. https//en.wikipedia.org/wiki/Petrochemical.

Wittcoff, H.A. and Reuben, B.G. (1980). *Industrial Organic Chemicals in Perspective, Part One: Raw Materials and Manufacture.* New York: Wiley.

Wittcoff, H.A. and Reuben, B.G. (1996). *Industrial Organic Chemicals.* New York: Wiley.

Wittcoff, H.A., Reuben, B.G., and Plotkin, J.S. (2013). *Industrial Organic Chemicals*, 3e. New York: Wiley.

Wohlleben, W. (2015). *Safety of Nanomaterials along their Life Cycle.* CRC Press.

Catalysis for Fuels. (2017). RSC Publishing.

Nomenclature

BP	Boiling point	°C
D	A darcy (or darcy unit) and millidarcy (md or mD) are units of permeability named after Henry Darcy.	
d^{15}	Specific gravity at 15°C	
E	Activation energy	J/mol
F	Molar flow rate	mol/s
G	Gibbs molar free energy	J/mol
GHSV	Gas hourly space velocity	$m^3/m^3 \cdot h$
H	Molar enthalpy	J/mol
ID	Internal diameter	m
K	Reaction equilibrium constant	
k	Reaction rate constant	s
λ	Air number	
M	Molar mass	g/mol
MP	Melting point	°C
MTD	Metric tons per day	
N	Feed of reactant	mol/time
n	Reaction order	
Nu	Nusselt number	
P	Pressure	Pa
Po	Poiseuille number	
ppm	Part per million	
R	Ideal gas constant	8.31 J/(mol·K)
Re	Reynolds number	
S	Molar entropy	J/(mol·K)
S	Selectivity	%
SN	Stoichiometric number	

SRT	Short residence time	s
SUS	Saybolt Universal Seconds, viscosity conversion (older SUS or SSU to current cSt. or mm^2/s)	
S/V	Surface-to-volume	cm^{-1}
T	Temperature	K
t	Temperature	°C
V	Reactor volume	
WHSV	Weight hourly space velocity	kg/kg·h
X	Conversion	%
Y	Yield	%
τ	Reaction time	[s]
ε	Expansion factor	

Abbreviations and Acronyms

AAGR	an annual growth rate
ABS	acrylonitrile butadiene styrene copolymer
ACC	American Chemistry Council
AGO	atmospheric gas oil
APU	auxiliary power unit
ATR	autothermal reformer
BDS	biodesulfurization
BPA	bisphenol A
BTX	benzene, toluene, xylene
BTEX	benzene, toluene, ethylbenzene, xylene
CAA	Clean Air Act
CAD	computer-aided design
CAGR	compound annual growth rate
CAR	Council for Automotive Research
CCR	Conradson carbon residue
CCR	continuous catalytic regenerative reformer
CD	catalytic distillation
CD	catalytic dehydrogenation
CDU	crude distillation unit
CFB	circulating fluidized bed
CFD	computation fluid dynamics
CHP	cumene hydroperoxide
CIS	Commonwealth of Independent States
CNG	compressed natural gas
^{13}C NMR	nuclear magnetic resonance
COD	conversion olefin to diesel
COS	carbonyl sulfide
CTO	coal to olefin
DCC	deep catalytic cracking
DEFC	direct ethanol fuel cell
DIR	direct internal reforming
DME	dimethyl ether
DMFC	direct methanol fuel cell

DNA	deoxynucleic acid
DOE	Department of Energy (USA)
DTBGE	di tertiary butyl glycerol ether
DSSC/SCA	Deep Scavenger steam cracking/steam cracking activation process
DS	downstream
EPA	Environmental Protection Agency (USA)
ETBE	ethyl tertiary butyl ether
ETL	ethylene to liquids
FC	Friedel-Crafts
FCC	fluid catalytic cracking
FCs	fuel cells
FFB	fixed fluidized bed
FTS	Fischer–Tropsch synthesis
FTSC	Fischer–Tropsch steam cracking
G	glycerol
GCR	gas-cooled reactor
GC/FID/MS	gas chromatography/flame ionization/mass spectrometric detection
GGE	gasoline gallon equivalent
GIR	gradual internal reforming
GR-N	copolymers of butenes and butadiene with acrylonitrile
GR-S	copolymers of butenes and butadiene with styrene
GTL	gas to liquid
HC	hydrocarbon
HDPE	high-density polyethylene
HP	high pressure
HTS	high-temperature water-gas shift
HVC	high-value chemicals
IB	isobutene
IC	ion chromatography
IEA	International Energy Agency
IFP	Institut Francais du Petrole
KSF	kinetic severity function
LCO	light cycle oil
LDPE	low-density polyethylene
LH_2	liquid hydrogen
LLDPE	linear low-density polyethylene
LNG	liquefied natural gas
LOE	light olefin etherification
LP	low pressure
LPG	liquefied petroleum gas
LPO	low-pressure oxo process
LT	low temperature
LTS	low-temperature water-gas shift
MCFCs	molten carbonate fuel cells
MDI	methylene diphenylene diisocyanate

MOGD	Mobil olefin to gasoline and distillate
MON	motor octane number
MPW	municipal plastic waste
MSW	municipal solid waste
MTBE	methyl tertiary butyl ether
MTBGE	mono tertiary butyl glycerol ether
MTO	methanol to olefins
MTOE	million tonnes of oil equivalent
MTPA	million tonnes per annum
NREL	National Renewable Energy Laboratory
PNNL	Pacific Northwest National Laboratory
NPW	novel process windows
OCM	oxidation coupling methane
OCT	olefins conversion technology
OE	oil equivalent
OOIP	original oil in place
PDH	propane dehydrogenation
PE	polyethylene
PEB	polyethylbenzene
PEMFCs	proton exchange membrane fuel cells
PET	polyethylene terephthalate
PFR	plug flow reactor
PM	particulate matter
PMDI	poly MDI
POX	partial oxidation
PP	polypropylene
PROX	preferential oxidation
PS	polystyrene
PSA	pressure swing adsorption
PVA	polyvinyl alcohol
PVC	polyvinyl chloride
R&D	research and development
RL	run length
RON	research octane number
RVP	Reid vapor pressure
SAN	styrene acrylonitrile copolymer
SAPO	silico-alumina phosphate
SB	styrene butadiene copolymer
SBR	styrene butadiene rubber
SC	steam cracking
SM	styrene monomer
SMB	simulated moving bed
SMDS	Shell Middle Distillation Synthesis
SMR	steam methane reforming
SO	selective oligomerization

SOFC	solid oxide fuel cell
SR	steam reforming
SR	straight run
SRR	semi-regenerative reforming
SWNT	single-walled carbon nanotube
TAEE	tertiary amyl ethyl ether
TAME	tertiary amyl methyl ether
TBA	tertiary butyl alcohol
TCM	trillion cubic meters
TDI	diisocyanatotoluene
TEA	triethanolamine
TEM	transmission electron microscopy
TIGAS	Topsoe Improved Gasoline Synthesis
TLE	trans-line exchanger
TPP	triphenylphosphine
TTBG	tri tertiary butyl glycerol ether
US	upstream
USEIA	United States Energy Information Administration
VD	vacuum distillation
VDC	volts direct current
VGO	vacuum gas oil
VI	viscosity index
VLPCR	very low pressure catalytic reforming with continuous catalyst regeneration
VOC	volatile organic compounds
VR	vacuum residue
WCR	water-cooled methanol reactor
XRD	X-ray diffraction
XPS	X-ray photoelectron spectroscopy
YSZ	yttrium stabilized zirconia
ZSM	Zeolite Socony Mobil

1

Chemical Technology

CHAPTER MENU

Petrochemistry: Petrochemical Processing, Hydrocarbon Technology, and Green Engineering,
First Edition. Martin Bajus.

1.1 Introduction

Chemical technology is the practical application of science to commerce or industry and is a multicomponent discipline that, in this context, deals with the application of chemical knowledge to the solution of practical problems. Technology is also a human action that involves the generation of knowledge and (usually innovative) processes to develop systems that solve problems and extend human capabilities.

Historically, the word *technology* is a modern term and rose to prominence during the Industrial Revolution, when it became associated with science and engineering. The word *technology* can also be used to refer to a collection of techniques, which refers to the current state of humanity's knowledge of how to combine resources to produce desired products, solve problems, fulfill needs, or satisfy wants; in includes technical methods, skills, processes, techniques, tools, and raw materials. The distinction between science, engineering, and technology is not always clear. However, technologies are not usually exclusively products of science, because they have to satisfy requirements such as utility.

Chemical technology is that branch of chemistry that formulates and solves the problems inherent in the exploitation of all raw materials, and produces from those materials, on an appropriate scale, many other products that do not occur naturally, that are scarce, or whose methods of preparation are capable of being improved upon. *Hydrocarbon technology* is a branch of chemical technology based on hydrocarbon raw materials, all derived from petroleum and natural gas. Petroleum is perhaps the most important substance consumed in modern society. And natural gas is increasingly being recognized as a hydrocarbon raw material, a preferred chemical feedstock for the twenty-first century.

The three major factors that have driven chemical technology development in recent years have been identified as energy, economic pressure, and environmental concern. These factors impinge directly on the activities of chemical technology, most clearly in the field of chemical processing. Hydrocarbon technology is made up of very large, global, integrated companies that can produce finished or semi-finished goods from hydrocarbons. In chemical products, the major impact is felt via the consumer: both the direct consumer who buys chemicals and the *intermediary* consumer, whose product provides use. In fact, the term *chemical* may not exist at some point in the future, and chemicals might just be intermediates inside a more complex production module. Chemicals, in some form, will probably always exist and will be continue to be marketed globally. There will always be a need for some of the chemicals used in the manufacture of products that are relatively small and do not lend themselves to integrated, mass manufacture.

Chemical technology has had a long, branched road of development. Processes such as distillation, dyeing, and the manufacture of soap, wine, and glass have long been practiced in small-scale units. The development of these processes was based on discoveries and empiricism rather than thorough guidelines, theory, and chemical engineering principles. This situation persisted until the seventeenth and eighteenth centuries. Only then were mystical interpretations replaced by scientific theories.

It was not until the 1910s and 1920s that continuous processes became more common, when disciplines such as thermodynamics, material and energy balances, heat transfer, fluid dynamics, as well as chemical kinetics and catalysis became (and still are) the foundations on which process technology rests. Allied with these are unit operations, including distillation, extraction, etc. In chemical process technology, various disciplines are integrated. These can be divided according to their scale:

- Scale independent
  - Chemistry, biology, physics, mathematics
  - Thermodynamics
  - Physical transport phenomena
- Micro level
  - Kinetics
  - Catalysis on a molecular level
  - Interface chemistry
  - Microbiology
  - Particle technology
- Meso level
  - Reactor technology
  - Unit operations
  - Scale-up
- Macro level
  - Process technology and process development
  - Process integration and design
  - Process control and operation

Of course, this scheme is not complete. Other disciplines such as applied materials science, information science, process control, and cost engineering also play a role.

In the development stage of a process or product, all necessary disciplines are integrated. The role and position of the various disciplines perhaps can be better understood from Figure 1.1, in which they are arranged according to their level of integration. Process development, in principle, roughly proceeds through the same sequence, and the x-axis also to a large extent represents the time progress in the development of a process. The initial phase depends on thermodynamics and other scale-independent principles. As time passes, other disciplines become more important, e.g. kinetics and catalysis on the micro level; reactor technology, unit operations, and scale-up on the meso level; and process technology, process control, etc., on the macro level.

Of course, there should be intense interaction between the various disciplines. To be able to quickly implement new insights and results, these disciplines should preferably be applied more or less in parallel rather than in series, as can also be seen from Figure 1.1. The plant is at the macro level. When one would like to focus on chemical conversion, the reactor is the level of interest. When one's interest changes to the molecules converted, the micro level is reached.

An enlightening way of placing the discipline of chemical engineering in a broader framework has been put forward by chemical engineers. Chemical engineering is a broad, integrated discipline. On one hand, molecules, having dimensions in the nanometer range and vibration time on the nanosecond scale, are considered. On the other hand, chemical plants may be half a kilometer in size, while the life expectancy of a new plant is 10–20 years.

Every industrial chemical process is designed to produce an economically desired product or range of products from a variety of starting materials (i.e. feed, feedstocks, or raw materials). Figure 1.2 shows a typical structure of such a process.

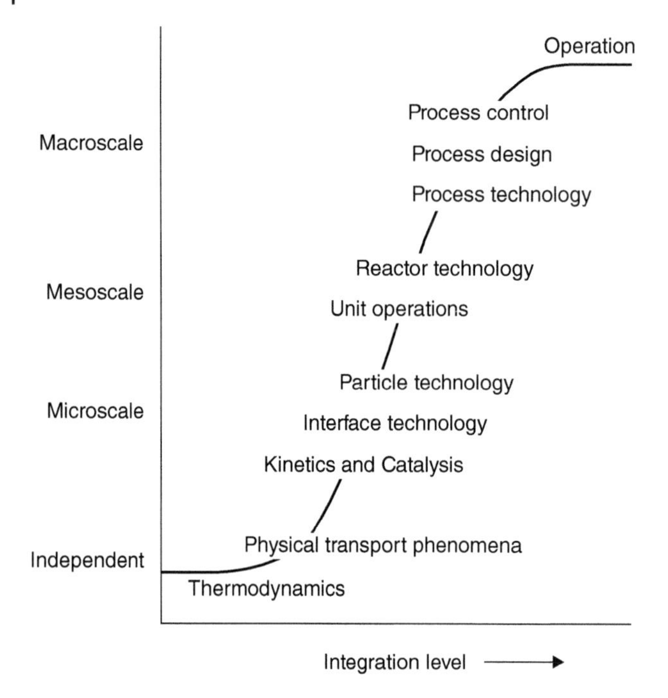

Figure 1.1 Disciplines in process development organized according to level integration.

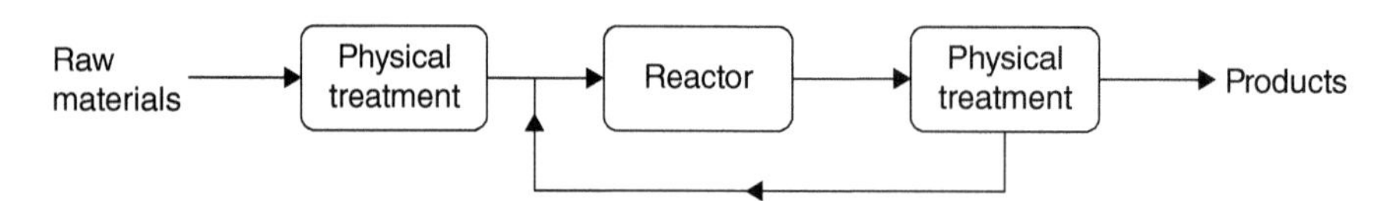

Figure 1.2 Typical chemical process structure.

The feed usually has to be pretreated. It may undergo a number of physical treatment steps, e.g. coal has to be pulverized, liquid feedstocks may have to be vaporized, water is removed from benzene by distillation prior to its conversion to ethylbenzene, etc. Often, impurities in the feed must be removed by chemical reaction, e.g. desulfurization of the naphtha feed to a catalytic reformer, making raw synthesis gas suitable for use in an ammonia converter, etc. Following the actual chemical conversion, the reaction products need to be separated and purified. Distillation is still the most common separation method; however, extraction, crystallization, membrane separation, etc., can also be used.

In this book, emphasis is placed on the reaction section, since the reactor is the heart of each process, but feed pretreatment and product separation will also be given attention. In the discussion of each process, the following questions will be answered:

- Which reactions are involved?
- What are the thermodynamics of the reactions, and what operating temperature and pressure should be applied?
- What are the kinetics, and what are the optimal conditions in that sense?

- Is a catalyst used, and if so, is it heterogeneous or homogeneous? Is the catalyst stable? If not, what is the deactivation time scale? What are the consequences for process design? Is regeneration required?
- Apart from the catalyst, what phases are involved? Are mass- and heat-transfer limitations important?
- Is a gas or liquid recycle necessary?
- Is feed purification necessary?
- How are the products separated?
- What are the environmental issues?

The type of the reactor and the process flow sheet determine the answers to these questions. Of course, the list is not complete, and specific questions may be raised for individual processes, e.g. how to solve possible corrosion problems in the production of acetaldehyde. Other matters are also addressed, either for a specific process or in general terms:

- What are the safety issues?
- Can different functions be integrated in one piece of equipment?
- What are the economics (comparison between processes)?
- Can sustainable technology be used?

1.2 Chemical Engineering

Chemical engineering is the branch of engineering that deals with the application of physical science (such as chemistry) to the process of converting raw materials (for example, petroleum) or chemicals into more useful or valuable forms. Chemical engineering largely involves the improvement of, design, and maintenance of processes involving chemical transformations for large-scale manufacture. Chemical engineers (process engineers) ensure that the processes are operated safely, sustainably, and economically.

Chemical engineering is applied in the manufacture of a wide variety of products. The scope of chemical technology includes manufacturing inorganic and organic industrial chemicals, ceramics, fuels and petrochemicals, agrochemicals (fertilizers, insecticides, herbicides) plastics and elastomers, oleo-chemicals, explosives, detergents and detergent products (soap, shampoo, cleaning fluids), fragrances and flavors, additives, dietary supplements, and pharmaceuticals. Closely allied or overlapping disciplines include wood processing, food, environmental technology and the engineering of petroleum, glass, paints and other coatings, inks, sealants, and adhesives.

Also, while the contemporary theoretical and practical structures of modern chemical technology tend to ignore the formal dichotomy in the subject, the emphasis at the seat of learning must still be on two traditional areas that are intimately related and indicate the nature of the materials and products being dealt with. These two areas are inorganic chemical technology and organic chemical technology. The scheme presented in Figure 1.3 summarizes the contents and aims of these branches of chemical technology. As indicated in the first section of the scheme, both inorganic and organic chemical technology have the primary aim (historically) of applying chemical procedures to the transformation of natural materials. This is the traditional distinction between the areas of applied chemistry and chemical technology.

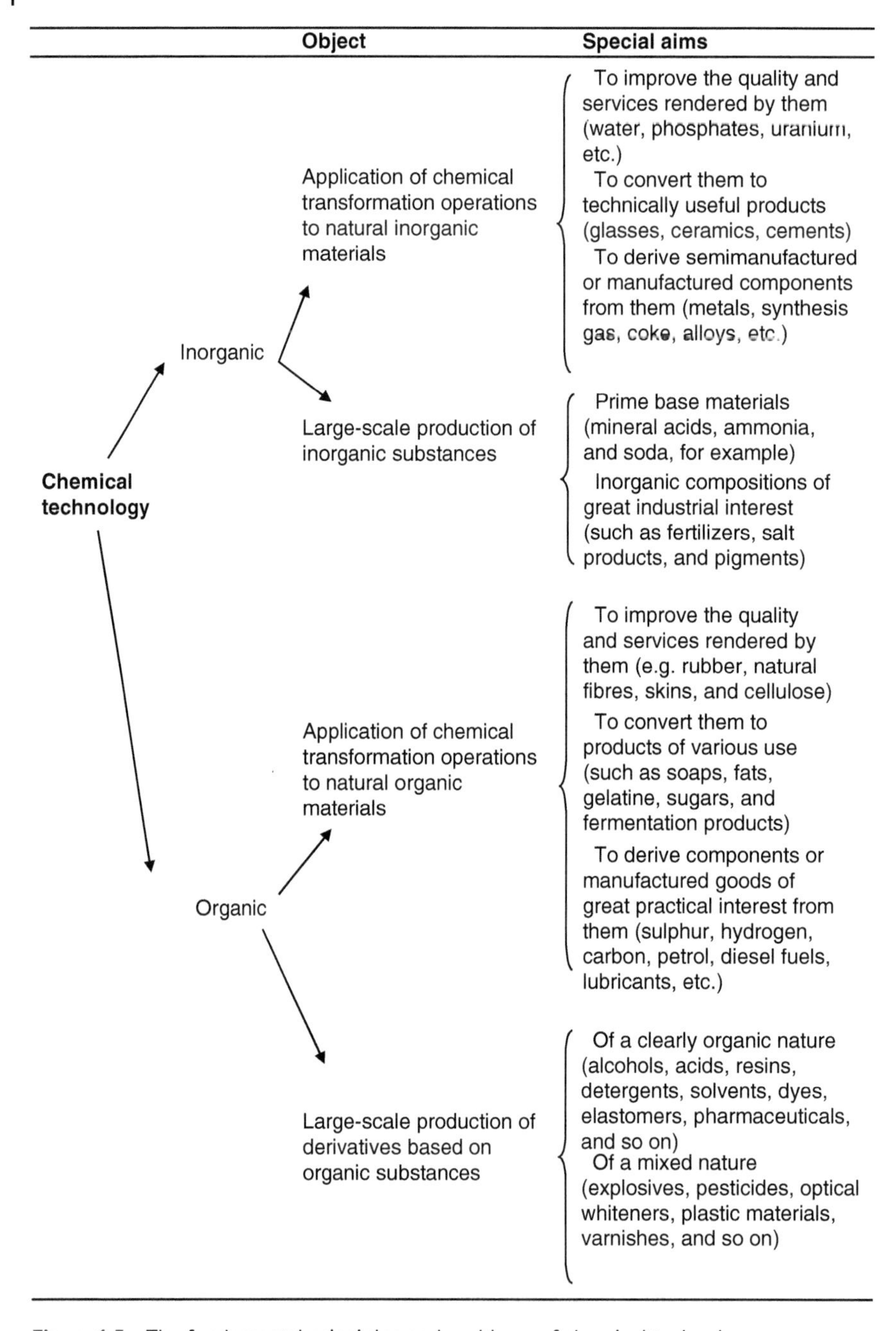

Figure 1.3 The fundamental principles and problems of chemical technology.

Nowadays, however, with the existence of so many substitute materials that are hybrids of those belonging specifically to these two traditional areas – and, more importantly, with the rapid integration of the basic procedures that lie at their foundation – a breakdown in this formal distinction has taken place.

Chemical engineers design processes to ensure the most economical operation, and the entire production chain must be planned and controlled for costs. A chemical engineer can simplify complicated showcase reactions while also gaining an economic advantage. Using a higher pressure or temperature makes several reactions easier; ammonia, for example, is simply produced from its component elements in a high-pressure reactor. On the other hand, reactions with a low yield can be recycled continuously (recycled to extinction, which no more is made), which would be complex, arduous work if done by hand in laboratory. It is not unusual to build 6-step or even 12-step evaporators to reuse vaporization energy for an economic advantage. In contrast, laboratory chemists evaporate samples in a single step.

The individual processes used by chemical engineers (e.g. distillation or filtration) are called *unit operations* and consist of chemical reactions, mass-transfer operations, and heat-transfer operations. Unit operations are grouped together in various configurations for the purpose of chemical synthesis and/or chemical separation. Some processes are a combination of intertwined transport and separation unit operations: for example, during reactive distillation, the product is formed as the still temperature is increased, and the product distills from the reaction mixture.

Three basic physical laws underlie chemical engineering design: (i) conservation of mass, (ii) conservation of energy, and (iii) conservation of momentum.

1.2.1 Conservation of Mass

The law of conservation of mass (principle of mass/matter conservation) is that the mass of a closed system (in the sense of a completely isolated system) remains constant over time. The mass of an isolated system cannot be changed as result of processes acting inside the system. But while mass cannot be created or destroyed, it may be rearranged in space and changed into different types of particles. This implies that for any chemical process in a closed system, the mass of the reactants must equal the mass of the products.

The change in mass of certain kinds of open systems – where atoms or massive particles are not allowed to escape, but other types of energy (such as light or heat) are allowed to enter or escape – went unnoticed during the nineteenth century because mass-change associated with the addition or loss of the fractional amounts of heat and light associated with chemical reaction was very small. Mass is also not generally conserved in open systems (even if only open to heat and work) when various forms of energy are allowed into or out of the system (for example, bond energy).

But mass conservation for closed systems proves to be true. The mass-energy equivalence theorem states that mass conservation is equivalent to energy conservation, which is the first law of thermodynamics. The mass-energy equivalence formula requires closed systems, since if energy is allowed to escape a system, mass will escape also.

1.2.2 Conservation of Energy

The law of conservation of energy states that the total amount of energy in an isolated system remains constant over time. A consequence of this law is that energy can neither be created nor destroyed; it can only be transformed from one state to another. The only thing that can happen to energy in a closed system is that it can change form, such as transformation of chemical energy to kinetic energy.

Conservation energy refers to the conservation of the total system energy over time. This energy includes the energy associated with the mass of the reactants, as well as all other forms of energy in the system. In an isolated system, although mass and energy (heat and light) can be converted to one another, both the total amount of energy and the total amount of mass of the system remain constant over time. If energy in any form is allowed to escape such systems, the mass of the system will decrease in correspondence with that loss.

1.2.3 Conservation of Momentum

The conservation of momentum is a fundamental law of physics, which states that the momentum of a system is constant if there are no external forces acting on the system. Momentum is a conserved quantity insofar as the total momentum of any closed system (a system not affected by external forces) cannot change.

One of the consequences of this law is that the center of any system of objects will always continue with the same velocity unless acted on by a force from outside the system.

In an isolated system (one where external forces are absent), the total momentum will be constant, which dictates that the forces acting between systems are equal in magnitude but opposite in sign due to the conservation of momentum.

1.2.4 Thermodynamics of Chemical Reactions

The two thermodynamic parameters to be considered when analyzing the thermodynamics of chemical reactions are enthalpy (**H**) and Gibbs energy (**G**). Generally speaking, the variations in enthalpy and in Gibbs energy associated with a chemical change must be assessed so as to draw the relevant conclusions for the heat balance and equilibrium advancement.

The Gibbs energy variation associated with a chemical reaction allows the position of the chemical equilibrium state to be situated between reactants and products for specified operating conditions. It also enables the position to be expressed by a value: the equilibrium constant.

Given the equilibrium reaction

$$aA + bB \rightleftarrows cC + dD \tag{1.1}$$

the equilibrium constant is related to Gibbs energy variation by the following formula:

$$(\Delta G^\circ{}_R)_T = (\Delta H^\circ{}_R)_T - T(\Delta S^\circ{}_R)_T = -RT \ln K^\circ \tag{1.2}$$

It can readily be shown that:

$$\frac{d(lnK^\circ)}{d(1/T)} = \frac{-(\Delta H^\circ_R)_T}{R} \tag{1.3}$$

By plotting ln K° versus 1/T, a curve is obtained that is practically a straight line. Its slope is – $(\Delta H^\circ{}_R)/R$ as long as $(\Delta H^\circ{}_R)$ can be assumed to be invariant within the temperature range under consideration. For an exothermic reaction $((\Delta H^\circ{}_R) < 0)$, the slope is seen to be positive, and consequently the value of the equilibrium constant decreases when the temperature increases: the equilibrium moves backward. Of course, the situation is just the opposite for endothermic reactions.

From the change of the Gibbs energy ($\Delta G°_f$, change of the standard free enthalpy of formation; $\Delta G°_r$, change of the standard free enthalpy of reaction), it follows that the stability of hydrocarbons of all homologous series decreases with increasing temperature, with the exception of acetylene (Figure 1.4). Table 1.1 contains the values of the standard free enthalpy of formation for some hydrocarbons. The tendency of hydrocarbons to decompose increases with increasing molecular mass, regardless of the homologous series. In the transition from alkanes and cycloalkanes to alkenes and aromatics, the thermodynamic stability increases. Similarly, the resistance of alkenes to decomposition increases in comparison with alkanes.

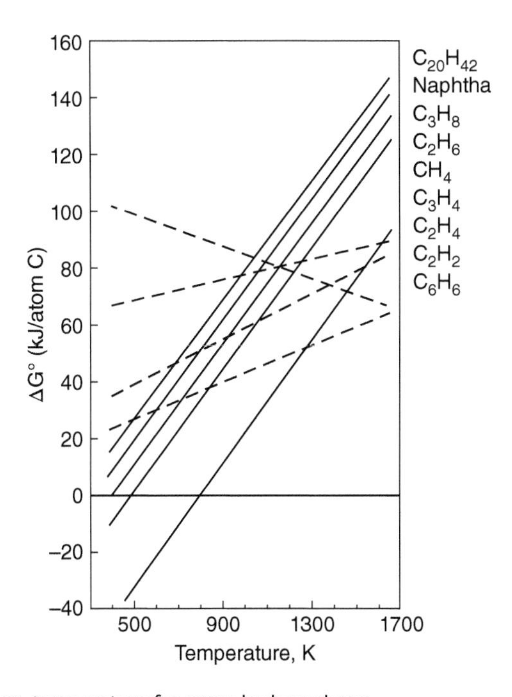

Figure 1.4 Gibbs energy vs. temperature for some hydrocarbons.

Table 1.1 The values of the Gibbs energy for some hydrocarbons.

ΔG_f[kJ mol^{-1}] Temperature, K	300	600	800	1000	1500
C_2H_2	210	190	180	170	145
C_3H_4	196	205	215	225	250
CH_4	−51	−23	−3	19	75
C_2H_6	−33	25	67	110	218
C_7H_{16}	10	220	365	515	890
C_2H_4	68	88	103	120	159
1-C_4H_8	72	150	207	264	407
1,3-C_4H_6	151	196	228	265	344
C_6H_6	130	183	220	260	360
C_6H_{12}	33	200	320	437	730

The stability of thiols, sulfides, and disulfides is comparable to that of alkanes. Thiophene and its homologs are more resistant to decomposition, in comparison not only with aromatics, but also with alkenes. Carbon disulfide and hydrogen sulfide form an exception: in the region of higher temperatures, the stability of hydrogen sulfide decreases (Table 1.2).

Chemical changes that occur during refining operations involve multiple reactants in multiple crisscrossing reactions. *Lumping*, i.e. grouping, by families of reactants and reactions is required. Lumping is very common in chemical kinetics and in determining equilibria between phases, but it is less commonly used in chemical equilibrium. It is often possible to situate the theoretical limits of a change at least qualitatively based on the behavior of a few model compounds that are representative of all the compounds and take part in a limited number of determining reaction paths.

The influence of temperature, pressure, and dilution parameters on the equilibrium position can then be examined for each of the reactions. This gives a quick, rough idea of the assumed equilibrium position according to the exo- or endothermic nature of the reaction and the overall increase or decrease of the number of moles during the change.

In fact, the problem is not so much defining the equilibrium position accurately. Rather, it is to work out which activation methods (thermal or catalytic) will help approach equilibrium reasonably quickly under satisfactory selectivity and stability conditions. To take the

Table 1.2 $\Delta G°_r$ (kJ/mol), hydrocarbons and sulfur compounds.

Temperature, K	300	400	500	600
Heptane	9.2	76.8	147.0	218.9
Methylcyclohexane	28.4	91.4	156.9	224.0
Ethylcyclopentane	45.6	105.1	167.3	231.1
Toluene	122.5	147.7	174.5	202.2
Ethylene	68.2	74.0	80.5	87.5
Diethyl disulfide	22.9	60.3	93.9	140.8
1-Butanethiol	11.6	48.2	83.3	123.2
Thiophene	126.8	133.2	137.3	148.8
Carbon disulfide	66.6	50.1	35.2	21.1
Hydrogen sulfide	−33.1	−37.1	−40.0	−42.2
Temperature, K	**700**	**800**	**900**	**1000**
Heptane	292.0	365.8	440.1	514.7
Methylcyclohexane	292.0	360.4	429.2	498.0
Ethylcyclopentane	296.1	361.7	427.8	494.1
Toluene	230.6	259.3	288.4	317.6
Ethylene	94.9	102.5	110.2	118.2
Diethyl disulfide	172.6	222.3	264.0	309.2
1- Butanethiol	261.3	204.1	245.1	288.8
Thiophene	151.3	164.8	172.8	181.7
Carbon disulfide	7.7	−16.1	−16.7	−17.4
Hydrogen sulfide	−43.9	−50.7	−45.6	−41.0

example of catalytic reforming, thermodynamics dictates that it is advantageous to operate at low pressure and high temperature (highly endothermic change, increased number of moles, and, in particular, significant hydrogen production by aromatization). All efforts have been focused on developing active, selective, stable catalysts that are compatible with these thermodynamic requirements. Moreover, as mentioned earlier, a thermodynamic constraint can be overcome in many refining processes by limiting conversion, creating disproportions in the initial mixture, diluting, eliminating one of the products as it is formed, changing process conditions, etc.

1.2.5 Chemical Kinetics

The mass balances and heat balances associated with the molar extent of reaction (or conversion) have been written, and the conditions of thermodynamic equilibrium have been examined. Now, the rate of change that incorporates the time dimension will be discussed. Before defining a formal expression of reaction rate, the main features of a chemical reaction will be analyzed:

- The nature and location of the reaction phase. In a reaction environment involving a single gas or liquid phase, the problem of location does not arise. In contrast, when the environment involves several phases, the chemical change is usually seen to occur either in one of the phases, at the interface between two phases, or in the vicinity of this interface (for instance, in a hydrotreating process; catalytic distillation).
- The means of activating the reaction and the active intermediate forms (Section 1.2.8).

1.2.5.1 Reaction Rate: Activation Energy

It is often said that the rate of a chemical reaction doubles whenever the temperature is raised by 10 °C. This statement is slightly misleading, since the margin of variation is in fact much wider. However, it does underscore the considerable influence of temperature on thermally activated reactions, as well as on some others. The influence of temperature makes itself felt via the variation in the reaction rate constant k.

Reaction rate constants, whose dimensions depend on partial orders of reaction, vary with temperature according to the Arrhenius equation:

$$\vec{k} = \vec{k}_0 \exp\left[-\frac{\vec{E}}{RT}\right] \tag{1.4}$$

or

$$\frac{\mathrm{d}(\ln \vec{k})}{\mathrm{d}(1/T)} = -\frac{\vec{E}}{R} \tag{1.5}$$

The term E that appears in Eq. (1.4) is called the *activation energy* and is generally expressed in J/mol or kJ/mol. R, the ideal gas constant, is equal to 8.31 J/(mole K). Written as in Eq. (1.5), the expression recalls a form known in thermodynamics as the *Clapeyron–Clausius equation*.

Likewise, the constant k is expressed as a function of temperature by

$$\frac{\mathrm{d}(\ln \overleftarrow{k})}{\mathrm{d}(1/T)} = -\frac{\overleftarrow{E}}{R} \tag{1.6}$$

The result is that

$$K_C = \frac{\vec{k}}{\overleftarrow{k}} = \frac{\vec{k}_o}{\overleftarrow{k}_0} \exp\left(\frac{\overleftarrow{E} - \vec{E}}{RT}\right) \tag{1.7}$$

$$\frac{d(\ln K_C)}{d(1/T)} = \frac{\overleftarrow{E} - \vec{E}}{R} \tag{1.8}$$

Since thermodynamics states that

$$\frac{d(\ln K°)}{d(1/T)} = -\frac{(\Delta H_R^{\circ})_T}{R} \tag{1.9}$$

considering thermodynamic relating K_c and K°, consequently

$$\Delta H_R \cong \vec{E} - \overleftarrow{E} \tag{1.10}$$

Therefore, as an initial approximation, the heat of reaction is equal to the difference between the reaction's activation energy in the forward and reverse directions. Generally speaking, the activation energy of a chemical reaction ranges between 40 and 200 kJ/mol. Any value outside this range should be considered questionable. A low observed value in particular is almost always indicative of diffusion limit skewing. The activation energy of thermally activated reactions is also frequently higher than that of catalyzed reactions. This is not surprising since one of the functions of a catalyst is precisely to lower the potential barrier that separates the reactants from the products.

It is often preferable to use the Arrhenius equation in a different form. k_0, called the *frequency factor*, is the value of the reaction rate constant that corresponds to an infinite temperature. The concept is therefore somewhat abstract. To overcome this conceptual difficulty, another reference temperature T_0 is often chosen, which can be any of the following:

- The initial temperature of a reaction carried out in a closed system
- The average temperature of the cooling fluid used to cool reaction vessel
- The feed inlet temperature in continuous reactor
- Any other appropriate temperature

Equation (1.11) is then written

$$k_T = k_{T_0} exp\left(\frac{E}{R}\left(\frac{1}{T_0} - \frac{1}{T}\right)\right) \tag{1.11}$$

1.2.6 Reactors

1.2.6.1 Conversion, Selectivity, and Yields

In a closed system, the conversion of a reactant A is designated by

$$X = \frac{n_{A°} - n_A}{n_{A°}} \tag{1.12}$$

Conversion is dimensionless and is frequently expressed in %:

$$\text{For selectivity } S = \frac{n_C}{n_{A°} - n_A} \tag{1.13}$$

$$\text{For yield } Y = \frac{n_C}{n_{A°}} \tag{1.14}$$

This gives the basic expression

$$Y = S.X \tag{1.15}$$

with:

n_A°: initial amount of species A_a (mole)

n_c: amount of species C_c (mole)

The case where the only values accessible are element mass concentrations is frequent in hydrocarbon technology. For differential reactor volumes (with plug flow), the balance is written as shown here:

$$N\,dn_A = r_A\,dV \tag{1.16}$$

with:

N: feed of reactant A in mole/time (reactor output)

N_{A°: amount of converted moles per mole of feed

r_A: total rate conversion of feed (mole/volume time) owing to the order of reaction

V: volume of the reactor

This allows us to obtain the basic equation for calculating the volume of a reactor:

$$\frac{V}{N} = \int_0^{n_A} \frac{dn_A}{r_A} \tag{1.17}$$

Reactors involving a single fluid phase present only a little difficulty in scaling up. Homogeneous reactions may be carried out using one of the following reactor systems:

- (Semi)-bath reactor (fine chemistry, biotechnology)
- Continuous tubular reactor (steam cracking)
- Continuous stirred-tank reactor/cascade of stirred-tank reactors (homogeneous catalysis).

1.2.6.2 Continuous Tubular Reactor

An example of a process employing a continuous tubular reactor involving one fluid phase is steam cracking of naphtha (see Chapter 7). The most important concern in the scale-up of a steam-cracking unit is heat transfer and, associated with this, the temperature profile. On a small scale, an electric furnace (Figure 1.5) is often used, whereas on an industrial scale, the furnace is always heated by a burner (Figure 1.6). To be able to translate laboratory results to the industrial unit, a kinetic model of the reaction system is necessary because the interaction of temperature and chemical reaction has to be predicted.

In order to be able to calculate the effective reactor volume, the temperature profile in the reactor must be known. The longitudinal temperature profile in the reactor is determined mainly by the characteristics of the heating elements, volume velocity, thermal capacities of the reactants and products, and endothermicity of the reaction. Of these factors, only volume velocity and heating current can be controlled. Calculations based on empirical correlations show that the temperature difference between the wall and the gas phase at 700 °C is about 5 °C. In the laboratory case, the reaction temperature is measured inside the reactor, and it is assumed that the radial temperature gradient is even smaller. In a tubular flow reactor, there is no sharp difference between the preheater and reactor sections. The equivalent volume of the reactor was determined as follows by Hougen and Watson.

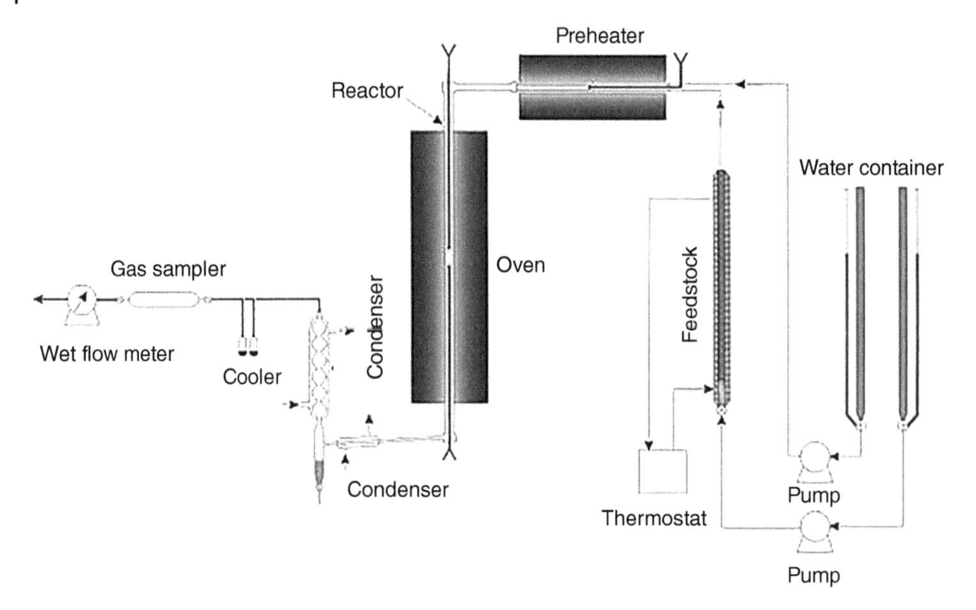

Figure 1.5 Laboratory tubular reactor.

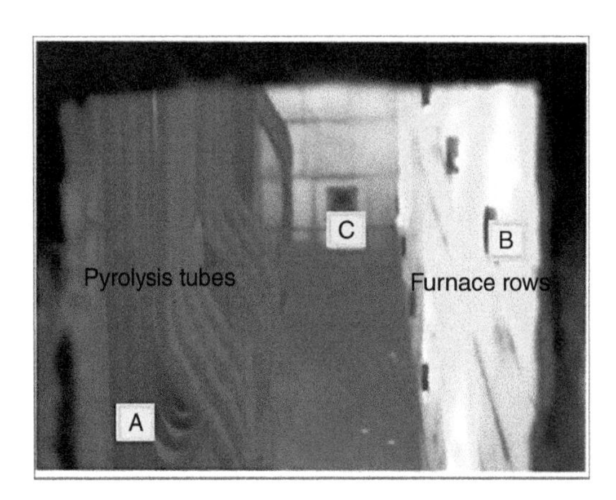

Figure 1.6 Industrial-scale reactor in the furnace.

A. Naphtha vapor flows through the inside of the tubes in the furnace.
B. Rows of furnace guns which burn methane to generate heat inside the furnace.
C. A peephole (eyehole) Inside a tubular reactor (pyrolysis furnace) being used for steam cracking naphtha. The temperature is about 1150 K.

All data are related to a reference temperature T_R, the *temperature of pyrolysis*. The equivalent reactor volume V_R is defined as the volume that gives at the temperature T_R the same conversion as the real reactor with its temperature profile

$$r_{T_R} \cdot dV_R = r_T \cdot dV \tag{1.18}$$

$$V_R = \frac{1}{\exp\left(-\frac{E}{RT_R}\right)} \cdot \int_0^V \exp\left(-\frac{E}{RT}\right) dV \tag{1.19}$$

The individual temperature data are measured stepwise. The calculated equivalent volume is substituted into the relation for the residence time:

$$\tau = \frac{V_\mathrm{R}}{(M_\mathrm{F} + M_\mathrm{S} + M_\mathrm{P})\frac{RT_\mathrm{R}}{P}} \tag{1.20}$$

1.2.6.3 The Reaction Order

The reaction order is determined according to the method developed by Kershenbaum and Martin. For low conversions (low temperatures), the following relation may be derived:

$$\Delta F = \overline{X}^n \left[\frac{QA}{R^n} \int_0^L \exp\left(\frac{-E}{RT(l)}\right) \left(\frac{P(l)}{T(l)}\right)^n \mathrm{d}l \right] \tag{1.21}$$

At a constant temperature and pressure profile, the term in parentheses in Eq. (1.21) does not change so that

$$\log(\Delta F) = n \log \overline{X} + \log(\text{constant}) \tag{1.22}$$

We performed a series of measurements at identical temperature and pressure profiles but with different starting charges of heptane. Relation (1.22) is shown graphically. The order determined from the graph is 1.01 at 680 °C, 1.02 at 700 °C, and 1.12 at 720 °C.

1.2.6.4 Rate Constant

Starting from the value of the reaction order, which is approaching one, we assumed for the determination of rate constants and activation energy that the conversion of heptane is governed by a reaction of the first order. For an irreversible reaction we may write

$$A_F + S_W \rightarrow v_1 R_1 + v_2 R_2 + v_3 R_3 + \cdots v_i R_i + \cdots + S_W \tag{1.23}$$

According to Eq. (1.23), the steam functions only as an inert diluent. A certain part of this diluent reacts, however, with reaction components according to Eq. (1.24). This reaction is negligible, since a large excess of steam is present. For a stationary reactor with plug flow, where an irreversible reaction of the first order proceeds, Levenspiel derived Eq. (1.25):

$$\mathrm{C} + \mathrm{H_2O} \rightleftarrows \mathrm{CO} + \mathrm{H_2} \tag{1.24}$$

$$k \cdot \tau = (1 + \epsilon) \ln \frac{1}{1 - x} - \epsilon \cdot x \tag{1.25}$$

ϵ represents the relative change of the volume in the system when passing from zero to a complete conversion:

$$\epsilon = \frac{V_{x=1} - V_{x=0}}{V_{x=0}} \tag{1.26}$$

The problem is that many different molecules are involved in the reaction and that naphtha does not always have the same composition. An accurate kinetic model can be derived from experimentation on a bench/mini-plant scale and pilot plant scale together with the introduction of the *lumping principle*: various molecules grouped together to form lumps of e.g. the *n*-alkanes, the *iso*-alkanes, etc. The model treats these lumps as if they were one species, when in fact the lumps consist of several species (*n*-alkanes: C_4, C_5, etc.) without extensive additional experimentation.

This lumping approach is also commonly used in modeling of, for instance, the kinetics of catalytic cracking and catalytic reforming. The problem is to determine to what extent lumping is tolerated (varying from considering every molecule separately to putting all molecules in one lump).

Nowadays, due to increasingly faster computers, lumping models can contain hundreds of different lumps. A good example is naphtha cracking, where models are used based on real elementary reaction steps. In fact, de-lumping has taken place. Another example is the combustion of methane. In this case, all reaction steps can be taken into consideration in a model. This leads to several thousand equations of the concentrations of components and associated reaction rate constants. It is not surprising that the best examples are non-catalyzed reactions. Also, for fluid catalytic cracking (FCC) and hydrotreating, de-lumping is practiced by applying sophisticated models, but this is much more difficult than for non-catalyzed reactions. It is expected that molecular modeling will give this field a push.

1.2.7 Industrial Catalysts

Industrial catalysts are at the heart of petrochemical and refining processes and determine the way they will evolve in the future. Catalytic processes are developing increasingly at the expense of thermal ones, and discovery of new catalysts spurs the development of new processes.

Rich in carbon and hydrogen, crude oils naturally contain numerous impurities: heteroatoms, sulfur, oxygen, and trace metals such as vanadium, nickel, and arsenic. Crude oils are made up of a number of individual compounds: alkanes, aromatics, cycloalkanes, and heterocyclic structures whose molecular weight varies from 16 to several thousands (asphaltenes). In contrast with the extreme diversity of crude oils and their components, there are increasingly stringent specifications on refinery and petrochemical products. The constant change in world demand results in greater constraints on discharge combined with more environmental issues.

1.2.7.1 The Place of Catalytic Processes in Hydrocarbon Technology

The first hydrocarbon technologies were based on physical fractionation processes (see Chapter 5: Table 5.2 and Figure 5.2). The first catalytic unit came into operation around 1927 (hydrogenation) and was followed by catalytic cracking processes (Houdry, 1937), aliphatic alkylation, hydrotreating and finally catalytic reforming (UOP, 1950). At the time, world refined oil product consumption was less than 0.6 billion tons per year. Consumption quickly quintupled and reached 3 billion tons per year (1962). At the same time, the refining industry's energy requirements grew tremendously. Today, the worldwide cost of refined catalysts is about €1.7 billion per year, i.e. less than 0.5% of the value of petrochemical and refined products.

1.2.7.2 Homogeneous Catalysts

In homogeneous catalysis, soluble catalysts are applied, in contrast with heterogeneous catalysis, where solid catalysts are used. One example of homogeneous catalysis that will be discussed is the alkylation of isobutane with alkenes, with acids such as HF and H_2SO_4 as catalysts. In this chapter, we will confine ourselves to homogeneous catalyst based on transition metals. These catalysts are becoming increasingly important and already have found a large number of applications. Homogeneous catalysis is applied in essentially all sectors of chemical technology (polymerization in particular) in the synthesis of bulk chemicals (solvents, detergents, plasticizers) and in fine chemistry.

It is useful to compare homogeneous and heterogeneous transition metal–based catalysis. In homogeneous catalysis, the reaction mixture contains the catalyst complex in solution. This means all of the metal is exposed to the reaction mixture. In heterogeneous catalysis, on the other hand, the metal is typically applied on a carrier or as a porous metal sponge-type material, and only the surface atoms are active.

Homogeneous catalysis is a highly innovative area. It is also typically an area where progress is accelerated when chemical engineers and chemists cooperate closely from the start. New developments in homogeneous catalysis are in large part related to new insights in organometallic chemistry. Table 1.3 shows examples of industrially important reactions that are catalyzed by transition metals, which will be dealt with in the next (Section 12.2.12).

Table 1.3 Reactions of industrial importance catalyzed by transition metals.

Reaction	ΔH°_{298} kJ/mol	
Wacker synthesis of acetaldehyde		
$H_2C{=}CH_2 + 1/2\ O_2 \longrightarrow H_3C\text{-}CH{=}O$ (ethene → acetaldehyde)	−244	(1)
Acetic acid by methanol carbonylation		
$CH_3OH + CO \longrightarrow CH_3COOH$ (methanol → acetic acid)	−138	(2)
Hydroformylation (OXO reaction)		
$CH_3CH{=}CH_2 + CO + H_2 \longrightarrow (CH_3)_2CHCH{=}O + CH_3CH_2CH_2CH{=}O$ (propene → isoaldehyde + normal aldehyde)	−125	(3)
Witten synthesis of dimethyl terephthalate		
$H_3C\text{-}C_6H_4\text{-}CH_3 + 3\ O_2 + 2\ CH_3OH \longrightarrow H_3COOC\text{-}C_6H_4\text{-}COOCH_3 + 4\ H_2O$ (p-xylene → dimethyl terephtalate)	−1305	(4)
Synthesis of terephthalic acid		
$H_3C\text{-}C_6H_4\text{-}CH_3 + 3\ O_2 \longrightarrow HOOC\text{-}C_6H_4\text{-}COOH + 2\ H_2O$ (terephtalic acid)	−1360	(5)

1.2.7.3 Heterogeneous Catalysts

In previously sections, implicitly, many processes based on heterogeneous catalysis have been discussed. This is no coincidence. Heterogeneous catalysis is a crucial technology and a workhorse in chemical reaction engineering. In this light, homogeneous, transition-metal catalysis, as discussed the previous section, can be considered a fancy horse. Usually, its application is on a smaller scale and more specific than heterogeneous catalysis.

In industrial catalysis, it is generally preferable from an engineering perspective to use heterogeneous catalysts in continuous processes where possible. The main advantages of heterogeneous compared to homogeneous catalysis are:

- Easy catalyst separation
- Flexibility in catalyst regeneration
- Less expensive

The refining industry's requirements are evolving: more and more petroleum is processed into high added-value products: motor fuels, petrochemical feedstocks, and lubricants. The transportation industry is a driving force behind this development (reformulation of gasoline, kerosene, and diesel oil), which has become possible because of the ongoing progress in processes and catalysts. The changing specifications for motor fuels (gasoline and diesel oil) and heavy fuel bear witness to this fact. The same constraints exist and are more prevalent for other fuel products. In the area of processes and catalysts, requirements involve (Table 1.4):

- Hydrofining processes
- Catalytic processes to convert the bottom of the barrel
- Processes to upgrade FCC, reforming, and hydrocracking products

Further needs include:

- Lower process operating cost and investments (especially for hydrocracking and residue hydroconversion processes) along with improvements in present time performance (for example, in hydrocracking middle distillate selectivity) (Table 1.4).
- New processes and/or innovative catalysts: for example, the development of new naphthene decyclization processes to produce isoparaffins from the hydrocracking gasoline cut or high-cetane diesel fuels from FCC light cycle oils (LCOs). A solid catalyst with superacid properties for aliphatic alkylation would be useful to replace conventional inorganic catalysts (H_2SO_4, HF, etc.).
- More efficient emission control (SO_x, NOx, particulates) and reduced discharge (retreating of spent catalysts, water pollution control).
- Improved hydrogen production.

This list takes future needs into account. With investment costs having quadrupled compared to 1970 costs, refinery energy consumption that could reach 13%, and quintupled production costs, the refinery of the twenty-first century will require high-performance non-polluting processes based on innovative catalysts. The general objective is and will remain to produce quality products at the lowest cost by means of technologies and processes with minimum emissions and discharge.

Table 1.4 Uses for refining and petrochemical catalysts.

Objective	Processes involved	Aims
Purify feeds and products	Hydrorefining	Improved sulfur, nitrogen, metal, and asphaltene elimination yield
Convert the bottom of the barrel	Hydroconversion of vacuum resids, then FCC or hydrocracking	More active and selective catalysts with lower production costs
Upgrade product quality (gasolines, diesel oil)	Catalytic reforming	Increased MON, RON, cetane number, and hydrogen yield
	FCC	Decreased benzene, sulfur, nitrogen, and aromatic contents
Reduce investments and operating costs	All existing processes, mainly resid hydroconversion and hydrocracking	Reduced operating pressure
		Increased activity, selectivity, and lifetime for catalysts
Improve existing processes	FCC and downstream processes	New decyclization catalysts for naphthenes and naphthenoaromatics (production of isoparaffins and of monoalkylnaphthenes)
	Aliphatic alkylation	New solid catalysts
Control emissions Decrease discharge	Claus process	Improved sulfur recovery and higher SO_x and NO_x conversion
	Purification of effluents and of FCC and thermal power station flue gases	Decreased particulate emission
Produce hydrogen	Partial oxidation of resids	Reduced soot formation
	Steam reforming of gases (CH_4, LPG)	Lower temperature
		Improved H_2 selectivity

1.2.7.4 Classifying Catalysts

The available body of catalyst classifications can be broken down two groups: *empirical*, based on acquired experience; and *scientific*, based on explicative correlations. They involve the active elements (active phases) of catalysts: hyperdispersed mono or multimetallic aggregates on a support; or oxide and sulfide active phases; or even a combination of these different phases. The example of empirical classification is given in Table 1.5. It distinguishes, on the one hand, active elements causing homolytic or radical activation (metals, sulfide, semiconductors). On the other hand, there are the active elements responsible for heterolytic (or ionic) activation of hydrocarbons (insulating oxides). Families of reactions correspond to these different types of solids.

Other, more specific empirical classifications are available for families of refining and petrochemical reactions. For example, the hydrogenating activity of metals, sulfides, and oxides is classified according to the target type of the hydrogenation reaction

Table 1.5 Example of solid catalyst classifications.

Product class	Reaction families	Examples
Metals Homolytic activation of H_2, CO, N_2, hydrocarbons, heteromolecules R—C • |•H R—C • |•C—R′	Hydrogenation Dehydrogenation Hydrogenolysis Oxidation	Fe, Co, Ni Ru, Rh, Pd Ir, Pt Cu, Ag
	NH_3 synthesis →	Fe
Sulfide and oxide semiconductors The same homolytic activation as with metals *(heterolytic activation is possible in same cases)*	Oxidation Reduction Hydrogenation Dehydrogenation Cyclization	CuO, ZnO, Cr_2O_3 V_2O_5, MoO_3 Fe_2O_3, CoO, NiO
	Desulfurization Denitrogenation	MoS_2, WS_2, Co_9S_3, Ni_3S_2 RuS_2
Acid and insulating oxides Heterolytic activation R–C+ | : C–R′ Mechanism by carbonium or carbenium ion	Hydration Dehydration Isomerization Oligomerization	Acid zeolites SiO_2-Al_2O_3 Al_2O_3 (+Cl or F) H_3PO_4, H_2SO_4
	Alkylation → Cracking →	H_2SO_4, HF, $AlCl_3$ SiO_2/Al_2O_3 Acid zeolites

(e.g. hydrogenation of aromatics, polyaromatics, alkylaromatics, alkenes, oxygenated compounds, etc.) or the type of unsaturated bond under consideration. Similar correlations exist in acid catalysis.

Explicative correlations, the basis of scientific classification, are generally fewer in number. They relate catalytic properties of the active elements alone or associated with the intrinsic properties of the active solid or with the interaction properties that the solid exhibits with the reaction medium, especially with the properties characterizing adsorption and desorption, which are fundamental in heterogeneous catalysis. As an example, the *d* type fraction (%) of the metal bond and average interatomic distance (exposed faces) of metals has been related to their intrinsic catalytic properties (for instance, in hydrogenation). The last example involves acid catalysis (catalytic cracking, isomerization, alkylation, etc.) where the essentially heterocyclic hydrocarbon activity step forms carboionic species. Analogously, with catalysis by complexes, the property taken into account is basically acidity, characterized in terms of site density or strength.

Adding promotors, sometimes only as traces, can radically modify active-phase catalytic properties. For example, in hydrorefining, adding Ni or Co enhances molybdenum's catalytic activity.

1.2.8 Conversion of Hydrocarbons: Active Intermediate Forms

The two mains ways of activating chemical reactions involving hydrocarbons are raising the temperature and implementing a catalytic substance. The two methods can also be combined. When the only activation method consists of increasing the temperature, the active intermediate forms are generally radicals. When catalytic substances are implemented, the active intermediate forms can be carbocations (also called carbonium or carbenium ions). The temperature method can also generate other active forms such as coordination organometallic complexes, carbanions, and enzymic ferments.

In hydrocarbon technologies, carbocations are involved in catalytic reforming, catalytic cracking, hydrocracking, alkylation, isomerization, polymerization, and etherification reactions. Radicals are brought into play in visbreaking, coking, steam cracking, and partial oxidation. Neutral adsorbed molecules are observed in hydrogenation, hydrodesulfurization, hydrodenitrogenation, and hydrodemetallization. Oligomerization of olefins uses soluble organometallic complexes. The nature of the active intermediate forms is related to that of catalysts that are (or are not) implemented and to the operating condition used.

1.2.8.1 Carbocations

As mentioned earlier, a large number of refining operations involve carbocations. They are indispensable in triggering reactions and are produced by using catalysts, usually of the solid, acid type. The catalyst's acid sites can be of two kinds: Lewis and Brönsted. A variety of treatments, particularly with steam, can be used to transform Lewis sites into Brönsted sites. The nature, strength, and distribution of sites can be highly variable from one catalyst to other. Generally speaking, the hydrocarbon molecule is adsorbed on an electron receiver site that it reacts with to form a carbocation (Eq. (1.27)):

$$-\overset{|}{\underset{|}{C}}-H + A^{+} \text{ receiver site} \rightarrow \underset{\text{carbocation}}{-\overset{|}{\underset{|}{C^{+}}}} + AH \qquad (1.27)$$

The carbonium ion formed in this way on the surface can then evolve in different ways. It can be:

- Desorbed, restoring the receiver site to its original status
- Desorbed, losing a proton that remains on the receiver, thereby forming an alkene
- Rearranged by migration of the positive charge along the chain (H-shift) to the most stable position (tertiary C^+ is more stable than secondary C^+, which is more stable than primary C^+)

The ion modified this way can in turn follow the paths indicated previously. It can:

- Be rearranged by migration of a chain element, for example a methyl group (CH_3-shift, PCP branching)

- React with another molecule to effect a charge transfer or an addition
- Be cut at the ß position with respect to the carbon carrying the positive charge.

These possibilities can be used to explain the behavior of hydrocarbons involved in reactions qualitatively (as well as quantitatively, to the extent that the kinetics specific to elemental steps can be accessed). The conventional reaction scheme of iso-alkane alkylation by alkenes is presented next as an example.

The initial carbocation is provided by propene reacting with an acid site:

$$H_3C{-}CH{=}CH_2 + H^+ \longrightarrow H_3C{-}\overset{+}{C}H{-}CH_3$$

Then there is a charge transfer to an isobutane molecule:

$$H_3C{-}\overset{+}{C}H{-}CH_3 + H_3C{-}\underset{\underset{\displaystyle CH_3}{|}}{CH}{-}CH_3 \longrightarrow H_3C{-}CH_2{-}CH_3 + (CH_3)_3\overset{+}{C}$$

This tertiary ion is more stable than initial secondary ion. It reacts with the propylene molecule:

$$H_3C{-}\overset{\overset{\displaystyle CH_3}{|}}{\underset{\underset{\displaystyle CH_3}{|}}{C^+}} + H_2C{=}CH{-}CH_3 \longrightarrow H_3C{-}\overset{\overset{\displaystyle CH_3}{|}}{\underset{\underset{\displaystyle CH_3}{|}}{C}}{-}CH_2{-}\overset{+}{C}H{-}CH_3$$

which gives the following by H-shift

$$H_3C{-}\overset{\overset{\displaystyle CH_3}{|}}{\underset{\underset{\displaystyle CH_3}{|}}{C}}{-}\overset{+}{C}H{-}CH_2{-}CH_3$$

and then, by CH_3-shift:

$$H_3C{-}\overset{+}{\underset{\underset{\displaystyle CH_3}{|}}{C}}{-}\overset{\overset{\displaystyle CH_3}{|}}{C}H{-}CH_2{-}CH_3$$

Afterward, there is another charge transfer with isobutane:

$$H_3C{-}\overset{+}{\underset{\underset{\displaystyle CH_3}{|}}{C}}{-}\overset{\overset{\displaystyle CH_3}{|}}{C}H{-}CH_2{-}CH_3 + H_3C{-}\underset{\underset{\displaystyle CH_3}{|}}{CH}{-}CH_3 \longrightarrow H_3C{-}\underset{\underset{\displaystyle CH_3}{|}}{CH}{-}\underset{\underset{\displaystyle CH_3}{|}}{CH}{-}CH_2{-}CH_3 + (CH_3)_3\overset{+}{C}$$

This provides the main product, 2,3-dimethylpentane, and the tertiary carbonium ion that allows the cycle to start over. As a whole, the reaction is written as follows:

$$H_2C{=}CH{-}CH_3 + H_3C{-}\underset{\underset{\displaystyle CH_3}{|}}{CH}{-}CH_3 \longrightarrow H_3C{-}\underset{\underset{\displaystyle CH_3}{|}}{CH}{-}\underset{\underset{\displaystyle CH_3}{|}}{CH}{-}CH_2{-}CH_3$$

If the very small amount of propane produced during the initiation step is disregarded, side reactions are possible, even if they are not thermodynamically promoted:

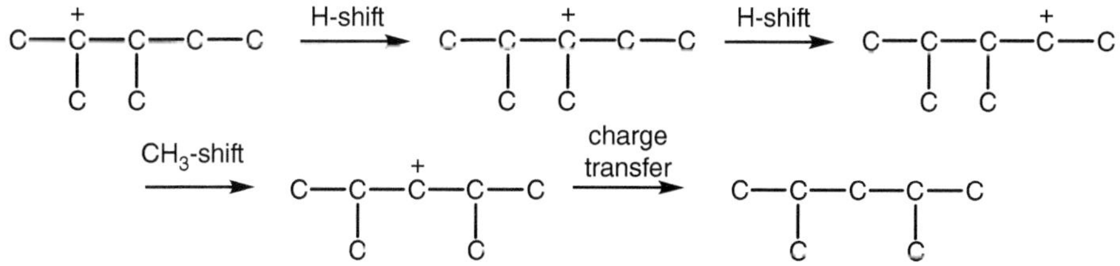

```
   +                          +                              +
C—C—C—C—C   --H-shift-->   C—C—C—C—C   --H-shift-->   C—C—C—C—C
  |  |                       |  |                       |  |
  C  C                       C  C                       C  C

                   +                 charge
--CH3-shift-->  C—C—C—C—C  --transfer-->  C—C—C—C—C
                  |   |                     |   |
                  C   C                     C   C
```

For these reactions to take place, it is indispensable for the alkane to be branched so that the relay carbocation is stable enough. Note also that the carbocations do not leave the surface of the catalyst. They are an adsorbed form of the relevant molecules, and all the changes occur in costant contact with acid sites.

1.2.8.2 Radicals

Radicals are formed by homolytic cleavage of a C—C or C—H bond, while carbocations are formed by heterolytic cleavage. The radicals are formed by preferential scission of C—C bonds as a result of the larger dissociation energy of C—H bonds. For the selective generation of radicals by thermal heating, thermal initiators with chemically different and more liable bonding are used: for example, C—O or C—N. Especially sensitive are some bonds like C—S, S—S, and S—H (Table 1.6).

In addition to selection of various types of precursor radicals, greater selectivity in thermal cracking of chemical bonds in organic compounds is reached by using unconventional thermal sources, which supply an intensive flow of infrared photons. In contrast to traditional methods, it is possible to generate reactants with specific populations of vibrational levels by interaction with infrared laser radiation. The interaction of reactants with infrared CO_2 laser radiation, which can be attuned to many characteristic vibrations, can lead to many new, uncommon chemical reactions that cannot be carried out by other techniques; the course of these reactions is made possible by primary activation of the reactants or specific conditions of these reactions. This is also the case for laser pyrolysis: to be more exact, laser-induced homogeneous pyrolysis (LIHP) of hydrocarbons.

The thermal decomposition of hydrocarbons going through radicals is chain-like and does not involve chain branching. It is remarkable because its course is not predictable/clear, with formation of a wide gamut of reaction products, small selectivity, and complex kinetics.

Basic radical reactions are studied on simple molecular systems, often requiring complicated physical and spectroscopic methods in combination with electron paramagnetic resonance. The lifetime of highly reactive radicals in pyrolysis is short, and their concentration is relatively low (Table 1.7).

Success can be achieved either by methods that indicate the level of free radicals directly in the stable state using the dynamic method, or by methods where rapid freezing on static or rotating cryostat isolates radicals from the hot zone.

Electron spin resonance (ESR) spectra of alkyl- and sulfur-containing radicals provide important information about the delocalization of the unpaired electron and the geometry of the species. Alkyl radicals have a smaller g factor than oxyl or peroxyl radicals, or radicals

Table 1.6 Bond dissociation energy of hydrocarbons and sulfur compounds.

C—C and C—H bonds	**kJ/mol**
$CH_3—CH_3$	360
$C_4H_9—C_3H_7$	314
$CH_2=CH—CH_3$	394
C—C in cyclopentane ring	293
C—C in cyclohexane ring	310
$C_2H_5—H$	410
$(CH_3)_3C—H$	373
$CH_2=CH—CH_2—H$	322
$C_6H_5—H$	427
C—H in cyclopentane ring	389
C—H in cyclohexane ring	389
C—S, S—S, and S—H bonds	kJ/mol
$CH_3S—CH_3$	323
$C_6H_5S—CH_3$	281
$C_4H_9—SH$	289
$RS—C_2H_5$	310
$RSS—C_2H_5$	226
$RSS—C_6H_5$	291
$CH_3S—SCH_3$	310
$C_6H_5S—SC_6H_5$	230
HS—H	379
RS—H	379
$C_6H_5S—H$	374

with an unpaired electron on sulfur (Table 1.8). The greater the delocalization of an electron on substituents, the smaller the absolute value of the splitting constant a (a_{x-h}).

An unpaired electron in a radical not only shows up in the electron resonance spectrum but also causes different vibrations between the atoms of the radical, so that infrared spectroscopy can also supply supplementary findings about the structure of the radical. The main primary process in the pyrolysis of 1-hexene, 2,5-dimethyl-1,5-hexadiene, and 1,5-hexadiene is fission, leading to the formation of an allyl radical that can be characterized by means of infrared spectroscopy.

Regarding the chemistry of radicals, the causes of their formation, and the course of their reaction – that is, in general, about the mechanisms of radical reactions – substantially less is known at present than about reactions taking place via carbocations. For a long time, these occupied one of the foremost places in investigations devoted to the clarification of reaction mechanisms. More research work was devoted to the study of carbocations than

Table 1.7 Concentration of radicals in the pyrolysis of ethane and propane.

Starting hydrocarbon Radical concentration	Ethane mole/l	Propane mole/l
$H^{\cdot}$	$10^{-8.3}$	$10^{-9.3}$
$CH_3^{\cdot}$	$10^{-8.3}$	$10^{-7.7}$
$C_2H_3^{\cdot}$	$10^{-6.6}$	$10^{-7.0}$
$C_2H_5^{\cdot}$	$10^{-6.7}$	$10^{-7.7}$
$C_3H_5^{\cdot}$	$10^{-7.9}$	$10^{-6.4}$
1-$C_3H_7^{\cdot}$	–	$10^{-9.3}$
2-$C_3H_7^{\cdot}$	–	$10^{-8.0}$
$C_4H_7^{\cdot}$	$10^{-6.6}$	$10^{-7.4}$
1-$C_4H_9^{\cdot}$	$10^{-10.3}$	–
2-$C_4H_9^{\cdot}$	–	$10^{-10.6}$

Experimental conditions:
Ethane: 1123 K, 50 mole% steam.
Propane: 1123 K, 40 mole% steam.

Table 1.8 ESR spectra of alkyl- and sulfur containing radicals.

Radical	g-factor	a_{X-H} (mT)
$C_6H_5^{\cdot}$	2.0023	2.75
$CH_3^{\cdot}$	2.0025	2.29
$CH_3CH_2^{\cdot}$	2.0025	2.22
$C_6H_5CH_2^{\cdot}$	2.0026	1.63
$CH_3SCH_2^{\cdot}$	2.0049	1.65
$RS^{\cdot}$	2.0106	–
$CH_3SO^{\cdot}$	2.0100	1.15 (3H)
$CH_3SO_2^{\cdot}$	2.0049	0.055

radicals, primarily because radical reactions very often are unclear, with the formation of complex reaction mixtures and low selectivity.

Another hindrance is the fact that radical reactions have a chain character with complicated kinetics, which considerably complicates direct measurements of the rates of the individual processes. A breakthrough in the study of radical reactions was brought about only after the introduction of new analytical and kinetic methods. The application of modern analytical, primarily chromatographic, methods makes it possible to carry out a quick quantitative analysis of pyrolysis products. This allows the unequivocal identification of the primary and secondary or even further products of the consecutive steps of thermal decomposition, which contributes to the determination of the rates of many elementary reactions.

Methods based on the analysis of reaction products are among the indirect methods of investigating radical reactions taking place in the pyrolysis of hydrocarbons.

The steps of radical processes are as follows:

- Initiation

$R-R' \rightarrow R\cdot + R'\cdot$ for alkanes and alkenes

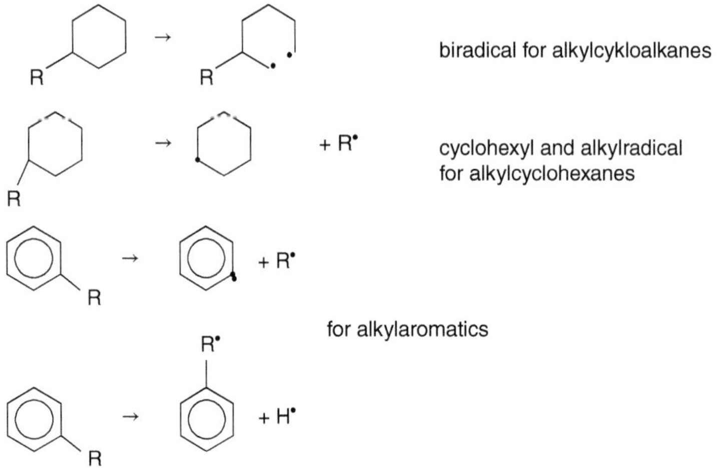

biradical for alkylcykloalkanes

$+ R^\bullet$ cyclohexyl and alkylradical for alkylcyclohexanes

$+ R^\bullet$

for alkylaromatics

$+ H^\bullet$

- Propagation
  $R\cdot + R' \rightarrow R + R'\cdot$
- Termination
  $R\cdot + R'\cdot \rightarrow R - R'$

Thermal cracking of *n*-heptane serves to illustrate this process, with the various steps shown in Figure 1.7.

Radical schemes differ widely from carbocations schemes in their results. In particular:

- The short $CH_3\cdot$, $C_2H_5\cdot$ radicals may generate large amounts of light gases.
- The cleavage always occurs in the chain for alkylaromatics, whereas it takes place flush with the aromatic ring for carbocations.
- Radicals do not lead to isomerization of the skeleton; there are no branched products other those already present in the feed.
- Thermal cracking of alkanes can produce 1-alkenes.

Speight and Pines have shown clearly the characteristic differences between the two reaction schemes when both are possible. Bajus et al. have analyzed the reaction mechanisms of the radical scheme for *n*-heptane and methylcyclohexane cracking in detail. When the theoretical rules deduced by Rice (1933) and modified by Kossiakoff and Rice (1943) are applied in the thermal decomposition of *n*-heptane, the cracking of four C_7H_{15} radicals proceeds according to the scheme in Figure 1.7.

1.2.8.3 Initiated Decomposition

The pyrolysis of hydrocarbon feedstocks to alkenes proceeds when the rate and selectivity of the conversion, even under optimal conditions of running the process, are relatively low. While methane is formed in substantial amounts in the primary reactions, heavy-liquid

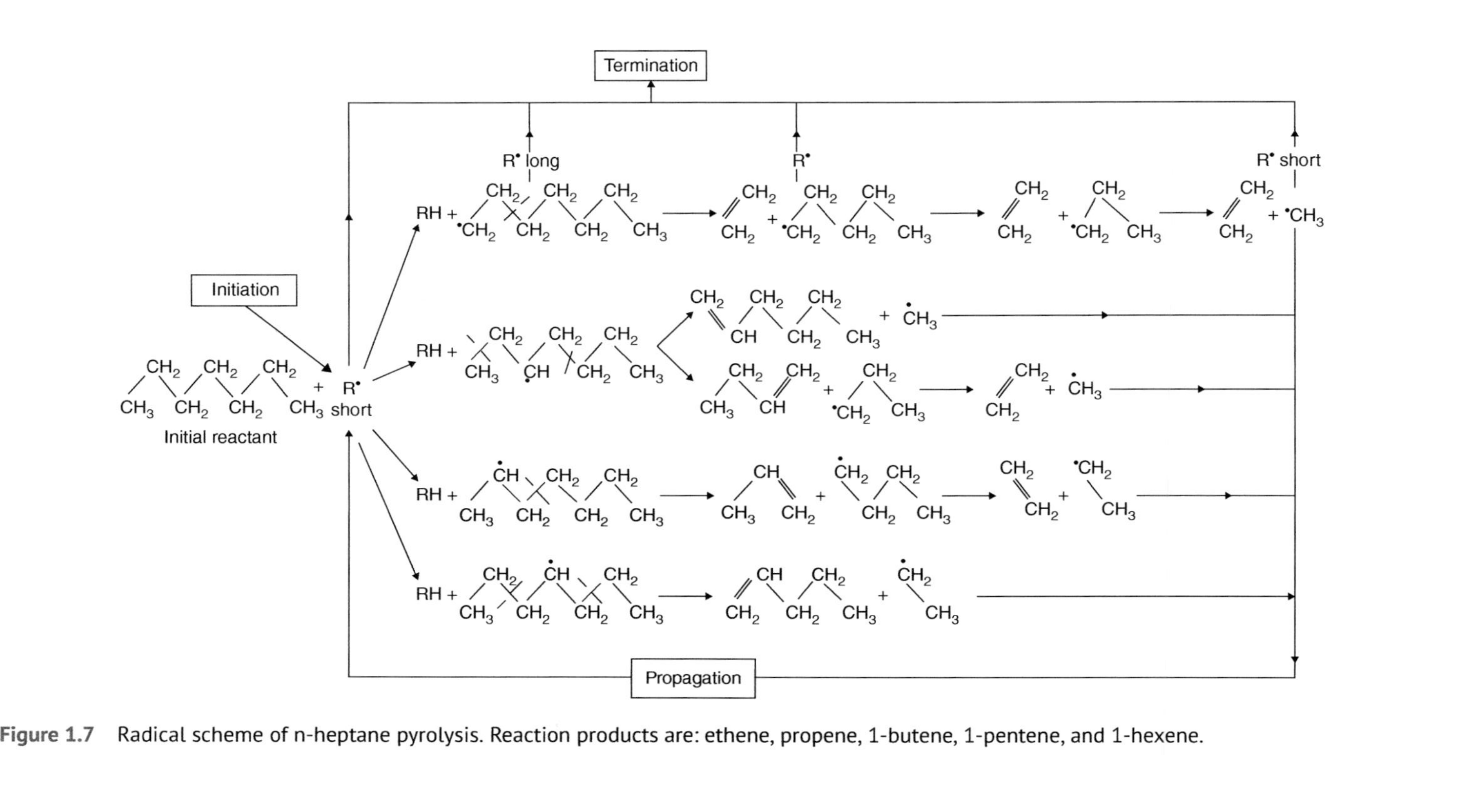

Figure 1.7 Radical scheme of n-heptane pyrolysis. Reaction products are: ethene, propene, 1-butene, 1-pentene, and 1-hexene.

pyrolysis fractions, pitch, coke, and oxides of carbon are formed as a result of secondary reactions, the course of which are made possible by a multicomponent, highly reactive alkene-aromatic system that is an appreciable catalytic effect of the inner surface of the reactor (see Section 7.4.1). One of the feasible ways of improving the production of lower alkenes is to carry out the pyrolysis process in the presence of substances that make it possible to lower the pyrolysis temperature, enhance the rate of radical conversion, increase the flexibility of the pyrolysis process, improve selectivity, and use feeds with different properties.

Presently, an intensive search is proceeding for such compounds (initiators, catalysts, activators, promotors) of a homogeneous and heterogeneous nature that favorably influence the pyrolysis process, as well as compounds (inhibitors, retarders, deactivators, passivators) that suppress the formation of unwanted pyrolysis products. As a result, several types of chemical substances have been studied; their broader application under running conditions is most often limited by their efficiency, availability, or cost.

Among the compounds that have the ability to influence the radical process of thermal decomposition of hydrocarbons are some inorganic and organic sulfur compounds. The effect of sulfur-containing additives on the pyrolysis process can be either favorable or unfavorable in various ways. One of the great advantages is that sulfur compounds are dosed into systems in small concentrations, several times smaller than in the use of other additives for the same purpose. Under pyrolysis conditions, sulfur compounds decompose mainly to hydrogen sulfide, which is readily removed from the pyrolysis gas by alkali washing. Several of the favorable effects of sulfur compounds can be seen irrespective of the conditions and type of feedstock. As far as negative effects are concerned, they are, primarily, corrosion of the exposed parts of the pyrolysis equipment. Part of the sulfur, which is concentrated in liquid products, deteriorates their quality and causes difficulties with further processing. In contrast, the use of sulfur compounds is very attractive not only because of the low cost, but primarily for their ability to influence the radical process of the thermal decomposition of hydrocarbons, especially the course of secondary reactions, which cause considerable difficulties.

1.3 Potential Steps Toward Greener Chemical Technology

Chemical and allied products technology has certain well-defined characteristics, which we will discuss in turn (Table 1.9). Most olefins (e.g. ethylene and propylene) will continue to be produced through steam cracking of hydrocarbons in the coming decade. In the uncertain commodity market, the chemical industry is investing very little in alternative technologies and feedstocks because of their current lack of economic viability, despite decreasing crude oil reserves and the recognition of global warming. From this perspective, some of the most promising alternatives are compared with the conventional steam cracking process, and the major bottlenecks of each of the competing processes are highlighted. These technologies emerge especially from the abundance of cheap propane, ethane, and methane from shale gas and stranded gas. From an economic point of view, methane is an interesting starting material, if chemicals can be produced from it. The huge availability of crude oil and the expected substantial decline in the demand for fuels imply that the future for proven technologies such as Fischer–Tropsch synthesis or methanol to gasoline is not

Table 1.9 Characteristics of chemical technology.

1.	Maturity and its consequences
2.	Participation in international trade
3.	Competition from developing countries
4.	Capital intensity and economies of scale
5.	Criticality and pervasiveness
6.	Freedom of market entry
7.	Stringent requirements of the CAA
8.	High R&D for ecologically oriented projects
9.	Dislocations and environmental impacts
10.	Feedstock recycling

bright. The abundance of cheap ethane and the large availability of crude oil have caused the steam cracking industry to shift to these two extremes, making room for the on-purpose production of light olefins: for example, by the catalytic dehydrogenation of propane.

1.3.1 Maturity

Maturity is highly prized in the individual lives of people but is feared in technology. No company could operate in the realm of chemical technology unless it fully understood all the ramifications of maturity, which expresses itself in overcapacity, intense competition, low prices, and low profitability. Ultimately, it leads to restructuring. All of these things have happened in chemical technology, particularly with commodity chemicals.

Maturity occurs because of market saturation, wide diffusion of technology, and low barriers to entry. In chemical technology, most of the available technologies are considered mature. Technological innovations in catalytic processes are thus expected from introducing substantial modifications to technology. New raw materials and catalysts are being used, and new process technologies, machinery, and equipment are being developed. Complying with new requirements and regulations requires fine-tuning existing processes. Work continues on optimizing all the process components, eliminating bottlenecks, and revamping existing plants. Processes that have a less severe environmental impact are being developed, which translates into lower pollutant emissions, less waste, and greater safety. There is also increasing demand for purer products for the production of high-quality polymers.

1.3.2 Participation in International Trade

The technology of the chemical industry tends to be very international. To a great extent, the same chemical companies and the same international service companies are found in most areas of the world; not surprisingly, the technology they employ is broadly the same regardless of the country they operate in. Fierce economic competition has been one of the key characteristic of the international petrochemical industry since its earliest days; this characteristic has been strengthened over the past 20 years by the global trend toward deregulation, competition, and privatization, creating a more pluralist industry that consists of many "players" in each segment of the industry and more transparent transfer prices between them.

Chemical technology in the United States has always had two advantages that make its products cheap not only at home but also in internationally. First, there is ample natural gas, which provides ethane and propane for steam cracking, and it is generally cheaper to crack gas than liquid naphtha and gas oil (Section 4.2). Second, the United States has ample supplies of propylene, not only because it is produced during steam cracking, but also because the huge capacity for catalytic cracking required by the gasoline industry means the few percent of propylene produced in this reaction translates into millions of tons of product.

Western Europe exports considerably more chemical products than the United States because companies like BASF, Lyondell Bassel, Degussa, Akzo Nobel, Evonik, and Sanofi are all higly exported-oriented. In 2009, world trade in chemicals amounted to 1871 billion euros, of which the European Union claimed 24% or 449 billion euros. In 2009, world chemical sales amounted to 1871 billion euros, of which the European Union 27 contributed 24% or 450 billion euros; 117 billion euros were exports outside the EU. Approximatelly 222 billion euros were intra-EU trade, and only 110 billion euros were domestic trade. Western Europe exported 73% of its production if one counts intra-European trade, or 26% if one excludes it.

American companies have preferred to ensure profits, tenuous though these sometimes are, by satisfying the local market and then assigning incremental production to world trade. American chemical technology exports products to more than 185 countries. In 2009, the United States exported $154.6 billion US, giving a negligible trade balance.

Japan, having based its chemical industry on relatively expensive imported naphtha and struggling with an expensive yen, cannot compete in the world market, although in the early 1990s it enjoyed sales with neighboring countries with expanding economies, primary China.

1.3.3 Competition from Developing Countries

But what about the future? Natural gas has been discovered in many places in the world, and of course many countries have petroleum. Many of them are eager to enter the chemical business because it promises greater added value than is possible when gas or oil is used for the energy production. Thus an impressive list of countries have built or are building chemical technology. The United States, Western Europe, and Japan have long-standing chemical technology. Newcomers include Saudi Arabia and other Gulf states; Canada; Mexico; Venezuela, Brazil, Argentina, and other Latin American countries including Trinidad and Chile; the former members of the USSR; and countries of the far east including Taiwan, Korea, China, Thailand, Indonesia, Malaysia, Philippines, and Singapore. Many of these countries have entered the chemical business to provide for their own needs. Taiwan and Thailand indicate that they will not be major exporters because they can consume most of their production locally. Korea, Saudi Arabia, Canada, and most other countries will, however, become formidable competitors in the international trade arena.

Japan and Western Europe reduced their trade of ethylene derivatives from 1978 to 2015 (estimated), while the Middle East and Canada gained. Japan lost its stake in trade of ethylene derivatives very early; indeed, this was true of virtually all Japanese chemicals because of expensive raw materials and strong currency. The trade that Japan enjoyed in 1993 was due largely to the needs of mainland China, where shipments from Japan had lower freight costs. Western Europe does not derive much of its huge chemical exports from ethylene derivatives, and much of what it enjoyed formerly has been taken over by Canada and Saudi Arabia. The United States still maintains a role, and it is believed that this will continue through the year 2005, particularly if gas continues to be available for steam cracking. Since Saudi Arabia only entered the chemical business of cracking ethane, it could only participate in ethylene derivatives trade. By 1995, however, another cracker was on stream for the cracking of naphtha and provided Saudi with the basic chemicals it needed to become a full-fledged competitor in the world's chemical arena.

Will Asia continue to be the growth market for chemical products? Certain parts will remain strong markets for imported chemicals, but Korea, Taiwan, and Thailand will mature. Countries such as China, India, and Indonesia that have high populations will continue to see strong growth. South America is a mix of newly industrialized and developing countries. While demand growth will be strong in South Asia and in the old Soviet bloc, those areas will be making the chemicals and derivatives they need, and their import requirements will be minimal. Africa, China, and the Indian subcontinent, because of their immense size, will be strong markets for imported goods. As an example, in 1998, only 700 000 tons of ethylene was consumed in Africa. However, by the year 2033, almost 10 million tons will be consumed there. This level of consumption represents only ethylene, not ethylene derivatives that will be imported. Africa may be the specialty petrochemical market of the future, with everyone wanting to export there.

1.3.4 Capital Intensity and Economies of Scale

Chemical technology is capital intensive. It produces huge quantities of homogeneous materials, frequently liquids or gases that can be manufactured, processed, and shipped most economically on a large scale. This was less so through the nineteenth century until World War II. The early petrochemical industry used general-purpose equipment and operated batch processes that required little capital investment but had high labor costs.

The petroleum refining industry was the first to convert to continuous operation on a large scale. Engineering developments for the petroleum industry were then applied to the petrochemical industry. Plant sizes escalated as dramatic economies of scale became possible. The capacity of a typical ethylene cracker rose from 32 000 tons per year in 1951 to 450 000 tons in 1972. In the early 1990s, plants with 680 000 tons per year capacity were built. Currently, there are few batch processors of any size in operation for commodity chemicals, and substantial economies of scale have become a characteristic of the modern petrochemical industry.

Many of the concepts just described have been demonstrated in the world's largest single-train plant, the Nova/Dow E-3 unit, in Alberta, Canada, to which previous references have also been made. This Stone & Webster design has been in service since 2000, and its original design capacity was 1.275 million tons per annum (MTPA) based on ethane feed; current operating rates approach 1.5 MTPA (Section 6.1). In 2014, the capacity of a typical ethylene cracker rose to 1.75 MTPA (Yanbu Petrochemical Co., Yanbu, Saudi

Arabia; Equistar Chemicals LP, Channelview, Texas, USA; INEOS Olefins & Polymers, Chocolate Bayou Works, Alvin, Texas, USA).

Total world ethylene capacity in 2014 was close to 146 MTPA. Propylene demand profile 2013 production estimate = 83 million metric tons (55% stm. crackers; 31% FCC splitters; 5% dehydro; 4% metathesis).

Economies of scale arise not only from improved technology but also from purely geometric factors. The capacity of a great deal of chemical equipment (e.g. storage tanks and distillation columns) varies with its volume, which is the cube of its linear dimensions. The cost, on the other hand, is the cost of a surface to enclose the volume and varies with the square of the linear dimensions. Consequently, cost is proportional to $(\text{capacity})^{2/3}$: this is called the square–cube law. It does not apply to all equipment. The capacity of a heat exchanger depends on its surface area, so the cost is proportional to $(\text{capacity})^{1}$, and there are few economies of scale. Control systems are not affected by capacity at all, so the cost is proportional to $(\text{capacity})^{0}$, and economies are infinite. It is claimed that for a modern petrochemical plant overall, cost is proportional to $(\text{capacity})^{0.6}$.

The size and complexity of a modern petrochemical plant demand high capital investment. Although other industries invest more capital per dollar of sales, chemical technology has the highest investment of current capital. That means the chemical industry invests more each year than other capital-intensive industries such as mining, where once equipment is bought, it remains in service for many years.

Capital intensity has a number of corollaries. The return on capital is relatively low. Because high capital investment reduces the labor force required, labor productivity (i.e. value added per employee) is high. Salaries contribute relatively little to costs (on the order of 2.0%), and employers worry less about pay increases than in labor-intensive industries such as food or apparel. Consequently, labor relations are unusually good.

The assets of a company correspond to the estimated value of the plant, land, and other capital good it owns. Such ratios as assets per employee, sales per dollar assets, and sales per employee are measures of both the capital and labor intensity of an industry. Generally, the petroleum refining industry has both the highest assets and the highest sales per employee. The chemical industry shows lower figures but still ranks high. The food and clothing industries are usually at the low end of the scale. Service industries with few assets and many employees, such as consulting, show both low assets and low sales per employee.

The move to specialty chemicals has altered these perceptions as far as overall chemical technology is concerned, although those sectors are not described in this book. Small, high-value, low-tonnage chemicals are frequently made by processes using computer-controlled equipment. Such equipment brings some of the advantages of continuous processing to batch processes.

1.3.5 Criticality and Pervasiveness

Chemical technology is critical to the economy of developed countries. In the nineteenth and first half of the twentieth century, a nation's industrial development could be gauged from its production of sulfuric acid, the grandfather of economic indicators. Today one uses ethylene production as a yardstick of industrial sophistication. An advanced economy cannot exist without a chemical industry; neither can a chemical industry exist without an advanced economy to support it and provide the educated labor force it requires.

Chemical technology is not replaceable. There is no other industry that could fulfill its function. It is pervasive and reflected in all goods and services. Not only is the chemical industry here to stay, but it is also a dynamic and innovative industry that has grown rapidly and on which the world will continue to rely in the future. Many of the problems concerning pollution and energy have been detected and monitored by chemical methods, and chemistry is playing a part in their solutions.

1.3.6 Freedom of Market Entry

Another characteristic of chemical technology is freedom of market entry. Anyone who wants to manufacture bulk petrochemicals may do so by buying so-called *turnkey* plants from chemical engineering contracting companies. Such companies have processes for preparation of virtually any common chemical and will build a plant guaranteed to operate for anyone who wishes to invest the money. This was the way many petroleum companies gained entry to the petrochemical business and also how many developing countries are laying the foundations of their own chemical technology.

The one requirement is a large amount of capital. To enter the basic chemical business requires a bare minimum of $1 billion. Large sums of money may be required for purposes other than capital investment. In the pharmaceutical industry, large sums are required for development, which in the early 2000s was on the order of $500–1000 million per drug. The detergent industry, on the other hand, requires money for massive amounts of advertising. These two industries underscore the importance of large cash flows to support research, development, and merchandising. Thus entry into the chemicals market is *free* in the sense in which economists use the word, but the expense is such that only governments, oil companies, and other giant enterprises can find the necessary capital.

What has been said about capital applies primarily to very large-volume, basic petrochemicals such as the seven basic petrochemicals or petrochemical groups and their first-line derivatives. Beyond that, there may be other barriers to entry such as lack of necessary technology or reluctance on the part of a patent holder to license technology. An example is DuPont's process for making hexamethylenediamine by the hydrocyanation of butadiene. This is probably the preferred process, but it is not available for license. On the other hand, the Dutch company AkzoNobel was able to enter the Aramid (DuPont's Kevlar) business, presumably by finding loopholes in the DuPont patents. Kevlar is used for fibers that are stronger, weight for weight, than steel; and DuPont believes that its future will be profitable. However, that future will be shared with others, since the patents were apparently not invincible.

Downstream operations also may provide a barrier to entry. Thus manufacturers of poly(methyl methacrylate) convert their product to acrylic sheets, which are then sold to molders. Potential manufacturers must decide whether they want to gain the expertise that participation and marketing the sheet requires, and whether the company's culture will allow participation in a business so far removed from basic chemical manufacture.

The low price of a product and correspondingly low profitability may deter entry. Furfural provides a classic example. When it was introduced many years ago, it was priced so low that it did not attract competition until well into the product's life cycle; and then major competition came not from US companies, but from China. Companies with products whose patents are about to expire often use this technique to discourage other manufacturers. Monsanto used it successfully with one of its herbicides that came out of a patent; the company was able to manufacture it cheaply in a depreciated plant using an optimized process. It was not worthwhile for other companies to invest fresh capital in order to compete. And, to Monsanto's surprise, the market increased as farmers used more of this cheap herbicide to control weeds in place of the more cumbersome processes of plowing or covering the ground with plastic. Union Carbide similarly exploited the technique with its pesticide carbaryl. By pricing it low, the company successfully avoided serious competition after the patent expired.

1.3.7 Stringent Requirements of the Clean Air Act (CAA)

Chemical technology is one of the most highly regulated areas of all the technology fields. The regulations are intended to protect and improve workers' and the nation's health, safety, and environment. The American Chemistry Council (ACC) has documented the industry's vigorous response to the need for pollution abatement.

Of all the regulations, the stringent requirements of the CAA have the most far-reaching economic impact on the industry, with cost projected to be $25 billion per year. A brief description of the acts that affect chemical technology is given in Table 1.10.

1.3.8 High R&D for Ecologically Oriented Projects

Chemical technology is research intensive. It hires many graduates – over 15% of all scientists and engineers in the United States – and most of them work in research and development (R&D) laboratories. Chemical technology's proportion of the total almost doubled from 1970 to 1992. But later in the 1990s and into the 2000s, the rate of growth of chemical R&D spending slowed. Research expenditures of some of the top chemical companies in 2009 are shown in Table 1.11. Companies with major pharmaceutical subsidiaries, such as Bayer, spend more on R&D than do the mainline chemical companies. Research-based pharmaceutical companies with few other interests spend 10–25% of sales on research, but the table is compiled, as far as possible, to exclude pharmaceutical sales. True speciality chemical companies have research budgets about a third of those of the pharmaceutical companies. At the other extreme, the major oil companies (e.g. Chevron-Phillips) are involved mainly with commodity chemicals and spend little on research. Various companies do not divulge their R&D spending and have been omitted from Table 1.11. It is claimed that private equity companies spend very little on R&D, and Huntsman exemplifies this. Ineos does not disclose its research spending.

How is the R&D budget spent? Research is a risky and expensive business. Finding the conditions that maximize the cost-effectiveness of an R&D budget preoccupies many pharmaceutical managers. Unfortunately, it is not a science, and success in the laboratory often depends on serendipity. Should a company rely on discoveries emerging from the interests of its researchers, or should it try to satisfy the pull of the marketplace?

Technology push was the initial approach. Research and development received a boost from government tax incentives during and immediately after World War II. Much of this research was related to finding new materials for which uses could be created. Thus, the period between 1940 and 1965 was a time of great discovery. Nylon, other synthetic fibers,

Table 1.10 Legislation affecting chemical technology.

The Paris Agreement	
An agreement within the United Nations Framework Convention on Climate Change (UNCCC) signed in 2016. Deals with greenhouse gas -emissions mitigation, adaptation, and finance. The agreement's long-term goals are to keep the increase in global average temperature to well below 2 °C above the preindustrial level; and to limit the increase to 1.5 °C, since this would substantially reduce the risk and effects of climate change.	
Environmental Protection Agency (EPA)	
Clean Air Act (CAA)	41 pollutants were to be controlled by 1995 and 148 more by 2003. Cost to industry estimated at $25 billion per year.
Federal Insecticide, Fungicide, Rodenticide Act (FIFRA)	Cleanup of hazardous and nonhazardous waste sites. Cost to industry: $9–60 billion in the decade 1990–2000.
Pollution Prevention Act (PPA)	
Resource Conservation and Recovery Act (RCRA)	
Safe Drinking Water Act (SDWA)	Sets standards for 83 chemicals in water. Ensures high-quality water.
Clean Water Act (CWA)	
Toxic Substances Control Act (TSCA)	Requires premanufacture notification to the EPA of tests on and effects of new products.
Food Quality Protection Act (FQPA)	
Superfund Amendments and Reauthorization Act (SARA)	Cleanup of hazardous sites. Mostly funded by taxes in industry.
Comprehensive Environmental Response Compensation and Liability Act (CERCLA)	
Food and Drug Administration (FDA)	
Federal Food, Drug, and Cosmetics Act	Sets standards for and evaluates tests of food, drugs, and cosmetics.
Good Manufacturing Practice	Center for Drug Evaluation & Research imposes a standard.
Department of Labor	
Occupational Safety and Health Act (OSHA)	Defines hazards in an attempt to prevent industrial accidents. Defines permissible exposure limits for 600 hazardous chemicals.
Department of Transportation	
Hazardous Materials Transportation Act	
Department of Justice	
Chemical diversion and Trafficking Act (CDTA)	Prevents the use of chemicals to make illegal drugs.
Other departments	
Poison Packaging Prevention Act (CPSC)	Reporting of production, handling, and storage of hazardous materials.
The Emergency Planning and Community Right-to-Know Act	
State laws and regulations	

(*continued*)

Table 1.10 (Continued)

1. The Emergency Planning and Community Right-to-Know Act	Reporting of production, handling, and storage of hazardous materials.
2. Clean Air Act (CAA)	41 pollutants must be controlled by 1995; 148 more by 2003. Cost to industry estimated at $25 billion per year.
3. Toxic Substances Control Act	Requires pre-manufacture notification to EPA.
4. Resource Conservation and Recovery Act	Cleanup of hazardous and nonhazardous waste sites. Cost to industry $9–60 billion in the decade 1990–2000.
5. Superfund	Cleanup of hazardous sites. Mostly funded by taxes in industry.
6. Clean Water Act	Ensures high-quality water.
7. Safe Drinking Water Act	Sets standards for 83 chemicals in water.
8. Chemical Diversion and Trafficking Act	Prevents use of chemicals to make illegal drugs.
9. Occupational Safety and Health Act (OSHA)	Defines hazards in an attempt to prevent industrial accidents. Defines permissible exposure limits for 600 hazardous chemicals.

the polymers that provide plastics, elastomers, coatings, and adhesives are examples of this development. Chemicals were also discovered, with numerous applications that made the industry generally more efficient, such as corrosion inhibitors, electronics chemicals, and food antioxidants.

In the mid-1960s, however, the concept changed to *demand pull*. What problems are there in the marketplace that require technical solutions? Market research to answer such questions became a discipline, and for the past 50 years, the industry has talked of *market orientation*. Examples of technology push include television, sulfonamides, and lasers. Examples of demand pull include hard-water-compatible detergents, jumbo jets, and automobiles with low-exhaust emissions. A catalytic cracking catalyst that gives increased amounts of isobutene (Section 5.5.1) is an obvious example of the result of a market-oriented research project, with isobutylene being required for the production of methyl tert-butyl ether (MTBE) for unleaded gasoline. Both kinds of research should be part of any large company's game plan, although there has been a marked trend to deemphasize the "blue-skies" research that leads to truly novel discoveries. A major area for research in today's world is monitoring and reducing pollution. One-fifth of new capital expenditure in the 1990s was for pollution abatement and control; approximately the same amount of the R&D budget of a large company is likely to be spent on ecologically oriented projects. Thus, first-generation research was blue-skies research. It required little participation by management, and the researchers were generally regarded as a group of people who were difficult to communicate with. It was only when a project reached the development and marketing stages that management was required. DuPont, General Motors, and IBM are examples of companies that made discoveries and brought them to the marketplace. Second-generation research involved going to the marketplace to find out what was needed and applying accounting processes to the monitoring of R&D projects.

Table 1.11 R&D spending, 2009.

Company	Chemical R&D spending ($ million)	Percent of chemical sales
Syngenta	512	6.1
DuPont	1378	5.3
DSM	548	5.0
Shin-Etsu Chemical	359	3.7
BASF	1930	3.5
Bayer	1198	3.5
PPG Industries	403	3.5
Dow Chemical	1492	3.3
Mitsui Chemicals	407	3.2
Arkema	190	3.1
Evonik	418	3.0
Eastman Chemical	137	2.7
AkzoNobel	471	2.4
Solvay	194	2.4
Clariant	138	2.3
Tosoh	147	2.2
Air Liquide	304	2.0
Lanxess	141	2.0
Huntsman Corporation	145	1.9
Rhodia	102	1.8
Borealis	110	1.7
DIC	132	1.6
Air Products	116	1.5
Celanese	75	1.5
Praxair	74	0.8
Linde	92	0.7
LyondellBasell	145	0.7
Chevron-Phillips	38	0.5
Braskem	29	0.4
Yara	14	0.1

Source: *Chem. Eng. News*, 26 July 2010.

This required strong participation by the marketing branch, but still little participation by top management.

Today, a third generation of research managers recognize that research should be a part of the organization rather than apart from it. Research and development should figure in corporate objectives and take its direction from these objectives in exactly the same way as any other business function. It should thus help the organization to achieve its overall goal.

Indeed, it should help to set goals by managing technology as opposed to only inventing and applying it.

Thus the R&D department has to determinate which technologies may by developed internally, which may be obtained through licensing, and which may be obtained through strategic alliances. This is very much like a "make or buy" decision in manufacturing. It was obviously better for many companies to license BP's ammoxidation technology than to try to work out an acrylonitrile process on their own. Similarly, when Himont wanted to develop highly sophisticated catalysts for propylene polymerization, it joined forces with Mitsui. Both companies had strong backgrounds in catalyst development, and jointly they could bring these to bear on the objective. DuPont decided to develop a superior maleic acid process, since this was an important raw material for a new process the company invented to prepare Spandex, its most profitable product in the late 1980s and early 1990s. Having discovered a new and, as DuPont perceived it, superior route, the company joined forces with Monsanto for the development of the process, because Monsanto was the largest maleic anhydride producer and had many years of experience. (Monsanto subsequently withdrew, because it planned to shed the maleic anhydride business, which it did by selling the business to Huntsman.)

A sensible R&D strategy avoids duplication, but a large amount of duplication takes place in the world's research laboratories. The patent literature discloses 25 processes for the manufacture of 1,4-butanediol and similar number for propylene oxide. At last 15 companies have worked on the homologation of methanol to higher alcohols, a process that none has commercialized. There are many other examples.

Of the $37 billion spent on chemical and drug industry research in 2009, about 10% was for basic research, 35% for applied research, and 55% for development. There has always been an academic argument that holds that basic research is the province of the university. This may well be so, but in any laboratory there may be need for theory that has not been developed but is necessary for the solution of a problem. The pursuit of this theory, which in essence is basic research, is appropriate for the industry that requires it in order to fulfil its objectives. This leads to many industry-academic collaborations and to industrially funded research in universities.

All the same, very little effort is being expended on new products. One reason for this is that the maturity of the industry does not offer as many opportunities for them (Section 1.3.1). Also, the many regulations that govern the chemical industry, particularly the Toxic Substances Control Act (Section 1.3.7) and the REACH regulations, require extensive and expensive testing before a product can be test-marketed.

The average spend on R&D overall in manufacturing industries in the United States in 2007 was 3.7%, and the chemical industry averaged a healthy-looking 7.9%, but that figure is affected drastically by the high spend of pharmaceutical companies. In the 2000s, it cost approximately $500–1000 million to develop a new pharmaceutical. The pharmaceutical companies have traditionally been willing to make such expenditures because of the lure of "blockbuster" drugs.

1.3.9 Dislocations and Environmental Impacts

An important concept in today's chemical technology is the ever-present possibility of dislocations. *Dislocations* are defined as events over which a given company has no control, but which markedly affect the company's business. This is particularly important for planners who find their scenarios askew because of a dislocation. In planning, one cannot forecast what a dislocation might be. Indeed, if it could be forecasted, it would not be a dislocation. But what must be anticipated in planning is that there will be dislocations, either for good or for ill.

A few examples illustrate the point. The advent of unleaded gasoline made lead tetraalkyls obsolete in the United States. The major manufacturer of these compounds was Ethyl Corp, with a reputed profit of $90 million annually. This figure rapidly declined to $20 million and would have been even lower had it not been for export sales. Obviously, Ethyl was a victim of a dislocation. This motivated Ethyl to use its skills to expand into a variety of specialized and semicommodity businesses, which allowed the company to recoup its profits. Thus, it became a large supplier of the bulk pharmaceutical ibuprofen, for sale to packagers that converted it into a consumer item. The synthesis involved organometallic chemistry developed for the unrelated area of α-olefin production.

The same unleaded gasoline dislocation proved to be a windfall for ARCO. ARCO's two-for-one process for the manufacture of propylene oxide and tert-butanol made available to the company large quantities of the latter, at the time the plant went on stream. However, there was only a very little use for it. The need for octane improvers in unleaded gasoline soon provided a market. Dehydration of *tert*-butanol and reaction with methanol gave MTBE, and in the early 1990s ARCO was the world's largest supplier of it. Between 1977 and 1993, production rose from virtually nil to about 13 billion pounds per year and reached 20 billion pounds by the end of the 1990s. Capital expenditure was conservatively estimated at $50 billion. There was then another dislocation. Although MTBE functions well as an octane improver, it is water soluble. Inspection and repair of hundreds of thousands of gasoline storage tanks was judged impracticable, and instead ethanol was proposed as a replacement for MTBE.

A third example also relates to unleaded gasoline. At least one petroleum company announced that it would achieve the desired octane number by removing lead and increasing the aromatics content of its gasoline. A few years later, the CAA specified that the aromatics content of gasoline must be decreased from about 35 to 25%. Thus, the CAA provided a second dislocation that negated that company's reaction to the earlier dislocation, provided by unleaded gasoline.

A fourth example: Phillips Petroleum Company never used the metathesis reaction (Section 8.3) to convert propylene into more expensive ethylene and 2-butene, which, in turn, may be dehydrogenated to butadiene. One might assume that it did not opt to carry out this interesting chemistry because of the widely held belief in the 1970s that within a 15-year period, declining US gas supplies would make naphtha and gas oil the major steam cracking feeds. Accordingly, large quantities of butadiene would become available. The United States has always imported butadiene from Europe because insufficient quantities were produced by the cracking of gas. But naphtha and gas oil never became major feedstocks in the United States because of another dislocation provided by Saudi Arabia's entry into the chemical business. Saudi Arabia decided to use only ethane in its associated gas, making large quantities of liquefied petroleum gas (LPG) (a propane/butane mixture) available at low world prices. The United States now uses LPG and has not found it necessary to switch to liquid feeds. And so, the United States still imports butadiene, a situation that metathesis might have helped to avoid.

Dislocations frequently result from advances in technology. The producers of propylene oxide by the chlorohydrin route suffered a serious dislocation when Arco announced its new process via tert-butyl hydroperoxide. Every manufacturer except one went out of business. Similarly, Monsanto's acetic acid process using methanol and CO closed down every producer of acetic acid that used acetaldehyde as starting material.

The answer to dislocations is the concept of *robustness*. A robust process is one that can accommodate a variety of dislocations. For example, some companies, uncertain of their feedstock supply, built steam crackers that could operate on gaseous or liquid feedstocks. Their plants cost more than a single feedstock would have, but were sufficiently robust to withstand dislocations in feedstock supply. The fact that the petrochemical business in the United States in 1993 remained relatively profitable was due largely to such flexible crackers. Western Europe suffered because only 5 of its 52 crackers were flexible.; Petrochemical investments ($72–185 billion) and project annoucements ($97–317 billion) in the United States are on the rise (cumulative, 2013–2017).

Finally, the chemical business is extremely dynamic. It is affected not only by what it does itself, such as creating new technology, but also by what others do around it. Modern managers keep abreast as much as possible with what the rest of the world is doing that might affect their business.

1.3.10 Feedstock Recycling

Today, plastics are very important materials and have widespread use in the manufacture of a variety of products including packaging, textiles, floor coverings, pipes, foams, and furniture components. Plastics are synthesized mainly from petroleum-derived chemicals. Engineering plastics, particularly thermosets, are also used in composite materials. Their excellent technological properties make them suitable for applications in cars, ships, aircraft, telecommunications equipment, etc. In recent years, new and important areas of application for plastics have emerged in medicine (fabrication of artificial organs, orthopedic implants, and devices for the controlled release of drugs), electronics (development of conductive polymers for semiconductor circuits, conductive paints, and electronic shielding), and computer technology (use of polymers with nonlinear optical properties for optical data storage).

Today, plastic materials are used in almost all areas of daily life. Accordingly, the production and transformation of plastics are major worldwide industries. The production of plastic has grown exponentially in just a few decades-from 1.5 milion tons in 1950 to 322 million tons in 2015 worldwide – and with it, the amount of plastic waste. The environmental impact is a matter of great public concern. The variation in properties and chemical composition between different types of plastic materials hinders the application of an integrated and general approach to handling plastic waste. The light weight of plastic goods and the fact that plastic waste is mainly found in municipal solid waste (MSW) mixed with other classes of residues are factors that greatly limit their recycling. As a consequence, the primary destinations of plastic waste are landfill sites, where they remain for decades due to their slow degradation. In Europe, energy recovery is the most-used way to dispose of plastic waste, followed by landfills. Some 30% of all the generated plastic waste is collected for recycling, and recycling rates by country vary a great deal.

At present, there are three main alternatives for the management of plastic waste in addition to landfilling: (i) mechanical recycling by melting and regranulation of the used plastics, (ii) feedstock recycling, and (iii) energy recovery. Mechanical recycling is limited by both the low purity of polymeric waste and the limited market for recycled products. Recycled polymers only have commercial applications when the plastic waste has been subjected to a previous separation by resin; recycled mixed plastics can only be used in undemanding applications. On the other hand, energy recovery by incineration, although an efficient alternative for the removal of solid wastes, is the subject of great public concern due to the contribution of combustion gases to atmospheric pollution. There has also been some controversy in the past about the possible relationship between dioxin formation and the presence of Cl-containing plastics in the waste stream.

Consequently, feedstock recycling appears as a potentially interesting approach, based on the conversion of plastic wastes into valuable chemicals useful as fuels or as raw materials for the chemical industry. The cleavage and degradation of the polymer chains may be promoted by temperature, chemical agents, catalysts, etc.

1.4 The Top Chemical Companies

Table 1.12 provides data about the 50 leading chemical companies in the world in 2018. For the global chemical industry, 2018 was another strong year, yet it showed signs of slowdown. Excluding PetroChina, which reported chemical sales only for 2018, chemical companies in the top 50 combined for $926.8 billion in chemical revenue, an increase of 13.4% from the same companies' sales the year before.

Leading the global Top 50 is a new company: DowDuPont, the result the 2017 merger between Dow Chemical and DuPont. It ends BASF's reign as the largest chemical company in the world, but only temporarily. DowDuPont has already split into three separate firms: Dow, DuPont, and Corteva Agriscience. Dow and DuPont will certainly appear in next year's rankings.

The global top 50 lost two other firms to deal-making this year. Linde AG and Praxair merged into a new firm called Linde PLC. Akzo Nobel sold its chemical business, Nouryon, to the private equity firm the Carlyle and the Singapore sovereign wealth fund GIC, and Nouryon alone didn't make the cut.

Five new firms did make the cut. The largest is PetroChina, which is breaking out chemical sales for the first time. Nutrien is the result of the merger between Potash Corporation of Saskatchewan and Agrium. Platinum-group catalyst suppliers Umicore and Johnson Matthey are on the list. US chemical maker Celanese rounds out the list at number 50, where it debuts.

Table 1.12 Sales and profitability of major chemical companies, worldwide, 2018.

Rank			Chemical	Change	
2018	2017	Company	sales ($ millions)	from 2017	Headquarters
1	2	DowDuPont	$85.977	37.6%	US
2	1	BASF	74.066	2.4	Germany
3	3	Sinopec	69.210	22.4	China
4	4	Sabic	42.120	12.0	Saudi Arabia
5	5	Ineos	36.970	2.1	UK
6	6	Formosa Plastics	36.891	13.8	Taiwan
7	7	ExxonMobil Chemical	32.443	13.1	US
8	8	LyondellBasell Industries	30.783	8.7	Netherlands
9	9	Mitsubishi Chemical	28.747	7.1	Japan
10	11	LG Chem	25.637	9.7	South Korea
11	12	Reliance Industries	25.167	37.3	India
12	–	PetroChina	24.849	n/a	China
13	10	Air Liquide	24.322	2.8	France
14	14	Toray Industries	18.651	8.7	Japan
15	15	Evonik Industries	17.755	4.2	Germany
16	16	Covestro	17.273	3.4	Germany
17	27	Bayer	16.859	49.0	Germany
18	18	Sumito Chemical	16.081	8.7	Japan
19	20	Braskem	15.885	17.7	Brazil
20	19	Lotte Chemical	15.051	4.2	South Korea
21	24	Linde plc	14.900	30.3	UK
22	21	Shin-Etsu Chemical	14.439	10.6	Japan
23	23	Mitsui Chemicals	13.432	11.6	Japan
24	22	Solvay	13.353	3.7	Belgium
25	26	Yara	12.928	13.8	Norway
26	33	Chevron Phillips Chemical	11.310	24.8	US
27	28	DSM	10.951	7.4	Netherlands
28	35	Indorama	10.747	21.2	Thailand
29	29	Asahi Kasei	10.654	8.1	Japan
30	30	Arkema	10.418	5.9	France
31	32	Syngenta	10.413	12.6	Switzerland
32	31	Eastman Chemical	10.151	6.3	US
33	34	Borealis	9.852	10.2	Austria
34	36	SK Innovation	9.719	14.4	South Korea
35	45	Mosaic	9.587	29.4	US

(continued)

Table 1.12 (Continued)

Rank		Company	Chemical sales ($ millions)	Change from 2017	Headquarters
2018	2017				
36	37	Huntsman	9.379	12.2	US
37	41	Wanhua Chemical	9.172	14.1	China
38	43	PTT Global Chemical	8.969	15.7	Thailand
39	39	Ecolab	8.964	11.0	US
40	38	Air Products & Chemicals	8.930	9.1	US
41	40	Westlake Chemical	8.635	7.4	US
42	25	Lanxess	8.505	−25.5	Germany
43	–	Nutrien	8.130	75.9	Canada
44	–	Umicore	8.113	27.1	Belgium
45	42	Sasol	8.110	4.2	South Africa
46	44	Tosoh	7.803	4.7	Japan
47	–	Johnson Matthey	7.579	16.1	UK
48	46	DIC	7.296	2.0	Japan
49	47	Hanwa Chemical	7.273	3.3	South Korea
50	–	Celanese	7.155	16.5	US

1.5 The Top Chemicals

Table 1.13 lists the 35 most important chemicals by volume manufactured in the United States in 2008. The rank order would be more or less the same in any developed country.

Sulfuric acid heads the list by large margin as befits its position as an economic indicator, despite its maturity suggesting slower growth. Though it has many applications, about 45% is used for phosphate and ammonium sulfate fertilizers. Of the first 15 chemicals, six are associated with the fertilizer industry – sulfuric acid, nitrogen, ammonia, phosphoric acid, nitric acid, and ammonium nitrate. Oxygen is used by the steel industry and for welding. Most of these chemicals are also used to make organic chemicals, but their main markets lie elsewhere. Chlorine has a number of uses including bleaching of paper, as a disinfectant, and as a component of organic compounds, most important of which is vinyl chloride, whose precursor is ethylene dichloride. Many chlorine compounds, however, are now considered ecologically undesirable, as is the use of chlorine for bleaching paper and disinfecting swimming pools. Its use between 1990 and 1991 decreased by half million and its growth rate presently is only around 0.5% per year.

The two most important organic chemicals, ethylene and propylene, occupy positions 3 and 6. Benzene, the third most important building block, occupies position number 16. The majority of the remaining chemicals are organic, and these form the backbone of the so-called heavy organic chemical industry. *Heavy organics* are defined as large-volume commodity chemicals such as ethylene and propylene, as opposed to specialty chemicals such

Table 1.13 Top inorganic and organic chemicals and fertilizers, United States, 2008.

Rank	Product description	2004 (thousands of metric tons)	2008 (thousands of metric tons)
1	Sulfuric acid, gross (100%)	35 954	32 381
2	Nitrogen	30 543	NA
3	Ethylene	25 682	22 554
4	Oxygen	25 568	NA
5	Lime	20 104	NA
6	Propylene	15 345	14 783
7	Chlorine gas	12 166	10 669
8	Ammonia, synthetic anhydrous	10 762	9571
9	Phosphoric acid (100% P_2O_5)	11 463	9216
10	Sodium carbonate	10 247	NA
11	Ethylene dichloride	12 163	8973
12	Sodium hydroxide, total liquid	9508	8111
13	Vinyl chloride	8596	7782
14	Nitric acid (100%)	6704	7245
15	Ammonium nitrate, original solution	6021	7114
16	Benzene	7675	5588
17	Ethylbenzene	5779	4104
18	Styrene	5394	4100
19	Superphosphates and other fertilizers (100% P_2O_5)	8073	5539
20	Urea (100%)	5755	5241
21	Hydrochloric acid	5012	3902
22	Cumene	3736	3386
23	Ethylene oxide	3867	2903
24	Ammonium sulfate	2643	2514
25	Phenol	2200	1990
26	1.3-Butadiene	2204	1633
27	Vinyl acetate	1431	1267
28	Sodium silicates	1131	1106
29	Acrylonitrile	1551	1018
30	Aniline	813	1009
31	Aluminum sulfate (commercial)	922	NA
32	Finished sodium bicarbonate	536	682
33	Sodium chlorate	658	607
34	Potassium hydroxide liquid	539	582
35	Hydrogen peroxide	1083	437

as dyes and pharmaceuticals. Some of the chemicals have only one, very large use. For example, the major use for ethylene dichloride (no. 11) is to make vinyl chloride. The major use for ethylbenzene (no. 17) is to make styrene. Cumene (no. 22) is converted to phenol and acetone. Many of the top chemicals are monomers for polymers. Comparison of the data with the 2004 data shows the maturity of the chemical industry. Indeed, if one looks at the year 2009, the decline in the US industry is much greater; but there is hope that 2009 was a one-off. A noteworthy change is a decrease in the propylene ratio.

Further Reading

Bajus, M. (1997). *Pet. Coal* 39 (2): 8–14.

Bajus, M. (2002). *Pet. Coal* 44 (3–4): 112–119.

Bajus, M. (2012). *Microchannel-Technol.* 54 (3): 294–300. http://pubs.acs.org/cen/coverstory/83/pdf/8328production.pdf.

Bajus, M. (2014). Shale gas and tight oil, unconventional fossil fuels. *Pet. Coal* 56 (3): 206–221.

Bajus, M., Veselý, V., Leclercq, P.A., and Rijks, J.A. (1979). *Ind. Eng. Chem. Prod. Res. Dev.* 18: 30.

Bajus, M., Veselý, V., Leclercq, P.A., and Rijks, J.A. (1979). *Ind. Eng. Chem. Prod. Res. Dev.* 18: 135.

Bajus, M., Veselý, V., Leclercq, P.A., and Rijks, J.A. (1980). *Ind. Eng. Chem. Prod. Res. Dev.* 19: 556.

Bajus, M., Veselý, V., Baxa, J. et al. (1981). *Ind. Eng. Chem. Prod. Res. Dev.* 20: 741.

Bajus, M., Veselý, V., Baxa, J. et al. (1983). *Ind. Eng. Chem. Prod. Res. Dev.* 22 (336).

Kershenbaum, L.L. and Martin, J.J. (1967). *AICHEI* 13: 580.

Kossiakoff, A. and Rice, F.O. (1943). *J. Am. Chem. Soc.* 65: 580.

Lederer, J. (2013). Restrictive possibilities of ecologic effects at the production and using of fuels and petrochemicals. In: Book of Abstracts. Scientific Seminary, Slovak University of Technology in Bratislava, 143.

Rice, F.O. (1933). *J. Am. Chem. Soc.* 55: 3035.

Tulio, A.H. (2017). Global top chemical companies. *Chem. Eng. News* 93 (30): 14–26.

World Petrochemical Conference (1998). Thirteenth Annual CMAI, Houston, Texas.

Zamostný, P., Karaba, A., and Bělohlav, Z. (2013). Mathematic models of hydrocarbon pyrolysis – study and comparison. In: Book of Abstracts. Scientific Seminary, Slovak University of Technology in Bratislava, 145.

2

Current Trends in Green Hydrocarbon Technology

CHAPTER MENU

2.1 Introduction

As we have moved into the twenty-first century, a number of important changes are taking place – in particular, hydrocarbon technologies in the oil industry tend to be evolutionary rather than revolutionary. Most research and development is oriented toward evolutionary improvement. The technologies of the oil industry tend to be more international. The technological development of the oil industry really drives economics, the environment, and safety in hydrocarbon technology. Hydrogen will become more of a focal point and a basic necessity in the industry of tomorrow. The future should also bring about a gradual change in the current competition between oil and natural gas as general-purpose fuels. Selective and efficient catalysts will be critical for this new trend. The importance of natural gas will rise as it becomes the fuel of choice for power generation and the feedstock for petrochemistry. Liquefied petroleum gas (LPG) is both environmentally and ecologically friendly, and it is considered to be one of the world's most important fuels. Gas to liquid (GTL) technologies, the route to liquid and higher-value products from natural gas, are obtained by conversion via synthesis gas.

Petrochemistry: Petrochemical Processing, Hydrocarbon Technology, and Green Engineering, First Edition. Martin Bajus.

2.2 Eco-Friendly Catalysts

Catalysts can play a significant role in the production of higher-quality fuel, chemicals as required by standards, which are going to be introduced step-by-step all over the world due to growing consciousness of the damage to human health and the environment caused by existing products. Fuel reformulation has been seeded by growing awareness of the damage mankind has caused to the ecosystem and itself. Fuel reformulation means that fuels are defined based on their chemical composition according to engine-technology-related standards rather than purely based on performance. These standards, which are becoming more and more stringent, can be met in different ways, mainly by using catalysts and process operating conditions.

Huge improvements toward the development of environmentally friendly processes have been achieved in the alkylation of aromatics with alkenes during the last four or five decades (Figure 2.1). In particular, many efforts have been devoted to the research of solid catalysts adequate to substitute for the mineral or Lewis acids and free bases traditionally employed as catalysts in acid or base catalyzed alkylations. Various solid catalysts based on different zeolites have been developed for the production of ethylbenzene and cumene up to the industrial scale. Solid acid catalysts operate in a mixed phase; the zeolitic catalyst is packaged into specially engineered bales to perform the chemical reaction process. Similar technologies and catalysts are offered for cumene processes by UOP (the UOP Q-Max process) and Mobil/Badger. The Versalis cumene process, formerly by EniChem, is based on a proprietary beta zeolite catalyst (PBE-1).

An alkylate defines a mixture of C_8 isomers, mainly trimethylpentane, with minor amounts of higher-order components obtained via the acid-catalyzed addition of isobutane to linear butenes. Today, two families of alkylation technologies are available: H_2SO_4 based and HF-based, proposed by UOP and Phillips. However, both of them suffer from severe environmental impacts.

Therefore, the potential for reformulated gasoline faces severe constraints in the production cycle, which, in turn, generates a strong impulse to search for solid alkylation catalysts. Many materials have been tested, starting with zeolites in the late 1960s. However, the catalysts that have reached the stage of pilot units are not based on zeolites. Instead,

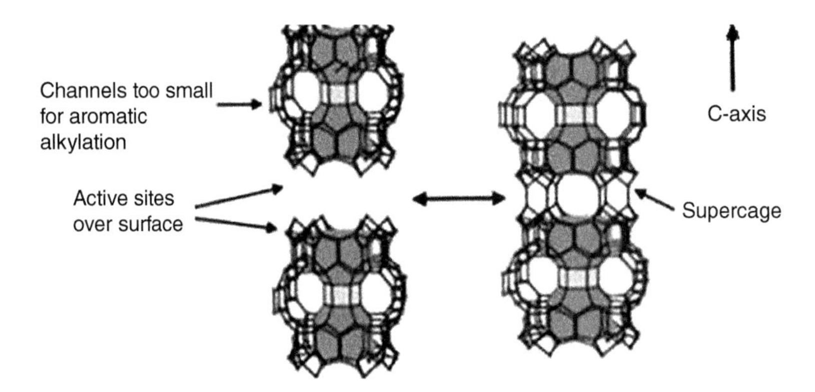

Figure 2.1 Alkylation zeolitic catalyst.

they are composites of mixed oxides. UOP (alkylene process) proposed $AlCl_3/Al_2O_3$ with a small amount of noble metal (Pt); Catalytica/Neste/ConocoPhillips selected a boron trifluoride on Al_2O_3, while CDTECH/Chevron developed SbF_5/acid-washed silica. All of them are supported, strong Lewis acids. Haldor Topsoe proposed the most innovative approach: supported or anchored trifluoromethansulfonic acid (trifluoric acid F_3CSO_3H) on silica.

As in other countries, cleaner or more environmentally friendly production of materials and chemical energies is becoming more and more important. Examples are the changes to solid acid from liquid acid catalysts, to catalytic oxidation from stoichiometric chemical oxidation, etc. Catalytic processes that use less-expensive and more abundant raw materials in a smaller number of reaction steps are always highly desired. Recently, this has been stressed particularly in the synthesis of fine chemicals, although the search for new functional compounds is also of great concern in this field.

After the recent development of two processes using heteropolyacids, Asahi Kasei has industrialized the production of cyclohexanol from benzene via partial hydrogenation and subsequent hydration. The two steps are catalyzed by ruthenium and a zeolite, respectively. The new process is superior to the conventional two-step oxidation process in several respects: selectivity, safety, corrosion, and energy consumption.

In the gas-phase Beckmann rearrangement of cyclohexanone oxime and synthesis of pyridine derivatives, there have been many attempts to replace the conventional process – which utilizes fuming sulfuric acid – with a gas-phase solid acid process. This long-term wish is about to be fulfilled by using silicalite-type catalyst that is almost non-acidic in the presence of methanol in the feed.

Asahi Kasei has developed a technology for highly selective partial hydrogenation of benzene to cyclohexene and succeeded in the commercialization of a new production process for producing cyclohexanol from benzene through cyclohexene (Figure 2.2). This process has been considered difficult for a long time, and several innovative technologies made the partial hydrogenation reaction possible. A catalyst that consists of specific metallic Ru particles exhibits excellent high selectivity of cyclohexene. The catalyst was obtained by reducing a Ru compound that contains a zinc compound; the use of a zinc or a strong acid as a co-catalyst exhibited a remarkable effect to enhance selectivity. Dispersing agents of metal oxides were found to extend the life of the metallic Ru catalyst, and some dispersing agents had the effect of enhancing selectivity in the partial hydrogenation reaction. One of the remarkable features of the reaction was the reaction field, which comprises four phases: vapor (hydrogen), oil, aqueous, and solid (Ru catalyst). The catalyst was used in the aqueous phase, and the reactants (benzene and hydrogen) were dissolved in the aqueous phase, where the reaction proceeded. Therefore, the products and reactants transferred between four phases through dissolution, diffusion, and extraction. Rendering quick transfer was a very important factor in enhancing reaction selectivity. The catalyst system and the reaction field described made the selectivity for cyclohexene very high: a yield of 60% for cyclohexene has been obtained.

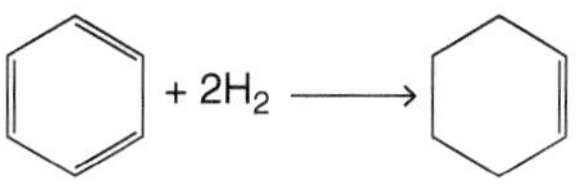

Figure 2.2 Partial hydrogenation of benzene to cyclohexene on a Ru-Zn/*m*-ZrO_2 nanocomposite catalyst.

$+ 2H_2 \longrightarrow$

Other promising routes in acid catalysis can be mentioned, such as an MCM-41/Nafion composite prepared from a sol-gel preparation of nanostructured silica in the presence of cethyltrimethylammonium and Nafion gel in monophasic conditions. This novel material was applied in the dimerization of α-methylstyrene to selectively produce the corresponding acyclic dimer.

When these modified materials were used to catalyze the condensation reaction between phenol and acetone, the highest selectivity in *p,p′*-bis phenol A isomer (I) versus *o,p′*-bis phenol A isomer (II) was obtained, because the reaction occurs inside the pores.

2.3 Hydrogen

Historically, refinery hydrogen consumption has increased as refiners increase the degree of conversion and process heavier and sourer crudes. However, the use of hydrogen is accelerating due to recently enacted environmental regulations that require refiners to produce cleaner-burning transportation fuels.

As refineries are reconfigured to produce clean fuels, there will be an increase in hydrogen demand, and refiners will be required to find new sources of hydrogen. Off-gas sources within refineries may provide a portion of these requirements, but more hydrogen will be needed than can be recovered economically. Expansion of existing steam reformers must be considered during any evaluation because of the potential to obtain hydrogen at an attractive incremental cost. The ultimate solution, however, may be new capacity using traditional steam-reforming technology; or, in specific situations, partial oxidation (POX) technology, which is also well-proven in plant operations. When new hydrogen demand is combined with bottom-of-the-barrel destruction, POX technology may have some very compelling synergies but must still provide hydrogen product at a competitive cost to be considered viable. The term *dry or CO_2 reforming* is short for CO_2 reforming of natural gas. In contrast to steam reforming of natural gas, the addition of CO_2 permits optimization of the synthesis gas composition for methanol production.

Development and utilization of more efficient energy-conversion devices are necessary for sustainable and environmentally friendly development in the twenty-first century. Fuel cells are fundamentally much more energy-efficient than internal combustion engines and can achieve as high as 70–80% system efficiency in integrated units including heat utilization, because fuel cells are not limited by the maximum efficiency of heat engines or IC engines dictated by the Carnot cycle.

Hydrogen would be an ideal fuel for fuel cells; but due to the lack of infrastructure for distribution and storage, processing of fuels is necessary for producing H_2 on-site for stationary applications or on-board for mobile applications. Hydrocarbons and alcohols can both be used as fuels for reforming on-site or on-board. Hydrocarbon fuels have the advantage of existing infrastructure of production and distribution, while alcohol fuels can be reformed at substantially lower temperatures. Further research and development are necessary in fuel processing for improved energy efficiency, size reduction of on-board hydrogen storage for energy (Section 13.4.1), and on-electrode catalysis related to fuel processing, such as tolerance to CO and sulfur components in reformate.

2.4 Alternative Feedstocks

Since the discovery and exploitation of big oil fields, the world has been supplied with cheap crude-oil-based fuels and chemicals. Crude oil is a fossil fuel and, hence, is limited by definition. Crude oil reserves will eventually be depleted, or their exploitation will become too costly. This opens the door for the use of alternative and cost-competitive feedstocks for the production of olefins. Since all feedstocks are limited, it is necessary to consider their best use and aim for the highest efficiencies. In addition, our common challenge to cut carbon emissions is an equally important criterion. CO_2 emissions lead to global warming, and more effort must be made to reduce emissions and mitigate the adverse effects of climate change. In this respect, biomass and waste streams are believed to be important for future use in the production of chemicals.

Biomass is a carbon-rich material that is mainly generated by photosynthesis and is widely available at relatively low cost. In contrast to fossil feedstocks, biomass can be considered virtually inexhaustible since it can be supplemented within a reasonable timescale. During photosynthesis, CO_2 is captured from the atmosphere; hence, if collection and processing are done in a sustainable manner, the use of this CO_2 will essentially not contribute to the increasing amounts of CO_2 in the atmosphere. The need for cleaner air and water and for environmental protection will eventually lead to the exploration of sustainable production possibilities, accelerated and activated by political decisions.

Furthermore, waste streams such as plastic solid waste and municipal waste need to be better managed in order to protect scarce environmental resources and prevent pollution. The amount of waste will grow as a result of the increasing global population and higher living standards; hence, waste is a potential feedstock to be used. Energy recovery for the production of heat and energy is the current focus, although chemical recycling is gaining momentum. These feedstocks are of a great benefit to the environment. Nevertheless, the enormous scale of current olefin-production facilities and the availability of biomass resources suggest that in the near future, renewables will be used to complement rather than to completely replace fossil resources.

Furthermore, the commercialization of technologies that convert these streams is limited by their poor economic viability. In addition, a large number of projects that primarily use classical steam cracking are going online in the coming decade and will probably put a halt to the importance of biomass or waste streams for the production of olefins. Therefore, inexpensive fossil feedstocks such as coal, natural gas, and shale condensates remain important sources of energy and chemicals and will remain dominant in the petrochemical industry. Thanks to technological advancements, shale gas has become a promising resource, causing its exploitation to increase exponentially since 2008. Also, cheap methane from stranded gas reserves will play an important role in the future. From an economic point of view, methane is a promising starting material for the production of chemicals.

Alternative fuels are often classified as fuels to replace conventional gasoline and diesel fuels, or fuels to extend conventional fuels with environmentally favorable reformulations. The following are possibilities:

- Replacements for gasoline and/or diesel fuel:
  - Methanol
  - Ethanol
  - Compressed natural gas
  - Liquefied petroleum gas
  - Electricity
- Gasoline and diesel fuel reformulation:
  - Varied gasoline volatility, T90, aromatic and olefinic hydrocarbon, sulfur, and oxygen content
  - Varied diesel sulfur and aromatic hydrocarbon content, natural cetane number, and cetane improver

On the other hand, direct SC of crude oil is also gaining importance. Using this process, petrochemical producers can skip the refining step and thus reduce their production cost. In 2014, ExxonMobil commissioned a world-scale facility in Singapore – the first of its kind – that produces 1×10^6 t/a of ethylene directly from crude oil. An announcement was also made regarding a joint venture by SABIC and Saudi Aramco. The use of coal as feedstock for chemicals or fuels is discouraged from an environmental point of view; the conversion of coal typically has very low carbon efficiencies and hence results in huge emissions of carbon dioxide (CO_2).

Researchers and engineers are expending considerable efforts to explore and optimize alternative production possibilities in an attempt to increase both efficiency and profitability. Nevertheless, more effort is absolutely necessary in order to take full advantage of sustainable or alternative feedstocks and technologies. Reliable fundamental multiscale models need to be developed and enhanced in order to explore unprecedented levels of efficiency, increase process maturity, and address major bottlenecks. In this respect, fast implementation with lower risk can be made possible. In addition to obtaining a fundamental understanding of the new processes, the end goals of these efforts are to face the fact that environmental resources are limited and to secure future needs in terms of energy and chemicals.

The use of fossil feedstocks as a fuel or for the production of chemicals will remain dominant in the near future, but this chapter focuses on promising olefin-production technologies that may challenge the current leading technology. Each of these alternatives benefits from the abundance of propane, ethane, and methane available from shale gas and stranded gas. Furthermore, the relevance of each pathway could be enlarged, as the shift toward lighter feedstock utilization in the steam cracking of hydrocarbons results in the decreased production of important co-products. Pathways that adopt renewables or waste streams will not be addressed here because they are believed to have rather low significance for the total quantity of olefins produced in the near future. The following technologies are of interest: catalytic dehydrogenation of light alkanes, oxidative coupling of methane (OCM), and syngas-based routes such as the Fischer–Tropsch synthesis (FTS) and methanol synthesis followed by methanol to olefins (MTO).

2.5 Alternative Technologies

The huge and continuously growing reserves of natural gas have stimulated its exploitation in terms of liquid fuels because of its intrinsic characteristic: high energy content, absence of heteroatoms, high hydrogen-to-carbon ratio in the raw material, geographical availability, and growing cost of reinjection and flaring (when dealing with oil-associated gas).

The abundance of cheap propane, ethane, and methane from shale gas and stranded gas will facilitate cost-competitive paths in the production of light olefins (Figure 2.3). In particular, the on-purpose production of propylene has grown as more and more steam crackers have shifted from naphtha feed to lighter shale condensates. This is especially true in the United States, where shale gas exploitation has grown exponentially, amplifying the issue of supply due to the strong growth in propylene demand compared with that of ethylene. Steam-cracker units cannot fill this gap due to the low propylene/ethylene ratio. In this respect, other production routes could be profiled as interesting alternatives to overcome this issue. Furthermore, technologies emerging from the possibility of valorizing methane into higher hydrocarbons or chemicals are promising, especially due to low methane prices and huge methane availability. However, high capital costs, low efficiencies, and low reliability of complex process sequences make commercialization very challenging and risky. In addition, some of these promising technologies still have room for improvement: to be industrially relevant, they should be economically comparable to current steam-cracker units and production capacities.

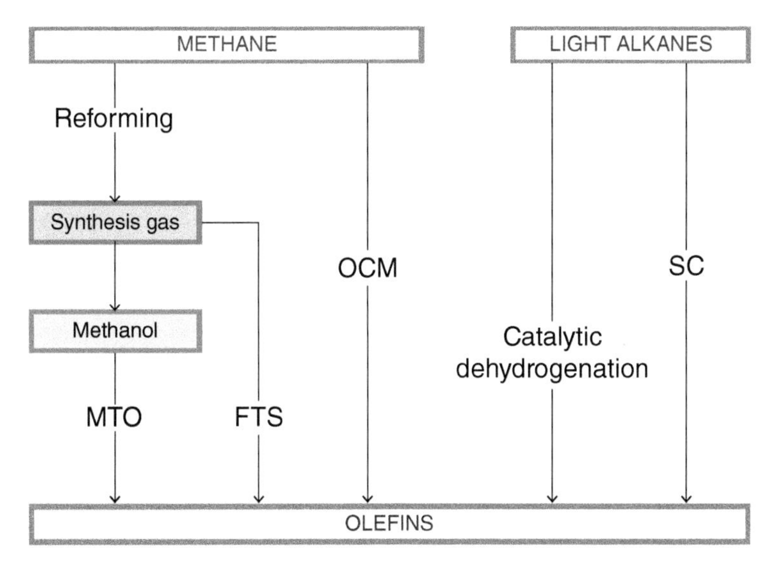

Figure 2.3 Different technologies of interest for the production of light alkenes from methane and light alkanes.

GTL technology has unfortunately become synonymous with Fischer–Tropsch technology. It is up to all of us in the gas conversion community to continue to point out the many business options and the rich chemistry in the conversion of natural gas to liquid fuels, fuel additives, and chemical feedstocks. Today, the route to liquid (or even solid) higher-value products from gas is through conversion via synthesis gas (syngas). Two basic types of liquid products can be manufactured: hydrocarbons (via FTS) and oxygenates such as methanol and dimethyl ether (DME). GTL is nowadays targeted at diesel because of the linearity of chains in the product mixture, which leads to a good cetane number.

The importance of these GTL technologies is the raw material: a liquid fuel is derived from natural gas instead of oil. This opens opportunities for natural resource exploitation, which will influence our future. Conversion of olefin to diesel (COD process) is in some ways bridging the oligomerization with the GTL technologies, aiming to produce high-quality fuel from natural gas; or, in more imaginative prospects, opening the door to a methane-based refinery.

2.6 Feedstock Recycling

The production of environmentally friendly materials and chemical energies (clean fuels) as well as environmentally friendly disposal and recycling systems are the main targets of current research. Although economically acceptable recycling is now very limited in quantity, the importance of recycling or waste treatment will grow rapidly. This particularly applies to polymer and plastics manufacturers. Quality improvements are also needed for transportation fuels, to meet stringent new regulations. This will necessitate significant alternations in oil refinery processes.

The severe limitations on the mechanical recycling of plastic wastes highlight the interest in and potential for feedstock recycling, also called *chemical* or *tertiary recycling*. It is based on the decomposition of polymers by means of heat, chemical agents, and catalysts to yield a variety of products ranging from starting monomers to mixtures of compounds (mainly hydrocarbons) with possible applications as a source of chemicals or fuels. The products derived from plastic decomposition exhibit properties and quality similar to those of their counterparts prepared by conventional methods.

A wide variety of procedures and treatments have been investigated for feedstock recycling of plastic and rubber wastes. These methods have been classified into the following categories:

- Chemical depolymerization by reaction with certain agents to yield starting monomers
- Gasification with oxygen and/or steam to produce syngas
- Thermal decomposition of polymers by heating in an inert atmosphere
- Catalytic cracking and reforming, during which polymer chains are broken down by the effect of a catalyst that promotes cleavage reactions
- Hydrogenation, during which polymers are degraded by combined actions of heat, hydrogen, and catalysts

2.7 Functionalization of Hydrocarbons

Early research on the functionalization of methane yielded only marginal results. On the basis of these efforts, it was easy to predict that chemistry would never be discovered to make methane the basic building block for chemical technology. However, the 1980s saw major advances in catalysis. Methane functionalization attracted intense research in the 1980s, which has accelerated to the present day. Data are accumulating so rapidly that we can do little more here than provide some insights into the approach. Three reactions achieve the goals: direct oxidation of methane to methanol and/or formaldehyde (see Section 12.2.4); the dimerization of methane to ethane, ethylene, or higher hydrocarbons; and the aromatization of methane (studied intensively today).

2.7.1 Partial Oxidation of Methane

The direct conversion of methane to methanol has been the subject of many academic research efforts during the past decade:

$$CH_4 + ½O_2 \rightleftarrows CH_3OH;\ \Delta H^{\circ}_{298} = -126\ \text{kJ/mol}$$

Such a process would be a spectacular event. In principle, efficiency would be increased enormously, compared to the conventional process of steam reforming followed by methanol synthesis. Furthermore, a real contribution to the reduction of the greenhouse effect would be realized. This is linked to the formation of CO_2, a major greenhouse gas, from methane in reformer feed and fuel: during steam reforming of hydrocarbons, CO_2 is formed by the water-gas shift reaction in the reformer. Moreover, combustion of fuel for heating the reformer furnace yields a large amount of CO_2. Both sources of CO_2 would be eliminated in a direct conversion process. Unfortunately, to date, the low yields achieved are a major obstacle to the commercialization of this route.

The primary alkane selective oxidation processes that exist or are under development are: (i) methane to formaldehyde or ethene, (ii) methane to vinyl chloride in the presence of HCl, (iii) ethane to 1,2-dichloroethane in the presence of HCl, (iv) ethane to acetic acid, (v) propane to acrylic acid, (vi) propane to acrylonitrile in the presence of ammonia, (vii) *n*-butane to maleic anhydride, (viii) isobutane to methacrylic acid, (ix) *n*-pentane to phthalic anhydride, (x) isobutane to *tert*-butyl alcohol, and (xi) cyclohexane to cyclohexanone and cyclohexanol. Some of these alkanes, e.g. *n*-butane and *n*-pentane, will be more available in the near future as environmental regulations impose increasingly stringent limits on the light-alkane content of gasoline. In addition, increasing demand for chemicals such as methyl *tert*-butyl ether (MTBE) favors the development of new synthetic routes. For example, the conversion of *n*-butane to isobutane and the conversion of isobutane to *tert*-butyl alcohol by directmonooxygenation might be of considerable industrial interest (*tert*-butyl alcohol is an intermediate in MTBE synthesis).

Out of the processes listed here, only the oxidation of *n*-butane to maleic anhydride has been commercialized. Butane is a particularly attractive material for this process because of its low cost, availability, and low toxicity. The fact of this commercialization refutes the commonly held notion that alkanes cannot be used as raw materials because of their low reactivity, which generally translates into low selectivity.

2.8 Biorefining

Until now, the major applications of biotechnology in the petroleum industry have been limited to microbiologically enhanced oil recovery, production of single-cell proteins, treatment of waste streams (waste waters or gases), and bioremediation of soils contaminated during the recovery, processing, and distribution of petroleum. There are no reports yet concerning the application of biorefining processes at the industrial scale.

Although most research and development efforts have been directed to biodesulfurization (BDS), other research projects have been initiated for removing other contaminants from petroleum, such as nitrogen and heavy metals. Projects concerning the transformation of heavy crudes into light crudes have also been reported. Depolymerization of asphaltenes is also envisaged. However, the most advanced area is BDS, for which pilot plants have been announced; the other areas of application for biorefining are still at the level of basic research.

It is possible that in the distant future, the use of biotechnology could also be extended to other areas of petroleum refining, making possible hydrocarbon cracking, isomerization, polymerization, or alkylation by biological catalysts. The introduction of biocatalysis in chemical technology is also expected. The success of biocatalysis will depend on the ability to enhance biocatalyst activity (i.e. rate and range of substrates) and stability under the conditions found in the petroleum refining industry, using genetic engineering strategies. In all cases, the development of commercial biorefining processes will depend on significant improvements in the cheap, abundant production of highly active and stable biocatalysts adapted to the extreme conditions encountered in petroleum refining. Important improvements will also have to be realized in the design of bioreactors and phase contact and separation systems.

As mentioned, BDS is currently the most advanced field of biorefining, although no commercial application has been announced yet. The results obtained for BDS may be generally applicable to other areas of biorefining, but the interdisciplinary participation of experts in biotechnology, biochemistry, refining processes, and engineering will be essential.

Further Reading

Bajus, M. (1997). Current trends and the process oil and petrochemical technologies for the future. *Pet. Coal* 19 (2): 8–14.

Bajus, M. (2000). Ropa a alternatívne energetické zdroje. *Ropa, uhlie, plyn a petrochémia* 42 (2): 24–28.

Bajus, M. (2001). Reformulované a alternatívne palivá-súčasnosť a budúcnosť. *Ropa, uhlie, plyn apetrochémia* 43 (2): 20–24.

Bajus, M. (2002a). Hydrocarbon technologies for the future, current trends in oil and petrochemical industry. *Pet. Coal* 44 (3–4): 112–119.

Bajus, M. (2002b). Alternatívne palivá. *Energia* 4 (1): 42–47.

Bajus, M. (2002c). Súčasnosť a budúcnosť alternatívnych palív. *Slovgas* 11 (5–6): 28–31.

Barrault, J., Pouilloux, Y., Clacens, J.M. et al. (2002). Catalysis and fine chemistry. *Catal. Today* 75: 177–181.

Brunel, D., Blanc, A.C., Galarneau, A., and Fajula, F. (2002). New trends in the design of supported catalysts on Mesoporous Silicas and their applications in fine chemicals. *Catal. Today* 73: 139–152.

Le Borgne, S. and Quintero, R. (2003). Biotechnological processes for the refining of petroleum. *Fuel Process. Technol.* 81: 155–169.

Misono, M. and Inui, T. (1999). New catalytic technologies in Japan. *Catal. Today* 51: 369–375.

Perego, C. and Ingallina, P. (2002). Recent advances in the industrial alkylation of aromatics: new catalysts and new processes. *Catal. Today* 73: 3–22.

Rossini, S. (2003). The impact of catalytic materials on fuel reformulation. *Catal. Today* 77: 467–484.

Song, C. (2002). Fuel processing for low-temperature and high-temperature fuel cells challenges and opportunities for sustainable development in the 21st century. *Catal. Today* 77: 17–49.

3

Clean Energy Technology

CHAPTER MENU

3.1 Rational Use of Energy

Every aspect of human life is affected by our need for energy. The sun is a central energy source for our solar system. The difficulty lies in converting solar energy into other energy sources and also to store that energy for future use. Photovoltaic devices and other means of using solar energy are intensively studied and developed but not at the level of our energy demands. Earth-based major installations using present-day technology are not feasible: the size of the collecting devices would necessitate the utilization of large areas of the planet. In addition, atmospheric conditions in most of the industrialized world are unsuitable to provide a constant solar energy supply. Perhaps a space-based collecting system that beams energy back to Earth can be established sometime in the future; but except for small-scale installations, solar energy will be of limited significance for the seeable future.

Petrochemistry: Petrochemical Processing, Hydrocarbon Technology, and Green Engineering,
First Edition. Martin Bajus.

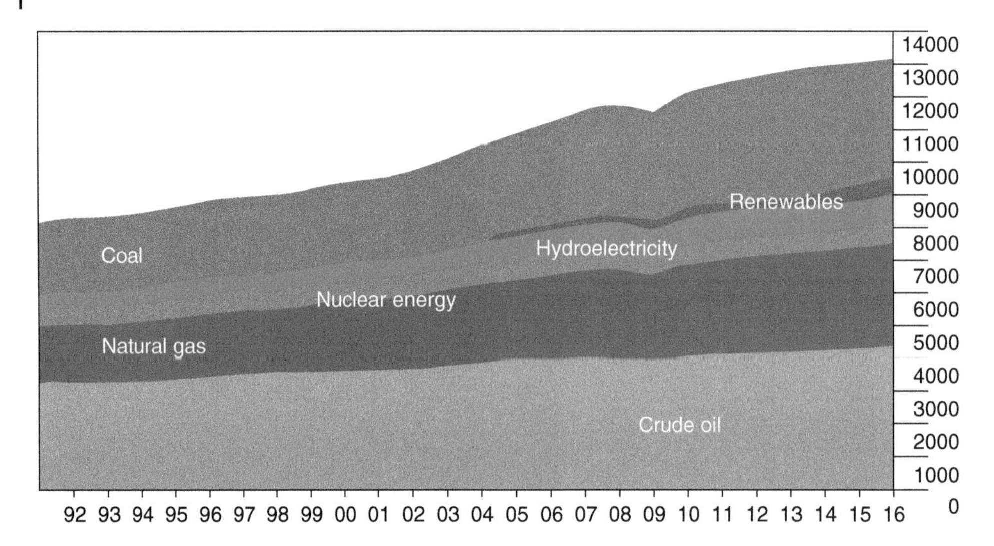

Figure 3.1 World primary energy consumption in 2016 (million tons oil equivalent).

Table 3.1 World consumption of energy in 2018.

	Mil. ton oil equivalent.	%
Crude oil	4662	33.6
Natural gas	3309	23.9
Coal	3772	27.2
Nuclear energy	611	4.4
Hydroelectricity	949	6.8
Renewable energy	561	4.1
Total	**13 864**	**100.0**

Unfortunately, the same must be said for wind, ocean waves, and other unconventional energy sources.

Our major energy sources are fossil fuels (i.e. oil, gas, and coal) and atomic energy (Figure 3.1). Fossil energy sources are, however, non-renewable (at least on our time scale), and their burning causes serious environmental problems (Table 3.1).

Increased CO_2 levels are considered to contribute to the greenhouse effect. The major limitation, however, is the limited nature of our fossil-fuel resources. The most realistic estimates have our overall worldwide fossil resources as lasting for not more than 200 or 300 more years, of which oil and gas would last less than a century. In terms of human history, this is a short period, and we will need to find new solutions.

The United States relies overwhelmingly on fossil energy sources, with only 8% of its energy coming from atomic sources and 4% from hydroenergy (Table 3.2); other industrialized countries utilize nuclear energy and hydroenergy to a much higher degree (Table 3.3).

Table 3.2 US energy sources (%).

Power source	1960	1970	1990
Oil	48	46	41
Natural gas	26	26	24
Coal	19	19	23
Nuclear energy	3	5	8
Hydro, geothermal, solar, etc.	4	4	4

Table 3.3 Power generated in industrial countries by non-fossil fuels, 2016.

	Non-fossil fuel power (%)		
Country	Hydroenergy	Nuclear energy	Total
France	12	75	87
Slovakia	12	63	75
Canada	58	16	74
Germany	4	34	38
Japan	11	26	37
UK	1	23	24
Italy	16	0	16
USA	4	8	12

However, in the last 20 years, concerns about safety and difficulties disposing of fission by-products have dramatically limited the growth of the otherwise clean atomic energy industry. One way to extend the lifetime of our fossil fuel energy reserves is to increase the efficiency of thermal power generation. Progress has been made in this respect, but heat efficiency even in the most modern power plants is limited. Heat efficiency increased substantially from 19% in 1951 to 38% in 1970. But then for many years 39% appeared to be the limit. Combined-cycle thermal power generation – a combination of gas turbines and steam turbines – allowed Japan to increase heat efficiency from 35 to 39% to as high as 43%.

Conservation efforts can also greatly contribute to moderating worldwide growth of energy consumption, but our planet's rapidly growing will put enormous pressure on our future. Figure 3.2 shows the relationship between energy consumption in exajoules per year and total world population in billion people.

Humankind's long-range energy future should focus on using nuclear energy to produce electric energy, which would increasingly free remaining fossil fuels as sources of convenient transportation fuels and as raw materials for synthesis of plastics, chemicals, and other substances. Eventually, in the not-too-distant future, we will need to make synthetic hydrocarbons on a large scale.

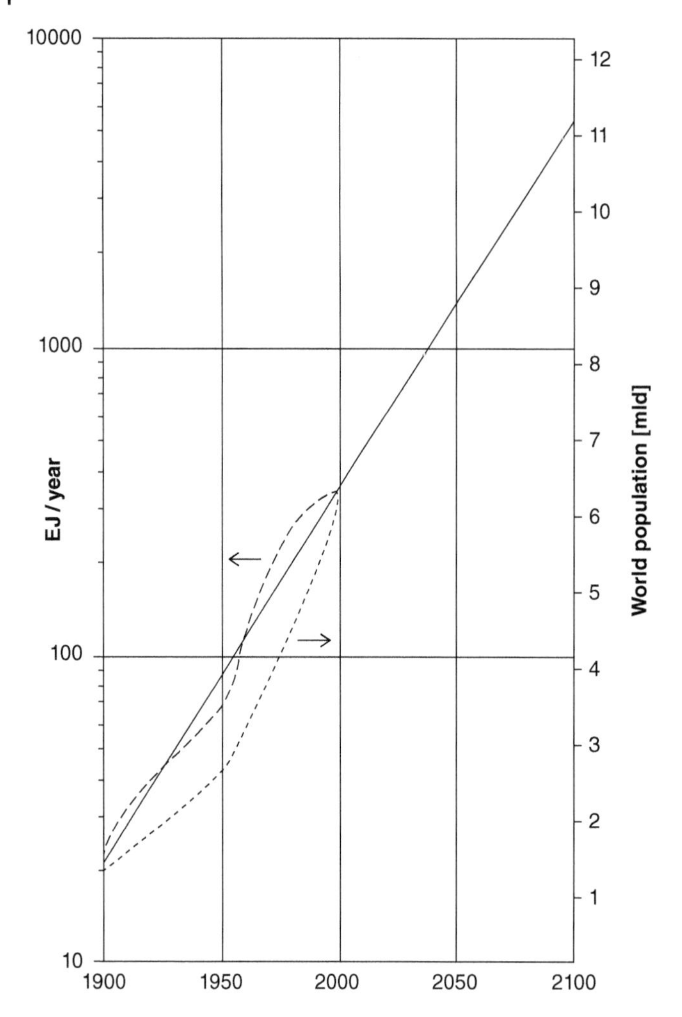

Figure 3.2 Dynamic growth of the world population (right) and energy consumption (left) in the period 1900–2000.

3.2 The Problem of Energy in Chemical Technology

Chemistry is one of the major industries of the world, and it is among the human activities that create wealth from large-scale operations. The chemical industry is the part of the industrial sector that uses the greatest amount of energy: in fact, in the highly industrialized nations (United States, Japan, Germany), the chemical industry consumes about 30% of the total average energy required by all of industry combined. More specifically, the average annual consumption of energy by the chemical industry in the United States is about 6870 GJ (1 GJ = 10^9 J). For this reason, taking into account the fact that many chemical processes furnish their own energy, it can be appreciated that the energy problem is of concern to the chemical industry. The situation is similar to the acute crisis that affected the chemical industry at the same time as the grave energy crisis in the early

1970s, which led to a recession; this was in direct contrast to the great expansion witnessed by this industry between the 1950s and 1970s, when energy was cheap everywhere and very readily available.

3.2.1 The Basics of Energy Management to Improve Economic Budgeting

Energy must be considered an essential element of the chemical industry, together with land, the workforce, capital, and raw materials. On the other hand, energy is not – as it may appear to be by drawing a simple parallel – a raw material, since it lacks the property of mass. It is treated as an independent entity, regarding both its provision and its application. Today, the energy costs of preparing many chemical technology products tend to be about 35% of their sale price, which means a great deal of attention has to be paid to *energy management*: that is, to a prearranged energy plan, the control of energy use, the improvement of yields, and the recovery of energy in the best way. Erroneous estimates of energy costs and of effective energy requirements constitute grave managerial errors from which major crises in economic management may follow.

While estimating the cost of energy may result in data that are quite objective and furnished by the market economy, energy requirements are to a large extent coupled with the capacity to recover energy in the productive process stages. The operations that require more than half of the total heat used in industrial chemistry are distillation, desiccation, and evaporation. On the other hand, these processes are most readily adapted to saving appreciable quantities of energy through the provision of good insulation, by using more operational units (evaporates, for example) that are linked together so units can successively utilize the heat produced by the preceding units, by equipping the apparatus with automatic control systems in order to reduce any fluctuations with timely correction of their causes, and by the compression of steam in order to revitalize it by raising its energy level.

In addition to indirect systematic energy resources of these types, there are also other occasional sources, which are related to the particular characteristics of the processes under consideration and the materials treated in those processes. These include more thorough draining of materials to be dried, reducing the drying medium to the bare minimum necessary, and improving contact between the medium and the material to be dried.

3.2.2 Types of Energy and Energy Sources for Chemical Technology

Basically, two types of energy are used in chemical technology: thermal energy and electrical energy. The proportions in which these are used are about 58 and 42%, respectively.

These percentages refer to the amounts of energy that are fed *ex novo* into chemical technology process plants from external energy sources. In practice, however, the amount of heat that is effectively employed is a higher percentage than noted, since the chemical processes themselves are concentrated sources of heat and also because electrical energy ends up by being degraded into heat that can be utilized to a greater or lesser extent.

The commonest use of *process heat* – that is, the heat generated for the two stated reasons in exothermic reactions and by operations carried out using electrical energy – is the production of steam with a certain energy content: for preheating a steam boiler, as latent heat, or for superheating the steam that has been produced.

Chemical technology directly exploits energy sources to procure the calorific energy it requires while utilizing electrical energy produced by external conventional sources for the same purpose. In terms of the amounts of each type that are used, the sources of electrical energy in order of importance are thermoelectric, hydroelectric, and finally, nuclear electric. At the moment (and, presumably, for a long time in the future), the amount of energy utilized in the chemical industry that is derived from biomass, wind, and other alternative energy sources is insignificant.

When the recycling of waste is not feasible or there is no market for the recycled product, incineration can be used to generate energy from waste combustion heat. Plastics are materials of high calorific value, and hence plastic wastes greatly contribute to the energy produced in incineration plants. Alternatively, they can be used as fuels in a number of applications: power plants, industrial furnaces, cement kilns, etc. Incineration of Cl-containing plastics has been the subject of great controversy due to possible formation and release of dioxins into the atmosphere. However, the relationship between PVC content in the waste stream and dioxin concentration has not been clearly demonstrated. In fact, it seems that the formation of dioxins depends mainly on the incineration conditions, rather than on the waste composition.

The energy sources that are directly exploited by chemical technology in order to supply itself with calorific energy are shown statistically in Table 3.1. As mentioned earlier, in the short term, it is not foreseen that there will be any appreciable contribution from alternative energy sources to the heat requirements of chemical technology if we exclude regional utilization of biogas, local exploitation of geothermal energy, and widespread but also minor use of solar energy for the acceleration of evaporation and drying processes.

It can be seen from analysis of the table that as with everything produced within the orbit of chemical processes, the heat that is exploited in chemical technology originates from the degradation of chemical energy – that is, the form of energy involved in the phenomenology of the bonding of atoms and molecules.

Chemical technology is inherently energy intensive: energy costs (including feedstock) average approximately 8% of the value added. For large-volume chemicals, these costs represent a much higher fraction.

Energy management includes energy conservation, but also encompasses utility system reliability; the intermeshing of process design with utility systems; purchasing, including plant locations for minimum energy cost; environmental impacts of energy use; tracking energy performance; and the optimization of energy against capital in equipment selection.

Chemical technology uses 21% of the energy consumed by the US industrial sector; three related process industries – paper, petroleum, and primary metals – combine for an additional 50% of industrial consumption. A separate breakdown of fuels and feedstocks for the chemical industry shows that the quantity of hydrocarbons used directly for feedstock is about as great as that used for fuel. Much of this feedstock is oxidized, accompanied by the release of heat; in many processes, by-product energy from feedstock oxidation dominates purchased fuel and electricity. A classic example is the reaction of propylene C_3H_6 and ammonia to make acrylonitrile, C_3H_3N. Here, two-thirds of the hydrogen is combusted just to satisfy stoichiometry. In addition, CO and CO_2 formation consumes about 15% of the feed propylene.

Two-thirds of the fuel used by US chemical technology in 1988 was natural gas, which is clean and easy to combust. Although relatively inexpensive at the wellhead, natural gas is costly to transport. Hence the chemical industry is concentrated in regions where natural gas is produced, keeping the average price paid by the US chemical industry for natural gas in 1988 to only 80% of the average US industrial price. Similarly, the movement of chemical commodity production to the Middle East is driven by the desire to obtain low-cost natural gas.

3.3 Waste Fuel Utilization

It is always preferable to minimize or upgrade by-products for sale as chemicals; however, when this is not feasible, the fuel value can still be recovered. Increased combustion of by-product gases and liquids was one of the principal components in the improvement in energy efficiency that occurred in the industry in the 1980s. An example of waste fuel utilization is the incineration of the off-gases from an acrylonitrile reactor followed by generation of high-pressure steam.

3.3.1 Electricity

Electricity, including the losses associated with production, represents 24% of the total energy used by the chemical industry. On a cost basis, electricity represents a higher share, at 29% of the energy bill including feedstocks. Increases in electrical costs have provided the driving force for increased cogeneration, i.e. the recovery of power as a by-product of other process plant operations. The historic cogeneration example is a steam turbine associated with a boiler plant. The relatively high cost of electricity has also led designers to focus on the efficiency of rotating equipment and has motivated a closer look at how processes can be controlled to reduce power, using such innovations as variable-frequency motor drives.

3.3.2 Energy Efficiency Improvements

Energy management is basically a game of economics played with a special set of technical rules. Saving millions of kilowatt hours or GJ is only communicable when converted into dollars, or into a ratio of dollars saved per dollar of incremental investment. The question of where in a process to focus on improvement is often answered by determining the biggest energy-cost items in a plant.

Efficiency improvement is driven by two distinct forces: the technological process, which is the long-term trend; and cost optimization, which is the short-term response to price swings. The baseline, long-term trend has been in the range of a 2–3% improvement per year. Improvement in the 1990s was 1–2% per year, reflecting a more mature industry. A 1–2% per year energy reduction will still yields large savings over time.

US chemical technology achieved an annual reduction of 4.2% in energy input per unit of output for the period 1975–1985. This higher reduction resulted from cost optimization: the trade-off of increased capital for reduced energy use driven by energy prices. In contrast, from 1985 to 1990, the energy input per unit of output was almost flat as a consequence of falling prices. The average price the US chemical industry paid for natural gas fell by one-third between 1985 and 1988.

Whereas energy conservation is an important component of cost reduction for the chemical industry, it is rarely the only driving force for technological change. Much of the increased energy efficiency comes as a by-product of changes made for other reasons, such as higher quality, increased product yield, lower pollution, increased safety, and lower capital. For example, process energy integration in design, enabled by computer simulation, saves energy as well as capital; substitution of variable-speed drives on motors for control valves saves energy as well as capital in the supply of power. One of the roles of energy management is making sure reduction in energy use is considered whenever processes are changed.

The refinement of processes occurred for production of low-density polyethylene, where a process requiring over 100 MPa was replaced by one taking place at 2 MPa. Such refinements may continue but are less than in the 1970s and 1980s. The introduction of biotechnology-derived processes is expected to cause a shift to lower-temperature, lower-pressure processing in chemical technology.

3.3.3 Energy and the Environment

The impact of energy usage on gaseous emissions has emerged as a primary environmental issue, and regulatory action has required emission reductions in NO_X and SO_2. Control of NO_X is achieved by limiting the temperature of combustion and limiting excess oxygen. Control of SO_2 requires either changing fuel or scrubbing flue gas. Because the preferred fuel of the chemical industry is already predominantly low-sulfur natural gas, the primary impact on the chemical industry of SO_2 regulation is expected to raise the price of electricity derived from coal.

Issues related to gases such as CO_2, which contribute to the global greenhouse effect, are also of increasing importance. Energy conservation directly reduces CO_2 emissions. The elimination of fugitive hydrocarbon emissions as a result of improved maintenance procedures is also a tangible step that the industry is taking.

3.3.3.1 Carbon and Greenhouse Emissions

Combustion is the sequence of exothermic chemical reactions between fuel and oxidant accompanied by the production of heat and conversion of chemical species. The release of heat can result in the production of light, usually in the form of a flame. Fuels of interest often include organic compounds (especially hydrocarbons) in the gas, liquid, or solid phase.

For the most part, combustion involves a mixture of hot gases and is the result of a chemical reaction, primarily between oxygen and a hydrocarbon (or a hydrocarbon fuel). In addition to other products, the combustion reaction produces carbon dioxide, steam, light, and heat. Combustion as the burning of any substance in gaseous, liquid, or solid form is a broad definition; combustion includes fast exothermic chemical reactions, generally in the gas phase but not excluding the reaction of solid carbon with a gaseous oxidant. Flames represent combustion reactions that can propagate through space at

Table 3.4 Formation of emissions: CO, NO_x, PM, and CO_2; and consumption of diesel fuel (CADC test).

	1	2	3	4	5	6	7
CO (g/km)	0.083	0.049	0.066	0.060	0.069	0.058	0.048
NO_x (g/km)	0.667	0.69	0.692	0.648	0.678	0.666	0.673
HC + NO_x (g/km)	0.675	0.694	0.694	0.649	0.683	0.667	0.676
PM (g/km)	0.043	0.043	0.033	0.029	0.030	0.029	0.028
CO_2 (g/km)	176	177	178	177	178	173	179
Fuel consumption (l/100 km)	6.64	6.70	6.83	6.80	6.88	6.64	6.89

1 Diesel in the car tank (on board).
2 Diesel in the external tank.
3 Diesel +5% wt. biocomponent.
4 Diesel +2% wt. di- + tri-tert-butyl ether.
5 Diesel +5% wt. di- + tri-tert-butyl ether.
6 Diesel +5% wt. biocomponent +1.0.25% wt. di- + tri-tert-butyl ether.
7 Diesel +4% wt. biocomponent +1% wt. di- + tri-tert-butyl ether.

subsonic velocity and are accompanied by emission of light. The flame is the result of complex interactions: chemical and physical processes whose quantitative description must draw on a wide range of disciplines such as chemistry, thermodynamics, fluid dynamics; molecules, atoms, and free radicals, all highly reactive intermediates of the combustion reactions, are generated.

For mixtures of higher ethers and diesel (and/or biodiesel) motor emissions, tests were performed (Table 3.4). From the results of the tests, it can be concluded that there is no negative influence of glycerol ethers on fuel properties. Furthermore, the tests indicated a positive influence on decrease of exhaust emissions, especially CO and PM.

The physical processes involved in combustion are primarily transport processes: transport of mass and energy systems in which the reactants flow, and transport of momentum. The reactants in a chemical reaction are normally a fuel and an oxidant. In practical combustion systems, the chemical reactions of major chemical species – carbon, hydrogen in the fuel, and oxygen in the air – are fast at prevailing high temperatures (greater than 930 °C), as the reaction rates increase exponentially with temperature. In contrast, the rates of transport processes exhibit much smaller dependence on temperature, and are therefore lower than those of the chemical reactions.

The greenhouse gases carbon dioxide, methane, and nitrous oxide are all produced during fuel oil combustion. Nearly all of the fuel carbon (99%) in fuel oil is converted to CO_2 during the combustion process. Although the formation of CO acts to reduce CO_2 emissions, the amount of CO produced is insignificant compared to the amount of CO_2 produced. The majority of fuel carbon not converted to CO_2 is due to incomplete combustion in the fuel stream.

3.3.3.2 Formation of Particulate Matter

When pseudo-hydrocarbon fuels are used in place of pure hydrocarbon fuels (such as hydrocarbons themselves, gasoline, diesel, and the like), combustion processes can (depending upon the properties of the fuel) emit large quantities of particles into the atmosphere.

Particles formed in combustion systems fall roughly into two categories. The first category, referred to as *ash*, comprises particles derived from non-combustible constituents (primarily mineral inclusions) in the fuel and from atoms other than carbon and hydrogen (heteroatoms) in the organic structure of the fuel. The second category consists of carbonaceous particles that are formed by pyrolysis of the fuel molecules.

Particles produced by combustion sources are generally complex chemical mixtures that often are not easily characterized in terms of composition. The particle sizes vary widely, and the composition may be strongly dependent on particle size.

3.3.3.3 CO_2 Emissions

Emissions of CO_2 from energy consumption increased by only 0.1% in 2016. During 2014–2016, average emissions growth was the lowest of any three-year period since 1981–1983. The good news is that carbon emissions were essentially flat in 2016; this was the third consecutive year in which we saw little or no growth in carbon emissions, in sharp contrast to the 10 years before that when emissions grew by almost 2.5% per year (Table 3.5).

The key question that arises is whether the experience of those three years signaled a decisive break from the past and a significant step toward the goals of the Paris Agreement (see Table 1.10), or whether it was largely driven by cyclical factors that are likely to unwind over time.

Novel technologies for olefin production need to be viable from an environmental perspective, as well as from a technical one. In this regard, it is important to evaluate the CO_2 emissions associated with different technologies discussed earlier. Figure 3.3 shows the total CO_2 emissions per ton of high-value chemicals (HVC) such as ethylene, propylene, and aromatics. These data were taken partly from a review by Ren et al. and partly from various figures published by the International Energy Agency (IEA). A distinction is made between CO_2 emissions resulting from the energy requirement of the process (i.e. fuel combustion) and the chemical CO_2 that is produced in the reaction. It is clear from Figure 3.3 that steam cracking is still the best-performing technology, even in terms of CO_2 emissions. The process produces almost no chemical CO_2, and the energy efficiency of the process has

Table 3.5 Carbon dioxide emissions.

Growth rate per annum	**Share**			
Million tons of carbon dioxide	2016	2016	2005–2015	2016
USA	5350	−2.0%	−1.1%	16.0%
China	9123	−0.7%	4.2%	27.2%
Russian Federation	1490	−2.4%	0.2%	4.5%
European Union	3485	*	−2.0%	10.4%
Slovakia	31	1.7%	−2.4%	0.1%
Czech Republic	105	2.5%	−1.9%	0.3%
Poland	299	2.8%	−0.6%	0.9%
Hungary	46	2.8%	−2.5%	0.1%
Total world	33 432	0.1%	1.6%	100.0%

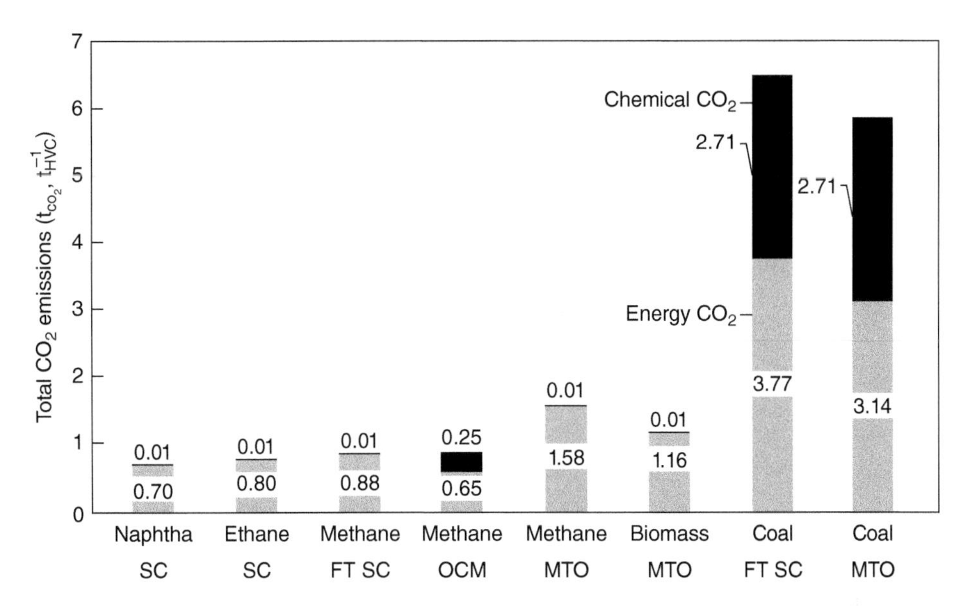

Figure 3.3 Total CO_2 emissions per ton of high-value chemicals (HVC) for different technologies.

been optimized in such a way that the energy CO_2 is very low in comparison with that of other techniques. Oxidation coupling of methane (OCM) looks very promising as well, as it has the lowest energy CO_2. However, because of the relatively low ethylene selectivity, the chemical CO_2 for this technology is still quite high. As expected, the coal-based techniques are major producers of CO_2, both energetically and chemically.

Carbon capture and storage or utilization techniques are a promising way to further cut carbon emissions from light-olefin-production technologies. Ethylene producers, which are considered large industrial emitters, will have to deal with the requirement (activated by political decisions) to reduce CO_2 emissions in the near future. These techniques are currently under rapid development; they are becoming promising methods to reduce carbon emissions and are therefore a crucial aspect of meeting CO_2 emission-reduction targets. In this context, it is important to distinguish between chemical-related, energy-related (process-related), and process-related (or chemical-related) emissions, in that the latter is inherent to the nature of the process. The use of coal as feedstock comes with high chemical carbon emissions.

This CO_2 stream is already separated; therefore, it could be further prepared for storage or utilization. However, this processing step comes with a higher cost and energy-efficiency penalty, as the CO_2 needs to be compressed and maintained within the supercritical envelope for storage. On the other hand, energy-related emissions are the result of the fuel-combustion process for heat or power generation. Hence, the applicability of these carbon-capture techniques is highly dependent on the process layout.

In the case of SC furnaces, two different carbon-capture techniques are of interest: oxyfuel combustion and the post-combustion process, both because of the high firing duty in the furnace. The latter can be seen as an add-on solution, with minor changes to the cracking furnace itself and, thus, a lower associated cost of implementation. However, the low CO_2

concentration in the flue gas leads to a high load for the processing unit. Oxyfuel combustion is based on separating oxygen from the air, leading to a more concentrated CO_2 stream and resulting in less-intensive post-processing steps. In addition, higher thermal efficiency in the radiant section can be expected, and there is no emission of thermal NO_x. A challenge remains the cost of separating oxygen from air. According to Weikl and Schmidt, a penalty on CO_2 emission of \$85 USD·t^{-1} would be necessary in order to arrive at a break-even point for the application of this carbon-capture technique. Such a penalty would result in a higher ethylene production cost, estimated at roughly \$130 USD·t^{-1} of ethylene.

3.4 Energy Technology

Energy management requires the merging of technologies such as thermodynamics, process synthesis, heat transfer, combustion chemistry, and mechanical engineering.

3.4.1 Thermodynamics

The first law of thermodynamics – which states that energy can be neither created nor destroyed – dictates that the total energy entering an industrial plant equals the total of all the energy that exits. Feedstock, fuel, and electricity count equally, and a plant should always be able to close its energy balance within 10%. If the energy balance does not close, there probably is a significant opportunity for saving.

The second law of thermodynamics focuses on the quality or value of energy. The measure of quality is the fraction of a given quantity of energy that can be converted to work. What is valued in energy purchased is the ability to work. Electricity, for example, can be totally converted to work, whereas only a small fraction of the heat rejected to a cooling tower can make this transition. As a result, electricity is a much more valuable and more costly commodity.

Unlike the conservation guaranteed by the first law, the second law states that every operation involves some loss of work potential, or energy. The second law is a very powerful tool for process analysis, because it tells what is theoretically possible and pinpoints the quantitative loss in work potential at different points in a process.

Typically, the biggest loss that occurs in chemical processes is in the combustion step. One-third of the work potential of natural gas is lost when it is burned with unpreheated air. Conventional analysis only points to recovery of heat from the stack as an energy improvement, but second law analysis shows that other losses are much greater.

The second law can also suggest appropriate corrective action. For example, in combustion, preheating the air or firing at high pressure in a gas turbine, as is done for an ethylene cracking furnace, improves energy efficiency by reducing the lost work of combustion. The combined cycle first fires fuel into a gas turbine and greatly increases the power extracted per unit of steam produced.

Gas turbine cogeneration is inherently relatively low in capital expenditures. The absence of heat-exchange surfaces in the gas turbine provides the basic capital advantage, and standardized equipment prepackaged as skid-mounted components adds to that capital advantage. When these factors are coupled with low-priced natural gas, a situation arises in which petrochemical plants have become exporters of cogenerated power to utilities. The gas turbine also has advantages that are firmly rooted in thermodynamics: it utilizes energy directly at a high-temperature level, without large driving forces for pressure drops and temperature differences.

Most gas turbine applications in the petrochemical industry are tied to the steam cycle, but turbines can be integrated anywhere there is a large requirement for fired fuel. An example is the use of the heat in gas turbine exhaust as preheated air for ethylene cracking furnaces.

The combined cycle is also applicable to dedicated power production. When the steam from a waste heat boiler is fed to a condensing turbine, overall conversion efficiencies of fuel to electricity in excess of 50% can be achieved. A few public utility power plants use this cycle, but in general utilities have been slow to convert to gas turbines.

3.4.2 Power Recovery in Other Systems

Steam is by far the biggest opportunity for power recovery from pressure letdown, but others exist, such as tail-gas expanders in catalytic crackers. An example of power recovery in liquid systems is the letdown of the high-pressure, rich absorbent used for H_2S/CO_2 removal. Letdown can occur in a turbine directly coupled to the pump used to boost the lean absorbent back to the absorber pressure.

3.4.3 Heat Recovery, Energy Balances, and Heat-Exchange Networks

The goal of heat recovery is to be sure that energy does the maximum useful work as it cascades to ambient. The *energy balance* is the summary of all the energy sources and all the energy sinks for a unit operation, a process unit, or an entire manufacturing plant; it is almost as important as the material balance for understanding how a process works. The energy balance is the basic tool for analyzing an operation for energy conservation opportunities. When incorporated into a computer program, the energy balance becomes the base for a model of the process. Operational changes, system configuration, and equipment alterations can be evaluated via the model.

Reactive distillation, also called catalytic distillation (Figure 3.4), is finding important new applications for the production of methyl tertiary-butyl ether (MTBE) and methyl acetate. In this process, the heat in the overhead stream of one distillation column(s) is utilized in the same system. Heat integration is generally low-risk technology, and its economics can be very attractive. Commercial installations have slowed in recent years in the United States due to low energy costs. Heat integration can be considered if appropriate temperature driving forces exist and integration of columns does not cause operational or control problems.

Membrane-based separation processes create separations by selectively passing (permeating) one or more components of a stream through a membrane while retarding the passage of one or more other components. Membrane processes generally do not require a phase change to make a separation. As a result, energy requirements are low, unless a great deal of energy is expended to increase the pressure of the feed stream in order to drive the permeating component(s) across the membrane. Especially in liquid separations, membrane

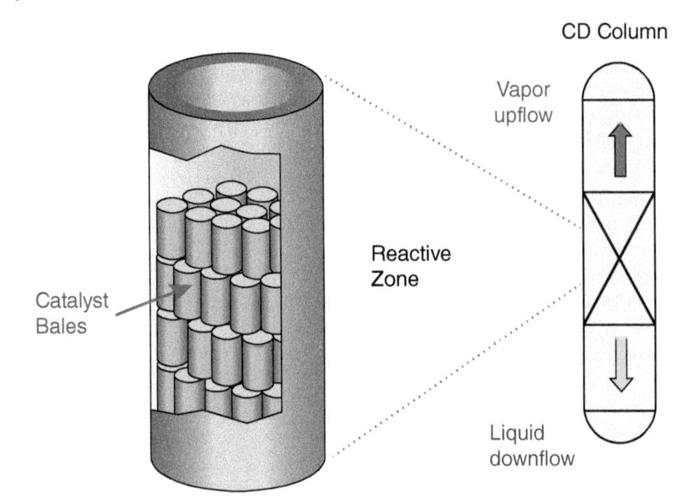

Figure 3.4 Catalytic distillation.

processes often show very low energy usage compared to other separation processes. The present impact of membrane technology on processes in the chemical, petroleum, and allied industries is exceedingly small; but there is reason to believe that this impact will grow in the future, driven mostly by improvements in membrane performance (both in selectivity and in permeability) as well as by the need to reduce both gaseous and liquid emissions from these plants. It is our contention that most of the new and expanded uses will be in the environmental area.

Heat exchanger network analysis, sometimes called *pinch technology*, is a special kind of model that has been developed into a sophisticated way of attacking heat-recovery problems. This type of analysis defines the optimum interchange between all the heat sinks and heat sources. It is similar to the concept used by furnace designers for many years for matching various heat sinks against the flue gas source. Using a temperature vs. enthalpy plot, the analysis can be done manually or via computer. It can even be set up to adjust the distillation sequence or operating pressures to improve energy efficiency and minimize capital.

3.4.4 Waste-Heat Boilers

In most chemical process plants, the steam system is the integrating energy system. Recovering waste heat by generating steam makes the heat usable in any part of the plant served by the steam system. Many waste-heat boilers are unique and adapted to fit a particular process. There is a long history of process waste-heat boiler failure resulting from inadequate attention to detail in design and the failure to maintain water quality. The high heat-transfer coefficients of boiling water are dependent upon clean surfaces. Designers should match the hardware as closely as possible to demonstrated designs, and operators should ensure that water treatment is monitored. Incinerators and gas turbines also involve heat-recovery boilers, and a number of fairly standard designs have evolved (Figure 3.5).

The gases that exit pyrolysis-furnace coils contain significant usable heat. Much of this heat can be recovered by use of transfer-line exchangers. At the cracking-coil outlet, the

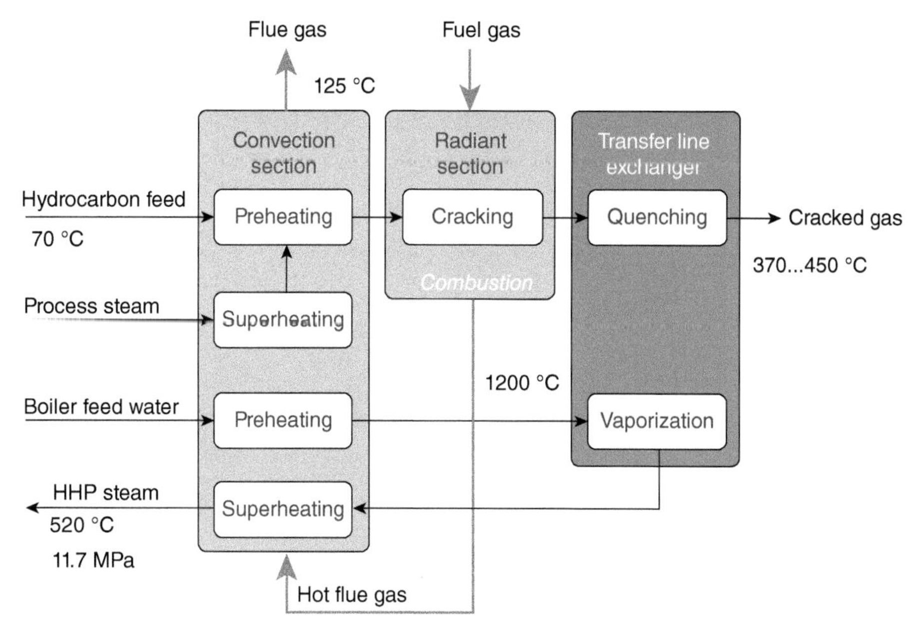

Figure 3.5 Integrated heating system of cracking furnaces.

primary and secondary reactions continue in the transfer line under adiabatic conditions. In order to avoid excessive transfer-line cracking, there is either a direct or an indirect quench downstream of the transfer line. Direct quenching features a more rapid temperature decrease than indirect quenching. However, indirect quenching permits heat recovery at a high temperature level suitable for generation of high-pressure (HP) steam. In older, energy-inefficient steam crackers, the furnace effluent was quenched directly with water (in the case of ethane or propane feed) or with a suitable oil (in the case of liquid feedstocks). Transfer-line exchangers, also called *quench coolers*, often utilized a heating medium such as Dowtherm until the late 1960s. Schmidtsche Schack | Arvox GMBH (Kassel, Germany) developed the first transfer-line exchanger (TLE) generating HP steam used for plant turbocompressors. Since then, the quench coolers of other designs generating HP steam suitable for feedstocks ranging from ethane to atmospheric gas oil (AGO) have become an integral part of steam crackers and have resulted in substantial improvements in operating costs.

3.4.5 Product-to-Feed Heat Interchange

Heat exchange is commonly used to cool the product of a thermal process by preheating the feed to that process, thus providing a natural stabilizing, feed-forward type of process integration. Product-to-feed interchange is common on reactors as well as distillation trains (Figure 3.6).

3.4.6 Combustion Air Preheat

Flue gas to air exchange, a type of product-to-feed heat exchange, is extremely important because of the large loss associated with the combustion of unpreheated air (Figure 3.7).

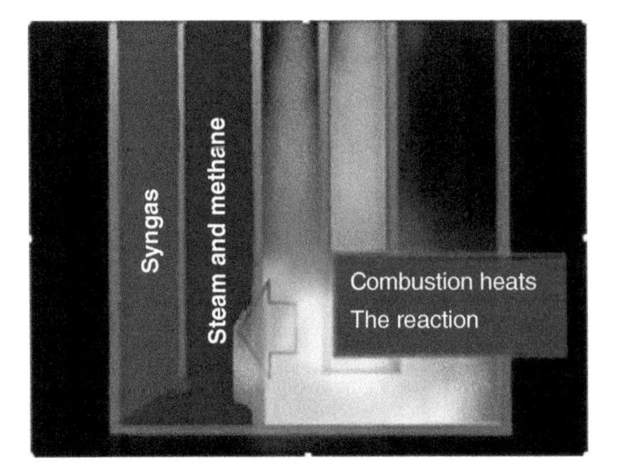

Figure 3.6 Product-to-feed.

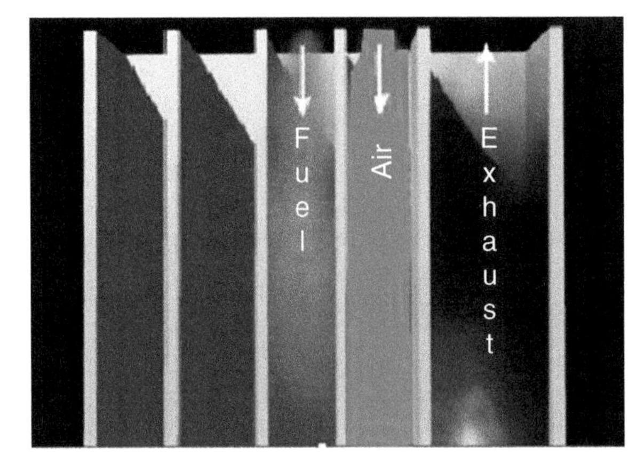

Figure 3.7 Air preheater.

This exchange process has generated fairly unique types of hardware, such as the Ljungstrom or rotary wheel regenerator. These are particularly useful for retrofitting because of the ability to move heat to a physically separated unit.

3.4.7 Heat Pumps

The use of heat pumps adds a compressor to boost the temperature level of rejected heat. This can be very effective in small plants that have few opportunities for heat interchange. However, in large facilities, a closer look usually shows an alternative for the use of waste heat. The fuel/steam focus of energy use has led to the application of heat pumps in applications where a broader examination might suggest a simpler system of heat recovery.

The word *energy* is commonly used in everyday life in a large number of sectors and covers very diverse concepts such as work, heat, cost, etc. Applied to technical and industrial fields, the definition and use of the word are restricted, but a wide variety of applications remain. Technicians monitor the flow of energy utilized in machines and equipment in a variety of forms from electricity to hot water. From another standpoint, operators are concerned with the quantities and costs of the two or three energy carriers they buy, electricity, and fuels. The relationship between technical activity and the economic result is not self-evident.

Today, after two oil price shocks and dramatic price fluctuations, the cost of all commercial energy carriers has fallen to low levels in processing industry balance sheets. Should it therefore be ignored?

A short introduction to rational management and use of energy will show that its function as a vital force is essential to satisfactory operation independent of its purchase price. A good understanding of energy implementation and the technology used are crucial for continued quality production. Energy also has a predominant impact on the natural environment, and operators must continue to be concerned with it, if only for this reason.

3.5 Energy Accounting

A production plant imports energy in various forms – for example, fuels or electricity – and can at the same time export surplus energy in the form of combustible residues, steam, and sometimes even electricity. In a simplistic approach, the cost of energy is taken into account comprehensively on the basis of bills paid for energy carriers purchased. The overall cost may be broken down and allocated to the products exiting several units interconnected on electricity or steam networks. This approach is quite correct from an accounting standpoint in order to establish an earnings-expenditures financial balance sheet.

However, from a technical standpoint, this type of procedure is totally unsuitable. It leaves the following questions pending:

- What is the relative weight of users in the process, and what form of energy do they receive?
- If primary energy is converted during operations in order to be distributed in the form of steam or hot oil, what additional cost should be taken into account?
- What technical and economic advantage will the use of a different form of energy afford if alternatives exist?
- Are there any seasonal variations in energy consumption and/or cost?

To better control energy costs, *energy accounting* must be set up on a technical basis, upstream from financial accounting. "There is no knowledge without measurement" is self-evident but needs to be remembered, since the measurement of energy fluids is often inaccurate in the field. Considerable investment has been made to better measure consumption of individual users ever since the first oil price shock. It can even be said that the first energy saving is accomplished by installing a meter. However, all measuring elements require maintenance and processing of the information they produce: two cost factors that must be taken into account in a development plan for a production unit. Industrial data processing and distributed control systems have dealt with the second factor and have transferred its cost to the real-time supervision item. Thus, a major obstacle of applying energy accounting has disappeared.

The more fine-tuned the analysis, the more measuring sensors there are, and the larger the mass of data to be processed. Obviously, some trade-off is necessary to adapt the scope of the analysis to the desired aim. Therefore, in setting up an energy accounting system, it is advisable to start with an evaluation of consumers and their classification according to the type of energy used and the technology implemented.

Table 3.6 Overview of energy users.

	Furnace	External heat	Internal heat	Steam heating	Discharged heat	Electricity	
						(GJ/h)	(MW)
Proc. furnaces	26						
Heat exchange:							
Exter. flows		–					
Inter. flows			214				
Steam				–			
Cooling							
Water/air					45		
Machines:							
Compressors						2.34	0.65
Turbines							
Pumps						0.54	0.15
Total machines						**2.88**	**0.80**
Reactors		(13)					
Process steam offers				10			
Total	**26**	**(13)**	**214**	10	45	2.88	0.80

Unit of measurement: GJ/h.

A process flow scheme labeled with all the energy exchanges, by inputs at battery limits or inside a process by exchanges between streams, will be the basis for a recapitulative table (see Table 3.6). The table gives an overview of all users by general categories and is the starting point for examining the correlation between technology and desired aim, by taking each user into account separately.

This is true provided the study is exhaustive! To be complete, the reactor exothermic energy (12.85) and the net difference between feed and product enthalpies must be added to the input items. Thus omissions or measurement inaccuracies and errors can be detected. It is often necessary to reconcile data in order to make the measured values more coherent. Accordingly, a mathematical model of the process is established, and the measured values are compared with the calculated ones. Simulation software is available for petroleum and petrochemical processes.

The largest item in the table should be noted: the heat exchanged internally, 214.07 GJ/h. This item totally escapes analysis; it is "non-commercial" energy but is in no way without importance.

An accounting plan can be set up on this basis to throw some light on the different facets of energy use. There are four types of balances:

- Technical balance based on enthalpy
- Physical balance based on energy type
- Primary energy balance based on equivalent fuel
- Monetary balance based on monetary cost

Further Reading

Ren, T., Patel, M., and Blok, K. (2006). Olefins from conventional and heavy feedstocks: energy use in steam cracking and alternative processes. *Energy* 31: 425–451.

Weikl, M.C. and Schmidt, G. (2010). Carbon capture in steam cracking. Paper presented at the 2010 Spring Meeting & Global Congress on Process Safety, San Antonio.

4

Sources of Hydrocarbons

CHAPTER MENU

Petrochemistry: Petrochemical Processing, Hydrocarbon Technology, and Green Engineering,
First Edition. Martin Bajus.

4.1 Introduction

Where do petrochemicals come from? Natural gas and petroleum are the main sources. From them come seven chemical building blocks on which a vast petrochemical industry is based: ethylene, propylene, the C_4 olefins (butenes and butadiene), benzene, toluene, the xylenes, and methane. The olefins – ethylene, propylene, butadiene, and the butenes – are derived from both natural gas and petroleum. The aromatics – benzene, toluene, and xylenes – are derived from petroleum and, to a very small extent, from coal. Methane comes from natural gas. Whether natural gas fractions or petroleum are used for olefins varies throughout the world depending on the availability of natural gas and the demand for gasoline. Both light and heavy naphthas are petroleum fractions that can be cracked to make olefins. They can also be used for gasoline. In the United States, demand for gasoline is higher than for other petroleum fractions; consequently, the price of naphtha has traditionally been high, and the petrochemical industry has preferred to extract ethane and propane for cracking to olefins from what has hitherto been abundant natural gas. Even with the decrease in natural gas reserves, the shift to liquids for cracking in the United States has been slowed by the availability of cheap liquefied petroleum gas (LPG).

In Western Europe and Japan, the demand for gasoline is lower because on a per capita basis there are half as many cars as in the United States, and they are smaller and are driven shorter distances. Thus, in Western Europe, more naphtha is produced than is required for gasoline. Also, natural gas is less abundant, even with the North Sea discoveries, and contains less ethane. Consequently, naphtha has been the important European raw material for the manufacture of olefins. This is equally true in Japan, which lacks resources of either natural gas or oil.

What once was a straightforward raw material supply situation was complicated by the discovery of natural gas in many parts of the world in the 1970s. Dramatic discoveries in Siberia mean that the Commonwealth of Independent States (CIS) possesses a large share of the world's natural gas reserves. Gas fields have also been discovered in Canada, New Zealand, the Arabian Peninsula, Thailand, Malaysia, Trinidad, and North Africa. In the Middle East, particularly Saudi Arabia, large quantities of associated gas (the gas that accompanies oil deposits, as opposed to that occurring in separate gas fields) are available. In many of these countries, particularly Canada and Saudi Arabia, this gas has become the basis for new chemical industries. The economic consequences of these developments will be discussed further.

Probably 90% by weight of the petrochemicals the world uses comes from petroleum and natural gas, and we therefore devote considerable space to them. In addition, we consider what might happen if natural gas and petroleum supplies are exhausted in the next 50–60 years. At at end 2018, the reserves-to-current production ratio was 50 years for oil and 51 years for natural gas. Nonetheless, it has been authoritatively predicted that the United

States will never again bring to the surface as much natural gas as it did in 1973. There is a need for strategies for the future. The glut of oil and gas in the 1980s and early 1990s clouds the inevitability of the development of future shortages.

A less important current source of chemicals is coal (carbochemistry). Coal was historically important, and much of the progress in chemical technology until World War II was motivated by the availability of coal. Indeed, the famous English chemist W.H. Perkin claimed to have founded organic chemical technology in 1865, when – while trying ineptly to synthesize quinine – he obtained a dye with a mauve color from coal tar intermediates. Perkin's dye was so important that its color gave its name to a period of history known in literature as the "mauve" decade.

The decline of coal coincided with the rise of petrochemicals. However, reserves of coal are much greater than those of oil. If petroleum becomes scarce, will coal come into its own again? Is it a realistic part of an alternative strategy? The so-called *C_1 chemistry* that developed in the 1960s and 1970s when petroleum shortages loomed large assumes that it will, if the industry is willing to provide the tremendous capital investment a switchover will require.

4.2 Natural Gas

Natural gas, depending on its source, contains – besides methane as the main hydrocarbon compound (present usually in >80–99%) – some of the higher homologous alkanes (ethane, propane, butane). In "wet" gases, the amount of C_2–C_6 alkanes is more significant (gas liquids).Typical compositions of natural gas of various origins are shown in Table 4.1.

Natural-gas liquids are generally of thermal value only but can be used for dehydrogenation to alkenes. Their direct upgrading to gasoline-range hydrocarbons is also pursued.

Natural gas as we know it is of biological origin (not unlike petroleum oil). Large gas reservoirs were discovered and utilized in the last century. Increasingly, deeper wells are drilled and deposits under the seas are explored and tapped. An interesting but unproven theory by Gold holds that hydrocarbons may also be formed by slow outgassing of methane from vast deep deposits dating back to the origin of our planet. Besides biologically derived oil and gas, "deep" carbon compounds trapped in the Earth's crust are subject to intense heat, causing them to release hydrocarbons that migrate toward the Earth's surface where they are trapped in different strata. Seepage observed at the bottom of the oceans and finds

Table 4.1 Composition of natural gas (wt.%).

Location	CH_4	C_2H_6	C_3H_8	C_4H_{10}
United States	89.5–92.5	5.1–2	2.1–0.7	1.6–0.5
Russian Fed.	98–99	0.5	0.2	—
Algeria	86.9	9.0	2.6	1.2
Iran	74.9	13.0	7.2	3.1
North Sea	90.8	6.1	0.7	0.1

of oil during drilling into formations (such as granite) where no "biogenic" oil was expected are cited as proof for "abiogenic" hydrocarbons. If abiogenic methane and other hydrocarbons exist (although most geologists presently disagree), vast new reserves would become available when improved drilling technology is developed to reach deeper into Earth's crust.

Other vast yet untapped reserves of natural gas (methane) are locked up as hydrates under the permafrost in Siberia. Methane gas hydrates are inclusion compounds of $CH_4.nH_2O$ composition. Their amount is estimated to equal or exceed known conventional natural-gas reserves. Their economical utilization, however, remains a challenge. Significant amounts ($\leq$500 million tons per annum [MTPA]) of natural methane are also released into the atmosphere from varied sources ranging from marshlands to landfills to farm animals. Methane in the atmosphere represents only a small component, although its increase can cause a significant greenhouse effect.

4.2.1 Definitions and Terminology

Natural gas is the gaseous mixture that arises from the decay of bio-organic material; the gas occurs alone or in conjunction with petroleum reservoirs and coal seams. It is predominantly methane, but also contains non-hydrocarbon compounds (Table 4.2), and it should not be confused with the gaseous products from the destructive distillation (often called *carbonization*) of wood and coal; these gaseous products are manufactured ashes (Table 4.3).

Non-associated natural gas is found in reservoirs in which there is no (or, at best, minimal) crude oil. Non-associated gas is often richer in methane but is markedly deficient in higher-molecular-weight hydrocarbons and condensate materials. *Associated* or *dissolved*

Table 4.2 Composition of natural gas.

Category	Component	Amount (%)
Paraffinic	Methane (CH_4)	70–98
	Ethane (C_2H_6)	1–10
	Propane (C_3H_8)	Trace–5
	Butane (C_4H_{10})	Trace–2
	Pentane (C_5H_{12})	Trace–1
	Hexane (C_6H_{14})	Trace–0.5
	Heptane and higher (C_7+)	Non-trace
Aromatic	Benzene (C_6H_{14})	Traces
Nonhydrocarbon	Nitrogen (N_2)	Trace–15
	Carbon dioxide (CO_2)	Trace–1
	Hydrogen sulfide (H_2S)	Trace
	Helium (He)	Trace–5
	Other sulfur and nitrogen compounds	Trace
	Water (H_2O)	Trace–5

Table 4.3 Classification of methane-containing gases.

1. *Natural gas*
   Associated with petroleum oil deposits, coal seams, or the decay of organic matter
2. *Manufactured gases*
   a) From wood – by distillation or carbonization – wood gas
   b) From peat – by distillation or carbonization – peat gas
   c) From coal – by carbonization – coal gas
      - By gasification
        (1) In air – producer gas
        (2) In air and steam – water gas
        (3) In oxygen and steam – Lurgi gas
      - By hydrogenation
   d) From petroleum and oil shale – by cracking – refinery gas
      - By hydrogenation – oil gas
      - By water gas reaction – oil gas
      - By partial oxidation – oil gas

natural gas occurs either as free gas or as gas in solution in crude oil. Gas that occurs as a solution with crude petroleum is *dissolved* gas, whereas gas that exists in contact with crude petroleum (as the *gas cap*) is *associated* gas. Associated gas is usually leaner in methane than non-associated gas but is richer in higher-molecular-weight constituents. *Gas condensate* contains relatively high amounts of higher-molecular-weight liquid hydrocarbons. These hydrocarbons may occur in the gas phase in the reservoir.

The nonhydrocarbon constituents of natural gas are (i) diluents such as nitrogen, carbon dioxide, and water vapor; (ii) contaminants such as hydrogen sulfide and/or other sulfur compounds. The diluents are noncombustible gases that reduce the heating value of the gas and are, on occasion, used as "fillers" when it is necessary to reduce the energy content of the gas. The contaminants are detrimental to production and transportation equipment in addition to being obnoxious pollutants. Thus, the primary reason for gas processing is to remove the unwanted constituents of natural gas.

Natural gas (Table 4.3) from different wells varies in composition and analyses (1); as a result of these variances in composition, several general definitions are applied to the different products. Thus, natural gas can be (i) *lean*, in which methane is the major constituent; (ii) *wet*, which contains notable amounts of the higher-molecular-weight hydrocarbons; (iii) *sour*, which contains hydrogen sulfide; (iv) *sweet*, which contains little, if any, hydrogen sulfide; (v) *residue gas*, which is natural gas from which the high-molecular-weight hydrocarbons have been extracted; and (vi) *casinghead gas*, which is derived from petroleum but is separated at the wellhead.

4.2.2 Origin

Natural gas (methane) is considered to originate in three principal ways: (i) the thermogenic process; (ii) the biogenic process; and (iii) the abiogenic process. The thermogenic process is the slow process of the decomposition of organic material in sedimentary basins, which usually requires some degree of heat. The biogenic process involves the formation of methane by the action of living organisms (bacteria) on organic materials. The abiogenic process, unlike the other two processes, does not require the presence of organic matter as the starting material.

It is generally believed that once formed, the direction of mobility of the gaseous hydro carbons in the earth is in an upward direction (i.e. toward the surface). However, it is more than likely that there are exceptions to this general rule. For example, hydrocarbons can also be envisaged as moving in downward or sideways directions from their place of formation (source rock) to their place of accumulation (reservoir rock). Irrespective of the direction of movement of hydrocarbons from the source rock, the movement causes displacement of some of the brine that originally filled the pore spaces of the sedimentary rock. This movement of hydrocarbons is inhibited when oil and gas reach an impervious rock that traps or seals the reservoir.

4.2.3 Occurrence

As stated in the introduction, natural gas occurs in the porous rock of the earth's crust either alone or with accumulations of petroleum. In the latter case, the gas forms a *gas cap*, which is the mass of gas trapped between the liquid petroleum and the impervious cap rock of the petroleum reservoir. When the pressure in the reservoir is sufficiently high, the natural gas may be dissolved in the petroleum and is released upon penetration of the reservoir as a result of drilling operations.

Natural gas, like petroleum, is located in the earth in reservoirs; but, just as conventional petroleum reservoirs can vary considerably in character, natural gas reservoirs can also vary considerably. However, for general purposes, natural gas reservoirs can be conveniently classified as *conventional* and *nonconventional* reservoirs. The latter reservoirs include formations such as tight sands, tight shales, geopressured aquifers, coal beds, deep sources, and gas hydrates.

4.2.4 Reserves

Reserves of natural gas are often classified into various categories. For example, *proven* reserves are those reserves of gas that are actually found (proven) by drilling, and the estimates have high degree of accuracy. On the other hand, the term *inferred* reserves is often used in addition to, or in place of, *potential* reserves. The term also usually includes those gas reserves that can be recovered by further development of recovery technologies. Thus, potential reserves of natural gas are the additional resources of gas believed (unsubstantiated) to be existent in the earth. Finally, the term *undiscovered* reserves is very speculative, and the data is regarded as having little value since they are open to questions about the degree of certainty. Recent estimates put the proven reserves of natural gas on the order of 141×10^{12} m^3 (of which some 260.1×10^{12} m^3 exist in North America).

4.2.5 Recovery

A *typical* gas production and processing system is an integrated system for gas production, collection, and processing that ultimately produces a purified gas product. However, the concept of site specificity may dictate the need for an *atypical* gas production and processing system. The produced gas is processed to remove higher-molecular-weight hydrocarbons and any liquid products. Any residue gas, rather than being sold, can be injected to maintain reservoir pressure. When the reservoir has been swept of the higher-molecular-weight materials so that retrograde condensation no longer can occur, the field is taken to full production.

A typical gas-processing plant produces residue gas and a variety of products such as ethane, LPG, and *natural gasoline*, which is a low-octane product of the gas recovery/processing system. Originally, gas-processing plants were used to remove gasoline components to be used as a blending stock for motor gasoline. Hence, the term *gasoline plant* was often inappropriately applied to gas-processing plants. Other fuel needs then caused a shift of focus to liquefied petroleum gas (propane, butanes, and/or mixtures thereof), as well as gasoline constituents. More recently, the extraction of ethane for petrochemical feedstocks has become an extremely important aspect of gas-processing operations.

4.2.6 Storage

The *storage* of gas is the process that matches the constant supply from long-distance pipelines to the variable demand of markets, which are subject to weather and other factors; as a result, its availability maximizes economic advantages. Storage facilities are usually classified as *market* or *field* storage. Market storage is near major consuming areas where the variable demands resulting from weather are serviced by a proper combination of pipeline and storage gas. In field storage, variable supply to the major market pipelines is supplemented by the availability of storage gas.

The most common method for underground storage is to use a previously producing gas or petroleum field. Gas is pumped into the old wells by means of compressors similar to those employed to move gas throughout the pipeline system. The gas is usually stored under the same pressure conditions that originally existed in the field.

The main commercial source of methane, ethane, and propane is natural gas. Selected properties of methane, ethane, and propane are summarized in Table 4.4. Compositions of typical natural gases are listed in Tables 4.5, and their properties are listed in Table 4.6. They are analyzed with conventional gas chromatographic columns.

4.3 Petroleum or Crude Oil

Petroleum or *crude oil* is a complex mixture of many hydrocarbons. It is characterized by the virtual absence of unsaturated hydrocarbons and mainly consists of saturated, predominantly straight-chain alkanes, small amounts of slightly branched alkanes, cycloalkanes, and aromatics (Table 4.7). Petroleum is generally believed to be derived from organic matter deposited in the sediments and sedimentary rocks on the floor of marine basins. The identification of biological markers such as petroporphyris provides convincing evidence for the biological origin of oil (see, however, the possibility mentioned earlier of abiogenic deep hydrocarbons). The effects of time, temperature, and pressure on the geologic transformation of organics to petroleum are not yet clear. However, considering the low level of

Table 4.4 Selected properties of methane, ethane, and propane.

	Methane CH_4	Ethane C_2H_6	Propane C_3H_8
Molecular weight	16.04	30.07	44.09
MP, K	90.7	90.4	85.5
BP, K	111	185	231
Explosivity limits, vol.%	5.3–14.0	3.0–12.5	2.3–9.5
Autoignition temperature, K	811	788	741
Flash point, K	85	138	169
Heat of combustion, MJ/mol	882.0	1541.4	2202.0
Heat of formation, kJ/mol	84.9	106.7	127.2
Heat of vaporization, kJ/mol	8.22	14.68	18.83
Vapor pressure at 273 K, MPa		2.379	0.475
Specific heat, J (mol-K) at 293 K	37.53	54.13	73.63
at 373 K	40.26	62.85	84.65
Density, kg/m^3 at 293 K	0.722	1.353	1.984
373 K	0.513	0.992	1.455
Critical point – pressure, MPa	4.60	4.87	4.24
Temperature, K	190.6	305.3	369.8
Density, kg/m^3	160.4	204.5	220.5
Triple point – pressure, MPa	0.0112	1.1×10^{-6}	3.0×10^{-10}
Temperature, K	90.7	90.3	85.5
Liquid density, kg/m^3	450.7	652.5	731.9
Vapor density, kg/m^3	0.0257	4.51×10^{-5}	1.85×10^{-8}
Dipole moment	0	0	0
Hazards	Fire, explosion, asphyxiation	Fire, explosion, asphyxiation	Fire, explosion, asphyxiation

oxidized hydrocarbons and the presence of porphyrins, it can be surmised that the organics were acted upon by anaerobic microorganism and that temperatures were moderate, <200 °C. By comparing the elemental composition of typical crude oils with typical bituminous coals, it is clear why crude oil is a much more suitable fuel source in terms of its higher H : C atomic ratio, generally lower sulfur and nitrogen content, very low ash content, probably mostly attributable to suspended mineral matter, vanadium, nickel (associated with porphyrins), and essentially no water content.

Finally, it is interesting to mention that recent evidence shows that even extraterrestrially formed hydrocarbons can reach the Earth. The Earth continues to receive some 40 000 tons of interplanetary dust every year. Mass-spectrometric analysis has revealed the presence of hydrocarbons attached to these dust particles, including polycyclic aromatics such as phenanthrene, chrysene, pyrene, benzopyrene, and pentacene of extraterrestrial origin (indicated by anomalous isotopic ratios).

Table 4.5 Composition of typical natural gases.

	Location				
	Salt Lake, Utah, USA	**Webb, Texas, USA**	**Cliffside, Texas, USA**	**Sussex, England**	**Lacq, France**
Component, vol.%					
Methane	95.0	89.4	65.8	93.2	70.0
Ethane	0.8	6.0	3.8	2.9	3.0
Propane	0.2	2.2	1.7		1.4
Butanes		1.0	0.8		0.6
Pentanes and heavier		0.7	0.5		
Hydrocarbons					
Hydrogen sulfide				1.0	15.0
Carbon dioxide	3.6	0.6			
Helium, nitrogen	0.4	0.1	25.6		
Helium			1.8		
Total	100	100	100	97.1[a]	90[a]

a) Components presents in trace quantities are not included.

Table 4.6 Physical properties of liquefied petroleum gas (LPG) components.

	Bp (at 101.3 kPa), °C,	**Vapor (at 37.8 °C), kPa**	**Liquid density, $d^{15.6}$ (at saturation pressure)**
Ethane	−88.6	–	354.9
Propane	−42.1	1310	506.0
Propylene	−47.7	1561	520.4
n-Butane	−0.5	356	583
Iso-Butane	−11.8	498	561.5
1-Butene	−6.3	435	599.6
cis-2-Butene	3.7	314	625.4
trans-2-Butene	0.9	348	608.2
n-Pentane	36.0	107	629.21

Table 4.7 Compositions (%) of typical light and heavy oils.

Fraction	**Light oil**	**Heavy oil**
Saturates	78	17–21
Aromatics	18	36–38
Resins	4	26–28
Asphaltene	Trace–2	17
Density at 15 °C, kg/l	0.831	0.886
Sulfur content, % wt.	1.1	2.84

Petroleum – a natural mineral oil – was referred to as early as in the Old Testament. The world *petroleum* means *rock oil* (from the Greek *petros* [rock] and *elaion* [oil]). It was found over the centuries seeping out of the ground – for example, in the Los Angeles basin (practically next door to where this was written) and what are now the La Brea Tar Pits. Vast deposits were found in varied places ranging from Europe, to Asia, to the Americas and Africa. In the United States, the first commercial petroleum deposit was discovered in 1859 near Titusville in western Pennsylvania, when Edwin Drake and Billy Smith struck oil in their first shallow (~20-m-deep) well. The well yielded 1.5 m^3 of oil per day. The area was known previously to contain petroleum, which residents skimmed from the surface of what was called Oil Creek.

The first oil-producing well opened up a whole new industry. The discovery was not unexpected, but it provided evidence for oil deposits in the ground that could be reached by drilling into them. Oil was used for many purposes, such as lamp illumination and even for medicinal remedies. The newly discovered Pennsylvania petroleum was soon also marketed to degrease wool, prepare paints, fuel steam engines to power light railroad cars, and so on. It was recognized that the well oil was highly impure and had to be refined to separate different fractions for various uses. The first petroleum refinery, a small stilling operation, was established in Titusville in 1860. Petroleum refining was much cheaper than producing coal oil (kerosene), and soon petroleum became the predominant source of kerosene as an illuminant. In the 1910s, the popularity of automobiles spurred the production of gasoline as the major petroleum product. California, Texas, Oklahoma, and, more recently, Alaska provided large petroleum deposits in the United States, whereas areas of the Middle East, Asia, Russia, Africa, South America, and, more recently, the North Sea became major world oil production centers.

The daily consumption of crude oil in the United States is about 61–64 thousand m^3. Most of this is used for the generation of electricity, as space heating, and as transportation fuel. About 4% of the petroleum and natural gas is used as feedstocks for manufacturing of chemicals, pharmaceuticals, plastics, elastomers, paints, and a number of other products. Petrochemicals from hydrocarbons provide many of the necessities of modern life, to which we have become so accustomed that we do not even notice our increasing dependence on them; and yet the consumption of petrochemicals is still growing at an annual rate of 10%. Advances in the petroleum-hydrocarbon industry, more than anything else, may be credited with the high standard of living we enjoy in the late twentieth century.

4.4 Coal and Its Liquefaction

Coals (the plural is deliberately used, as coal has no defined, uniform nature or structure) are fossil sources with low hydrogen content. The *structure* of coals means only structural models depicting major bonding types and components relating changes with coal rank. Coal is classified – or ranked – as lignite, subbituminous, bituminous, or anthracite. This is also the order of increased aromaticity and decreased volatile matter. The H : C ratio of bituminous coal is about 0.8, whereas anthracite has H : C ratios as low as 0.2.

From a chemical – as contrasted to geologic – viewpoint, the coal formation (coalification) process can be grossly viewed as a continuum of chemical changes, some microbiological and some thermal, involving a progression in which woody or cellulosic plant materials (the products of nature's photosynthetic recycling of CO_2) in peat swamps are converted to coals during many millions of years and increasingly severe geologic conditions. Coalification is grossly a deoxygenation-aromatization process. As the *rank* or age of the coal increases, the organic oxygen content decreases and the *aromaticity* (defined as the ratio of aromatic carbon to total carbon) increases. Lignite is a young or brown coal containing more organic oxygen functional groups than do subbituminous coals, which in turn have a higher carbon content but fewer oxygen functionalities.

The principal type of bridging linkages between clusters are short aliphatic groups $(CH_2)_n$ (where $n = 1$–4), different types of ether linkages, and sulfide and biphenyl bonds. All except the latter may be considered scissible bonds, as they can readily undergo thermal and chemical cleavage reactions.

The main approaches employed in converting coals to liquid hydrocarbons revolve around breaking down the large, complex structures generally by hydrogenative cleavage reactions and increasing the solubility of the organic portion. Alkylation, hydrogenation, and depolymerization, as well as combinations of these reactions followed by extraction of the reacted coals, are major routes taken. This process can provide clean liquid fuels such as gasoline and heating oil.

Three types of direct coal-liquefaction processes have emerged to convert coals to liquid hydrocarbon fuels. The first is a high-temperature solvent extraction process in which no catalyst is added. The liquids produced are those that are dissolved in the solvent or solvent mixture. The solvent usually is a hydroaromatic hydrogen donor, while molecular hydrogen is added as a secondary source of hydrogen.

The second, catalytic liquefaction process is similar to the first, except that there is a catalyst in direct combination with the coal. $ZnCl_2$ and other Friedel-Crafts catalysts including $AlCl_3$, as well as BF_3 phenol and other complexes, catalyze the depolymerization-hydrogenation of coals, but usually forceful conditions (375–425 °C, 10–20 MPa) are needed. Superacidic HF-BF induced liquefaction of coal involves depolymerization-ionic hydrogenation at a relatively modest 150–170 °C.

The third coal-liquefaction approach is direct catalytic hydrogenation (pioneered by Bergius) in which a hydrogenation catalyst is intimately mixed with the coal. Little or no solvent is employed, and the primary source of hydrogen is molecular hydrogen in the latter case.

The ultimate depolymerization of coal occurs in Fischer–Tropsch chemistry wherein the coal is reacted with oxygen and steam at about 1100 °C to break up, or gasify, the coal into carbon monoxide, hydrogen, and carbon dioxide. A water-gas shift reaction is then carried out to adjust the hydrogen : carbon monoxide ratio, after which the carbon monoxide is catalytically hydrogenated to form methanol or to build up liquid hydrocarbons.

4.5 Shale Gas and Tight Oil: Unconventional Fossil Fuels

The remainder of this chapter (Sections 4.5–4.7) deals with unconventional fossil fuels. After reviewing the application of unconventional fossil fuels such as shale gas and tight oil, a brief introduction is given that covers some frequently asked questions, key topics, terminology, and an overview of the work performed to date in the field is given.

On the supply side, the most noticeable phenomenon remains the American shale revolution in 2012, when the United States recorded the largest oil and natural gas production increases in the world and saw the largest gain in oil production in its history.

World primary energy production expansion across all types of energy forms plays an increasingly significant role. Renewables, shale gas, tight oil, and other new fuel sources in aggregate are growing at 6.2% p.a. and will contribute 43% of the increment in energy production to 2035.

The total recoverable shale gas in the United States was revised downward to 18.6 trillion cubic meters in 2013, and recoverable shale gas in Canada was revised upward to 16.1 TCM.

Although the prospects for shale gas production are promising, there remains considerable uncertainty regarding the size and economics of this resource. Many shale formations are so large that only a portion of the entire formation has been extensively production-tested. Most of the shale gas wells have been drilled in the last few years; as a result, there is considerable uncertainty regarding their long-term productivity. Another uncertainty is the future development of well-drilling and -completion technology, which could substantially increase well productivity and reduce production costs.

Tight oil is conventional oil that is found within reservoirs with very low permeability. The oil contained within this reservoir rock typically will not flow to the wellbore at economic rates without assistance from technologically advanced drilling and completion processes. Commonly, horizontal drilling coupled with multi-stage fracturing is used to access these difficult-to-produce reservoirs.

4.5.1 Introduction

A *hydrocarbon* is an organic compound consisting of carbon and hydrogen only. The inclusion of any atom other than carbon and hydrogen disqualifies the compound from being considered a hydrocarbon. The majority of hydrocarbons found naturally occur in petroleum (crude oil) and natural gas, where decomposed organic matter provides an abundance of many individual varieties of hydrocarbons.

Hydrocarbon fuels (gas, liquid, and solid) are those combustible or energy-generating molecular species that can be harnessed to create mechanical energy. Most liquid fuels in widespread use are derived from fossil fuels. Petroleum-based hydrocarbon fuels are well-established products that have served hydrocarbon technology and consumers for more than 100 years.

However, time is running out, and these fuel sources, once considered inexhaustible, are now being depleted at a rapid rate. In fact, there is little doubt that the supplies of crude oil are being depleted with each year that passes. In spite of all the arguments, it is not clear how long it will take to reach the bottom of the well; but based on current estimates of reserves, it should be assumed that the time frame for depletion to occur is within the next 50 years.

Unconventional natural gas and oil resources are considerable. At the present time, they are only produced marginally, due to technical and ecological difficulties and production costs.

The production cost of shale gas and shale oil varies with the gas and oil content per unit volume of reservoir rock. The production cost is highest for shale gas and oil from deep aquifers, for which the gas and oil content is particularly low.

Important developments can therefore be anticipated in the twenty-first century. Their velocity will depend on the scale of demand for shale gas and shale oil and on the technical breakthroughs that will help to achieve lower costs.

Questions and topics discussed will be as follows:

- What is shale gas or shale oil?
- What is tight shale gas or tight shale oil?
- What is tight light oil or tight oil shale?
- What is the difference between shale gas and oil shale?
- How are shale gas and shale oil produced?
- Where is shale gas found in North America?
- Tight and shale gas environmental concerns
- European shale developments: an update on the UK, Poland, Slovakia, and Ukraine
- The effect of shale gas on other industries
- Unconventional gas operator experiences
- New developments and drilling technologies

4.5.2 Glossary and Terminology

Shale gas refers to natural gas that is trapped within shale formations. *Shales* are fine-grained sedimentary rocks that can be rich sources of petroleum and natural gas. Over the past decade, the combination of horizontal drilling and hydraulic fracturing has allowed access to large volumes of shale gas that were previously uneconomical to produce. The production of natural gas from shale formations has rejuvenated the natural gas industry in the United States. Unconventional natural gas deposits are difficult to characterize overall, but in general they are lower in resource concentration, are more dispersed over large areas, and require stimulation or some other extraction or conversion technology. Extremely large natural gas in the place volumes are represented by these resources, and the United States has produced only a fraction of its ultimate potential.

Shale gas is defined as natural gas from shale formations. The shale acts as both the source and the reservoir for the natural gas. Older shale gas wells were vertical, while more recent wells are primarily horizontal and need artificial stimulation such as hydraulic fracturing in order to produce. Only shale formations with certain characteristics will produce gas. The most significant trend in US natural gas production is the rapid rise in production from shale formations. In large measure, this is attributable to significant advances in the use of horizontal drilling, well stimulation technologies, and refinement in the cost-effectiveness of these technologies. Hydraulic fracturing is the most significant among these.

Shale oil (also known as *tight oil* or *light tight oil*) is a petroleum *play* (defined shortly) that consists of light crude oil contained in petroleum-bearing formations of low permeability, often shale or tight sandstone. Economic production from tight oil formations requires the same hydraulic fracturing and often uses the same horizontal well technology used in the production of shale gas. Shale oil should not be confused with oil shale, which is shale rich in kerogen.

Thermal mode leads to the formation of thermal gas from organic water present in some sedimentary levels; this organic matter is incorporated in the sediments at the time of deposition. Debris from organisms that accumulate at the water-sediment interface is degraded by living organisms. In an anaerobic environment, degradation is slow and incomplete: the residues accumulate in the sediments to form complex macromolecular structures and debris that resist biodegradation. The overall mass is insoluble in organic solvents and constitutes *kerogen*. The accumulation of kerogen takes place nearly exclusively in fine-grained sediments, especially clays, for two reasons: (i) the hydrodynamic properties of the organic debris are similar to those of fine-grained minerals; and (ii) it is easy to create anaerobic conditions in these sediments.

Tight *oil shale* formation are heterogeneous: they vary widely over relatively short distances, and thus even in a single horizontal drill hole the amount recovered may vary, as may recovery within a field, or even between adjacent wells. This makes evaluation of plays and decisions regarding the profitability of wells on a particular lease difficult. Production of oil requires some natural gas; oil cannot be produced from a portion of a formation that contains only oil. Formations that formed under marine conditions contain less clay and are more brittle, and thus are more suitable for fracking than formations in fresh water, which may contain more clay. Formations with more quartz and carbonate are more brittle.

Unconventional natural gas and oil technology (deposits) are very often characterized by the following terms:

- *Annulus:* The space between two concentric objects, such as between the wellbore and casing or between casing and tubing, where fluid can flow.
- *Aquifer:* The subsurface layer of rock or unconsolidated material that allows water to flow within it. Aquifers can act as sources of groundwater, both usable fresh water and unusable saline water.
- *Casing:* Steel pipe placed in a well and cemented in place to isolate water, gas, and oil from other formations and maintain hole stability.
- *Carbonates:* Sedimentary rocks that are rich in calcium or magnesium carbonate, such as limestone and dolomite. The dissolution spaces (*vugs*) associated with these types of rock can contain oil or gas.
- *Completion:* The activities and methods of preparing a well for the production of oil and gas.
- *Flowback:* The flow of fracture fluid back to the wellbore after the hydraulic fracturing treatment is completed.
- *Formation:* A number of rock units that have a comparable lithology, facies, or other similar properties. Formations are not defined on the thickness of the rock units they consist of, and the thickness of different formations can therefore vary widely.
- *Horizontal drilling:* A drilling procedure in which the wellbore is drilled vertically to a kick-off depth above the target formation and then angled through a wide 90° arc such that the producing portion of the well extends horizontally through the target formation.
- *Hydraulic fracturing (aka fracking):* A method of improving the permeability of a reservoir by pumping fluid such as water, carbon dioxide, nitrogen, or propane into the reservoir at sufficient pressure to crack or fracture the rock. The opening of natural fractures or the creation of artificial fractures helps to create pathways by which oil can flow to the wellbore.

- *Multi-stage fracturing:* The process of undertaking multiple fracture stimulations in the reservoir section where parts of the reservoir are isolated and fractured separately.
- *Permeability:* The ability of fluids or oil to pass through rock. The higher the permeability number, the greater the amount of fluid or oil that can flow through the rock. Permeability is measured in a unit called *darcies*. Conventional reservoirs may have permeabilities in the tens to hundreds of millidarcies (mD) or occasionally darcy range. Unconventional or tight reservoirs usually have permeabilities in the micro to nanodarcy (one millionth of a millidarcy) range.
- *Play:* The extent of a petroleum-bearing unit within a formation.
- *Porosity:* The free space within fine-grained rock that can store hydrocarbons.
- *Propping agents/proppants:* Non-compressible material, usually sand or ceramic beads, that is added to the fracture fluid and pumped into open fractures to prop them open once the fracturing pressures are removed.
- *Reservoir:* Rock that contains potentially economic amounts of hydrocarbons.
- *Reclamation:* The act of restoring something to a state suitable for use.
- *Stimulation:* Any process undertaken to improve the productivity of the hydrocarbon bearing zone (e.g. formation fracturing).
- *Sweet spot:* The specific area within a reservoir where a large amount of gas is accessible.
- *Vug:* A small cavern or cavity within a carbonate rock.

4.5.3 Energy in 2018

Global primary energy consumption grew rapidly in 2018, led by natural gas and renewables. Nevertheless, carbon emissions rose at their highest level in seven years.

Primary energy consumption grew at a rate of 2.9% in the last year, almost double its 10-year average of 1.5% per year, and the fastest since 2010. Energy consumption growth was driven by natural gas, which contributed more than 40% of the increase. All fuels grew faster than their 10-year averages, apart from renewables, although renewables still accounted for the second-largest increment to energy growth. China, the United States, and India together accounted for more than two-thirds of the global increase in energy demand, with US consumption expanding at its fastest rate in 30 years.

The annual average oil price rose to $71.31 per barrel, up from $54.19/barrel in 2017. Oil consumption grew by an above-average 1.4 million barrels per day (b/d), or 1.5%. China (680,000 b/d) and the US (500,000 b/d) were the largest contributors to growth. Global oil production rose by 2.2 million b/d, or 2.4%. Almost all of the net increase was accounted for by the United States, with its growth in production (2.2 million b/d) a record for any country in any year. Elsewhere, production growth in Canada (410,000 b/d) and Saudi Arabia (390,000 b/d) was outweighed by declines in Venezuela (–580,000 b/d) and Iran (–310,000 b/d). Refinery throughput rose by 960,000 b/d, down from 1.5 million b/d in 2017. Nevertheless, average refinery utilization climbed to its highest level since 2007.

Natural gas consumption rose by 195 billion cubic metres (bcm), or 5.3%, one of the fastest rates of growth since 1984. Growth in gas consumption was driven mainly by the United States (78 bcm), supported by China (43 bcm), Russia (23 bcm), and Iran (16 bcm). Global natural gas production increased by 190 bcm, or 5.2%. Almost half of this came from the United States (86 bcm), which (as with oil production) recorded the largest annual growth seen by any country in history. Russia (34 bcm), Iran (19 bcm), and Australia (17 bcm) were the next largest contributions to growth. Growth in inter-regional natural gas trade was 39 bcm or 4.3%, more than double the 10-year average, driven largely by continuing rapid expansion in liquefied natural gas (LNG).

LNG supply growth came mainly from Australia (15 bcm), the United States (11 bcm), and Russia (9 bcm). China accounted for around half of the increase in imports (21 bcm).

Coal consumption grew by 1.4%, double its 10-year average growth. Consumption growth was led by India (36 million tons of oil equivalent [MTOE]) and China (16 MTOE). OECD demand fell to its lowest level since 1975. Coal's share in primary energy fell to 27.2%, its lowest in 15 years. Global coal production rose by 162 MTOE, or 4.3%. China (82 MTOE) and Indonesia (51 MTOE) provided the largest increments.

Renewable power grew by 14.5%, slightly below its historical average, although its increase in energy terms (71 MTOE) was close to the record-breaking increase of 2017. Solar generation grew by 30 MTOE, just below the increase in wind (32 MTOE), and provided more than 40% of renewables growth. By country, China was again the largest contributor to renewables growth (32 MTOE), surpassing growth in the OECD (26 MTOE) for the first time. Hydroelectric generation increased by an above-average 3.1%, with European generation rebounding by 9.8% (12.9 MTOE), almost offsetting its steep decline in the previous year. Nuclear generation rose by 2.4%, its fastest growth since 2010. China (10 MTOE) contributed almost three-quarters of global growth, with Japan (5 MTOE) the second largest increase.

Electricity generation rose by an above-average 3.7%, buoyed by China (which accounted for more than half of the growth), India, and the United States. With regard to fuel, renewables accounted for a third of the net increase in power generation, followed closely by coal (31%) and then natural gas (25%). The share of renewables in power generation increased from 8.4 to 9.3%. Coal still accounted for the largest share of power generation at 38%.

Cobalt and lithium production rose by 13.9% and 17.6%, respectively, both well in excess of their 10-year average growth rates. Cobalt prices rose 30% to their highest levels since 2008, while lithium carbonate prices increased by 21% to new highs.

4.5.4 Energy Outlook 2035

According to the BP Energy Outlook (https://www.bp.com/en/global/corporate/energy-economics/energy-outlook.html), world primary energy production will grow at 1.5% p.a. from 2012 to 2035, matching consumption growth. Growth is concentrated in the non-OECD, which accounts for almost 80% of the volume increment. There is growth in all regions except Europe. Asia Pacific shows both the fastest rate of growth (2.1% p.a.) and the largest increment, providing 47% of the increase in global energy production. The Middle East and North America are the next largest sources of growth, and North America remains the second largest regional energy producer.

There is expansion across all types of energy, with new energy forms playing an increasingly significant role. Renewables, shale gas, tight oil, and other new fuel sources in aggregate are growing at 6.2% p.a. and will contribute 43% of the increment in energy production to 2035. The growth of new energy forms is enabled by the development of technology and underpinned by large-scale investments.

By 2035, the growth of tight oil (0.78 million tons per day [MTPD]), biofuels (0.26 MTPD), and oil sands (0.45 MTPD) alone will account for 60% of global growth and all of the net increase in non-OPEC production. Tight oil will account for 7% of global supplies in 2035, while biofuels and oil sands will obtain market shares of 3 and 5%, respectively. North America will dominate the expansion in unconventionally with 65% of global tight oil and with Canada responsible for all the world's oil sands production. In the United States, the increase in tight oil production coupled with declining demand will continue the dramatic shift in import dependence. Imports are set to decline from a peak of well over (1.63 MTPD), or 60% of demand in 2005, to just (0.136 MTPD), or less than 10% demand in 2035.

Global gas supply is expected to grow by 1.9% p.a. or 4.82 Bm^3/d over the outlook period, reaching a total of 13.92 Bm/d by 2035. Shale gas will be the fastest-growing source of supply (6.5% p.a.), providing nearly half of the growth in global gas. Gas supply growth will be concentrated in the non-OECD (3.53 Bm^3 or 2.1% p.a.), accounting for 73% of global growth. Almost 80% of non-OECD growth will be from non-shale sources. OECD supply growth (1.5% p.a.) will come exclusively from shale gas (5.1% p.a.), which will provide nearly half of OECD gas production by 2035.

Shale gas supply is dominated by North America, which will account for 99% of shale gas supply until 2016 and for 70% by 2035. However, shale gas growth outside North America will accelerate and by 2027 will overtake North American growth. China is the most promising country for shale growth outside North America, accounting for 13% of world shale gas growth; together, China and North America will account for 81% of shale gas by 2035.

Perhaps the most dramatic adjustments to shale gas are seen in trade flows. The United States is set to shift from a net importer of gas today to a net exporter in 2018, with net exports reaching 0.30 Bm^3/d by 2035. It will become a net LNG exporter by 2016, reaching a total net LNG export volume of 0.31 Bm^3/d by 2035.

US domestic gas production has been revitalized by the shale gas "revolution." US shale gas output will exceed the highest level ever achieved by conventional gas production in the United States by 2029. By 2035, shale gas production will be just short of US total gas output in 2012. How will the market accommodate this shale gas shock? All the elements of the supply and demand balance will adjust, responding to relative price movements. Conventional supply will decline faster than it would have in the absence of shale; gas will gain share in the various segments of US energy markets, in competition with other fuels; and US gas will gain share in the international market in competition with other suppliers.

US shale gas output will grow by 4.3% p.a. between 2012 and 2035, enabling US gas production to rise by 45%. This is causing a series of adjustments in energy markets; some are already evident, and others will develop over time. Next, gas will gain market share in the industrial sector, from 39% in 2012 to 42% by 2035. And finally, gas will start to penetrate the transport sector. Gas is the fastest-growing fuel (18% p.a.) in a sector where overall demand is falling (−0.9% p.a.). By 2035, gas will account for 8% of US transport sector fuels, almost matching biofuels.

4.6 Shale Gas

Natural gas is playing a growing energy role. The scale of its reserves and its environmental advantages favor its use for fast-growing activities such as precision industries and generation of electricity. Although the shale gas potential of many nations has been significant, as of 2013, only the United States, Canada, and China produce shale gas in commercial quantities, and only the United States and Canada have significant shale gas production. There is a high abundance of ethane in shale gas, and more ethane is produced than US ethane crackers can consume. This excess opens the doors for ethane export, as many ethylene producers outside of the United States want to take advantage of the low ethane price. The margins are large enough that even shipping ethane and converting it to ethylene becomes profitable (Figure 4.1). Pipelines and export terminals are ready for the future export of ethane, as the projections estimate that 8×10^6 t of ethane will be exported by 2022.

Ethylene producers located mainly in India, Brazil, Canada, and several European countries will import ethane to feed their crackers (Section 7.7). To date, no shipment to China has been confirmed. Olefin production in China still mainly depends on naphtha cracking; hence, shifting toward ethane could increase China's competiveness. China is undergoing two significant developments that are driven by a fast-growing dependence on imported oil: coal-based chemical production and propane dehydrogenation (PDH, Section 8.1). These developments affect not only the domestic Chinese chemical market, but also the global market. By 2018, ethylene and propylene capacity from coal to olefins (CTO), methane to olefins (MTO), and PDH plants will account for nearly 40% of the country's olefin capacity. However, naphtha crackers will produce the largest fraction of the total ethylene production capacity.

Table 4.8 is based on data collected by the Energy Information Administration (EIA) of the United States Department of Energy. Numbers for the estimated amount of recoverable shale gas resource are provided alongside numbers for proven natural gas reserves.

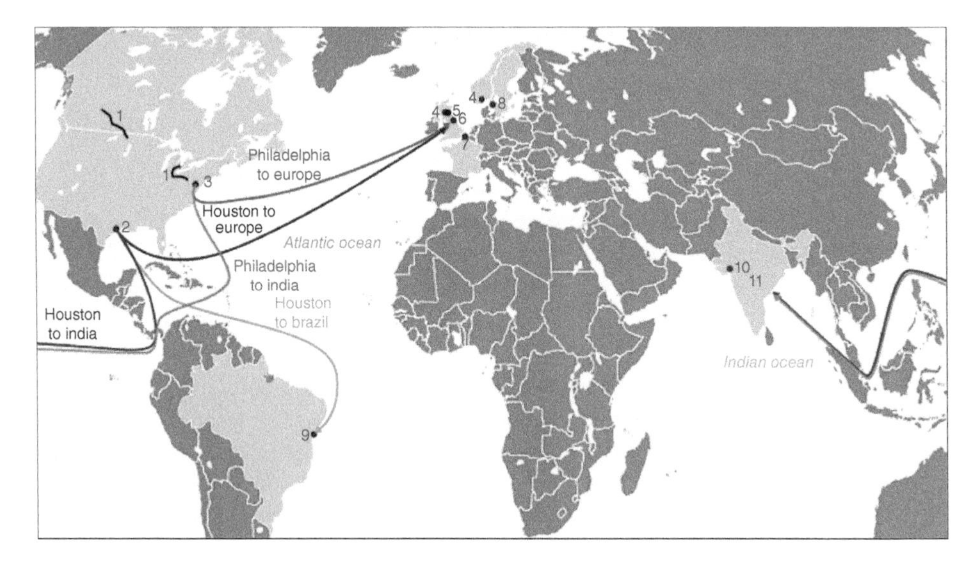

Figure 4.1 Pipelines and shipping routes for the export of ethane to ethylene-producing facilities outside of the United States.

Table 4.8 Shale gas by country.

Country	Estimated recoverable shale gas (trillion cubic meters)	Proven natural gas reserves of all types (trillion cubic meters)
China	31.22	3.47
Argentina	22.46	0.34
Algeria	19.80	4.45
United States	18.62	8.90
Canada	16.07	1.90
Mexico	15.26	0.48
South Africa	13.58	–
Australia	12.24	1.20
Russia	7.98	47.26
Brazil	6.86	0.39

The US EIA had made an earlier estimate of total recoverable shale gas in various countries in 2011, which for some countries differed significantly from the 2013 estimates. The total recoverable shale gas in the United States, which was estimated at 24.14 trillion cubic meters in 2011, was revised downward to 18.62 trillion cubic meters in 2013. Recoverable shale gas in Canada, which was estimated to be 10.86 TCM in 2011, was revised upward to 16.07 TCM in 2013.

Shale gas is found in shale plays, which are shale formations containing significant accumulations of natural gas and which share similar geologic and geographic properties.

A decade of production has come from the Barnett Shale play in Texas. Experience and information gained from developing the Barnett Shale have improved the efficiency of shale gas development the country. Another important play is the Marcellus Shale in the eastern part of the United States. US geophysicists and geologists identify suitable well locations in areas with potential for economical gas production by using surface and subsurface geology techniques and seismic techniques to generate maps of the subsurface.

Conventional gas reservoirs are created when natural gas migrates toward the Earth's surface from organic-rich source formation into highly permeable reservoir rock, where it is trapped by an overlying layer of impermeable rock. In contrast, shale gas resources form within organic-rich shale source rock. The low permeability of shale greatly inhibits the gas from migrating to more permeable reservoir rocks. Without horizontal drilling and hydraulic fracturing, shale gas production would not be economically feasible because the natural gas would not flow from the formation at high enough rates to justify the cost of drilling.

Shale gas was first extracted as a resource in Fredonia, New York, in 1821, in shallow, low-pressure fractures. Horizontal drilling began in the 1930s, and in 1947 the first well was fracked in the United States. George P. Mitchell is regarded as the father of the shale gas industry, by making it commercially viable in the Barnett Shale by getting costs down to $4 per million British thermal units. Mitchell Energy achieved the first economical shale fracture in 1998 using slick-water fracturing. Since then, natural gas from shale has been the fastest-growing contributor to total primary energy in the United States and has led many other countries to pursue shale deposits. According to the International Energy

Agency (IEA), shale gas could increase technically recoverable natural gas resources by almost 50%.

4.6.1 Geology

Because shales ordinarily have insufficient permeability to allow significant fluid flow to a wellbore, most shales are not commercial sources of natural gas. Shale gas is one of a number of unconventional sources of natural gas; others include coalbed methane, tight sandstones, and methane hydrates. Shale gas areas are often known as *resource plays* (as opposed to *exploration plays*). The geological risk of not finding gas is low in resource plays, but the potential profits per successful well are usually also lower.

Shale has low matrix permeability, so gas production in commercial quantities requires fractures to provide permeability. Shale gas has been produced for years from shales with natural fractures; the shale gas boom in recent years has been due to modern fracking technology to create extensive artificial fractures around wellbores. Horizontal drilling is often used with shale gas wells, with lateral lengths up to 3000 m within the shale to create maximum borehole surface area in contact with the shale (Figure 4.2).

Shales that host economic quantities of gas have a number of common properties. They are rich in organic material (0.5–25%) and are usually mature petroleum source rocks in the thermogenic gas window, where high heat and pressure have converted petroleum to natural gas. They are sufficiently brittle and rigid enough to maintain open fractures.

Some of the gas produced is held in natural fractures and some in pore spaces, and some is adsorbed onto the organic material. The gas in the fractures is produced immediately; the gas adsorbed onto organic material is released as the formation pressure is drawn down by the well.

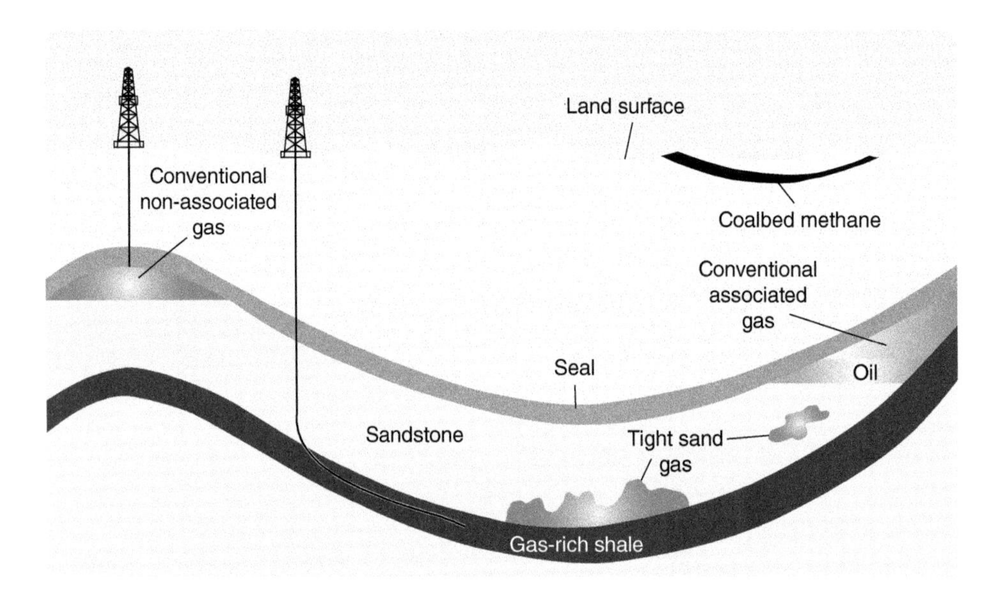

Figure 4.2 Schematic geology of natural gas resources.

4.6.2 Formation of Natural Gas Reservoirs

4.6.2.1 General

A natural gas reservoir occupies the intergranular porosity or the cracks of a rock that is called a *reservoir rock* or simply a *reservoir*. The permeability of this rock must be sufficient to obtain a gas flow enabling profitable production. Gas reservoirs are classed as *conventional* or *unconventional*. The former, resulting from large gas accumulations, can be exploited using current techniques. In the latter, the gas is stored under specific conditions and requires special and costly techniques to be produced.

A reservoir that contains gas also contains water, a fluid that is omnipresent in sediments, and very frequently oil. All or part of the gas may be dissolved in the water or oil, depending on the thermodynamic conditions prevailing in the reservoir and the chemical species present. If the gas phase is present, its density is lower than that of the water or oil, and it tends to move upward as a result of buoyancy. Hence, it can only remain in the reservoir if prevented from escaping by a barrier.

This barrier is either a permeability barrier, usually consisting of shale or salt, or a hydrodynamic barrier, in which case the movement of water counteracts the buoyancy. The reservoir-barrier combination constitutes a trap. Figure 4.3 gives an example of an unconventional trap in which gas formed in the deep parts of a basin (Alberta Basin, Canada) advances upward due to buoyancy. However, the permeability of the formation in which the gas advances is low, and the movement of the gas is extremely slow due to the existence of high capillarity forces that oppose its motion. In this figure, the scale marked in % R_m corresponds to the increase in the vitrine reflectance as a function of depth. It is assumed that the gas begins to be formed in significant quantities from an R_m of 1.3%. The gas located above this limit was therefore most likely formed deeper in the basin.

Most barriers, except for salt, are incapable of completely stopping the upward movement of gas. This gives rise to leaks, and the process is called dysmigration. Naturally, dysmigration is rarely perceptible on the human timescale. At the geological timescale, however, a gas reservoir appears as a temporary underground repository that only lasts as long as it is replenished at a higher rate than its losses.

The reservoir rock participates in the geological evolution of the basin. It may be dragged deeper by subsidence or rise toward the surface because of erosion of the overlying sediments, causing a change in the thermodynamic conditions prevailing therein. It may also be deformed by tectonics, cemented by mineral recrystallization, or even destroyed by erosion if it reaches the surface. Natural gas reservoirs are therefore complex objects. Each of them has its own individual history.

4.6.2.2 Unconventional Reservoir

Unconventional reservoirs are not distinguished by any particular configuration, and their formation is based on the same principles. They include the following types:

- *Reservoirs of tight sandstones or relatively impermeable sandstones*: The permeability of these sandstones is so low as to make it impossible to obtain a sufficient gas flow by simple drilling. The largest tight sandstone reservoirs, found in the United States, contain gas without associated oil.

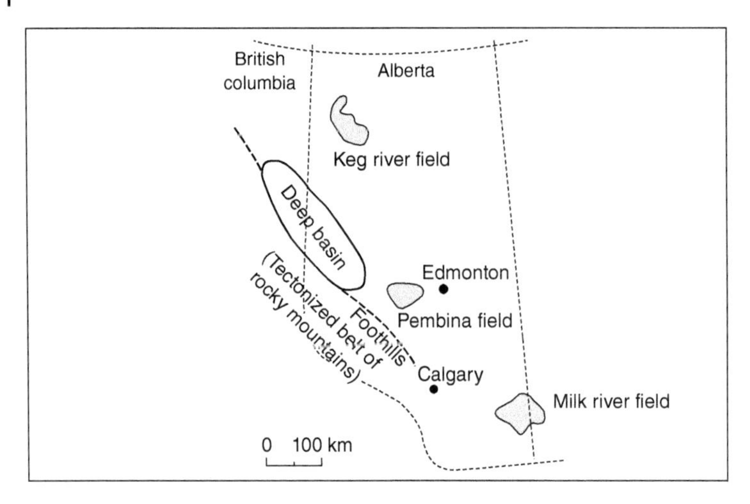

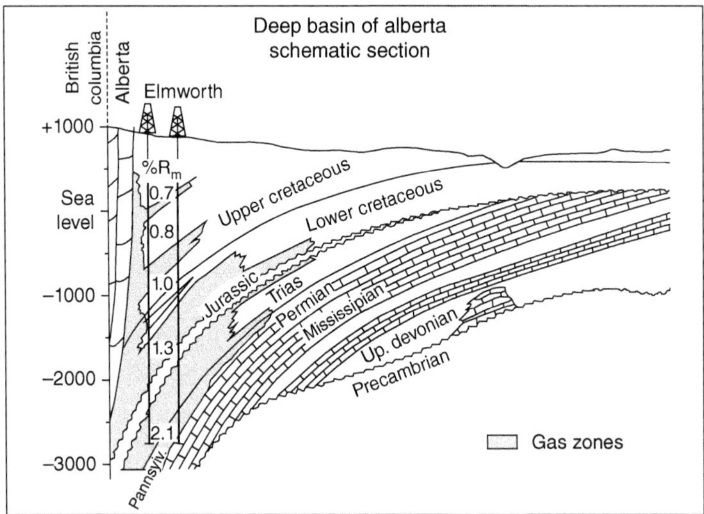

Figure 4.3 Deep-basin gas reservoir in Alberta (Canada).

- *Fractured shales*: These are shales containing millimetric cracks that are filled with gas. The cracks are supplied by the surrounding shale, but at a very low rate on the human timescale. These shales are in fact source rocks still containing a part of the gas formed therein. A classic example is the Albany Shales in the United States. These reservoirs also contain gas without associated oil.
- *Coal mine gas*: It is well known that gas or *firedamp* is found in many coal seams. This gas is rich in methane and often contains high proportions of carbon dioxide and nitrogen. It is formed during the burial of the coal, which is its source rock, and where it remains trapped. It is liberated by the decompression caused by mining operations. It is generally removed by ventilation, but its recovery is sometimes profitable.
- *Deep aquifer gas*: In these reservoirs, gas is found dissolved in water. One example is provided by the deep aquifers of the Mexican Gulf coast in the United States, already mentioned. The gas is pure methane, and the quantities dissolved range from 10 to

$20\,Nm^3/m^3$ of water. This region features projects for the geothermal tapping of deep aquifers, and the recovery of the methane present in the water is considered to be an important factor for the profitability of the operation.

This main unconventional sources of natural gas are:

- Low-permeability sandstone reservoirs
- Fractured shales
- Coalbed methane
- Methane dissolved in geopressurized aquifers
- Hydrate reservoirs

Unconventional sources of natural gas are considerable but have been tapped very little so far for economic reasons.

Gold and Sauter postulated the presence of large quantities of methane of inorganic origin in the deep layers of the Earth's crust. Tests performed to detect deep methane failed to yield positive results, and it has already been reported in that Gold's theory appears highly speculative today. However, as pointed out by Nederlof, even if the organic origin of the oil and gas reservoirs discovered so far in the sedimentary basins has been fully established, this does not necessarily mean that methane of a different origin cannot be found in the far deeper zones of the Earth's crust or the mantle.

4.6.2.3 Low-Permeability Gas Reservoirs

Large amounts of natural gas exist in tight sandstone reservoirs, characterized by permeability lower than 30 μD and low porosity (7–12%). The gas saturation of the porous medium is generally low, about 50% versus about 80% in usual reservoirs.

The combination of these factors implies low well productivity. A similar situation is found when the pressure in the reservoir is relatively low (3–4 MPa), with permeability ranging between 30 and 100 μD.

Low permeability may stem from two factors:

- The mineralogical composition of the porous medium. The presence of a mixture of shales and fine sediments leads to the formation of a dense, nonporous medium.
- The depth of the reservoir, which causes compaction of the porous medium.

Takach et al. investigated the stability of methane at the temperature and pressure conditions implied by extreme depth. They concluded that methane remains stable up to a depth of at least 12 km, but in the presence of carbonates and water, the carbon dioxide content may increase.

The production of tight reservoirs requires the application of relatively complex and costly methods. The depth of some reservoirs raises another problem and increases the cost of drilling operations.

The main solution employed to improve productivity is hydraulic fracturing. This helps to fracture the porous medium and thus to boost the productivity of the wells, which communicate with the fractures created.

Another promising technique for improving well productivity is horizontal drilling. It substantially increases the length of the drain and thus boosts well production by a factor of up to four or five in favorable situations.

4.6.2.4 Fractured Shales

Shale Formations Containing Gas Shales are usually considered impervious and can form a cap rock for a gas reservoir. In some shales, however, relatively high content of organic matter originally present have led to the formation of natural gas contained in natural fractures of the rock.

The Devonian shales in the eastern United States are the best known. These shales were formed from sediments deposited 350 million years ago in a shallow sea, which covered a large portion of the eastern United States. The kerogen present in the rock was transformed during its burial into methane and bitumen, and gas accumulated in the formation fractures. Reservoirs of the same type most likely exist in the rest of the world. For them to contain natural gas, it is first necessary for the geological conditions allowing the formation of methane from kerogen to be satisfied. The tectonic evolution of the formation should also have created sufficient porosity by natural fracturing of the rock.

Gas Production from Fractured Shales Natural gas production from fractured shales raises tremendous problems. As in the case of tight sandstone reservoirs, horizontal drilling and hydraulic fracturing are the most promising techniques to improve the productivity of these reservoirs.

Hydraulic fracturing has the disadvantage of creating an essentially unidirectional fracture. This is why another fracturing method, based on the use of a slow-action explosive, has been investigated. It generates a set of radial fractures, ensuring better communication between the well and the natural fractures in the rock.

The explosive helps to produce a very short pressure peak. Multiple fractures appear if the pressure buildup takes place in a time interval between 0.05 and 5 ms, depending on the diameter of the well, and the shortest times correspond to the smallest diameters.

Horizontal drilling also helps to improve well productivity. Horizontal drilling tests have been conducted in a shale formation with positive results, and this technique appears to be promising for these reservoirs. Horizontal drilling can also be combined with hydraulic fracturing.

4.7 Tight Oil

Crude oil, also known as *petroleum* or *fossil fuel*, is found in some rock formations deep below the earth's surface. Crude oil forms the foundation for the petroleum industry and is relied upon for fuels as well as feedstocks for the petrochemical industry.

Oil is commonly defined as either *heavy* or *medium-to-light* grade depending on the density of the hydrocarbons and its ability to flow. *Heavy oil* generally refers to crude oil that is too viscous for pipeline transport without dilution, or oil that is mined in the oil sands in Northern Alberta. *Conventional oil*, which is referred to as *light* or *medium* in grade, is found in reservoir rocks that have enough permeability (the ability for a fluid to move through a rock formation) to allow the oil to flow to a vertical or horizontal well.

Tight oil is conventional oil that is found within reservoirs with very low permeability. The oil contained within these reservoir rocks typically will not flow to the wellbore at economic rates without assistance from technologically advanced drilling and completion processes. Commonly, horizontal drilling coupled with multi-stage fracturing is used to access these difficult-to-produce reservoirs.

Oil is trapped within the open spaces in the rock (called *porosity*). This porosity may be in the form of the small spaces between grains in sandstone or as small, open vugs, or cavities within carbonates (limestone or dolomite type rocks). For the reservoir to flow oil to a wellbore, the rock must have some form of permeability, either in interconnected pathways between pore spaces or in natural fractures found in the rock. The percentage of pore volume, or void space within the rock, is generally less than 30%, and in tight oil reservoirs is commonly less than 10%. The amount of oil stored within a reservoir is directly related to the porosity of the reservoir and other geological characteristics.

Crude oil has a number of characteristics or properties that allow it to be classified into different types. One of the main properties of oil is its density. The higher the density, the more resistant it is to flowing in the reservoir. A measure of a fluid's resistance to flow is termed *viscosity*. Most tight oil produced is of the medium to light variety, with lower viscosity.

The world has relied extensively on the production of oil for many years and continues to be dependent on it as the primary source of transportation fuels. As countries continue to utilize oil resources, there is a natural decline in production as resources are depleted due to easy access.

Extensive oil and gas resources are known to be present in tight oil reservoirs; however, they require additional technology to enable them to be produced. Tight oil is of high quality but commonly found in regions where reservoir properties inhibit production using conventional drilling and completion techniques. The oil itself requires very little refinement, and in many cases existing surface infrastructure can be utilized, reducing both surface impact and capital investment.

Industry and government often report the original oil in place (OOIP) resource for regions or geological formations that are believed to have, or are proven to contain, oil and gas potential. The OOIP is simply the amount of oil that is trapped within the reservoir underground. This amount is often many times larger than the actual amount of oil industry is capable of recovering. With the application of horizontal drilling and multi-stage hydraulic fracturing, industry is successfully extracting additional oil from these reservoirs.

Tight oil is found throughout Canada's known oil-producing regions, as well as numerous basins in the United States. The industry has explored for and developed conventional oil reservoirs for many years. As these resources have diminished, companies have expanded their search to look at new sources of oil, such as shale oil and other tight oil reservoirs. Drilling and completion technologies used to produce unconventional resources are also being applied to increase oil production from some conventional oil reservoirs where recovery has been low (Figure 4.4).

4.7.1 Types of Tight Oil Plays

In the oil and gas industry, the types of oil and gas deposits are generally classified into different categories called *plays*. Plays are differentiated based on geology and the

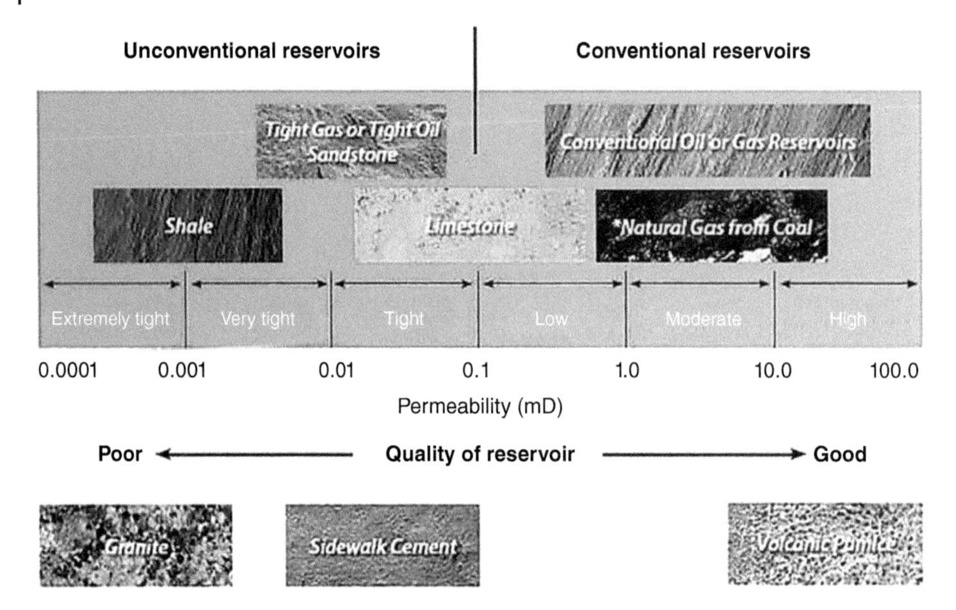

Figure 4.4 Conventional and unconventional reservoirs.

technology required to produce the oil. There are numerous play types associated with tight oil, including conventional oil plays and unconventional oil plays.

In some existing oil fields, the fringe regions or *halos* surrounding the areas of historical production are known to contain oil. The reservoir properties in the halo are not as favorable as those within the previously developed area. Applying new technologies, such as horizontal drilling, allows oil to be recovered from the halo. Examples of this type of play are the Cardium and Vikong formations in Western Canada.

4.7.1.1 Geo-Stratigraphic Play

This type of play describes a geological formation known to contain significant oil resources over a large geographic region. This type of play also requires the use of advanced technology to yield economic oil production. The Middle Bakken Formation, which occurs in parts of Saskatchewan, North Dakota, and Montana, is an example of this type of play. It contains oil that has been sourced from the overlying and underlying organic rich shale units.

4.7.1.2 Shale Oil Play

In shale oil plays, the rock material is predominantly organic-rich shale, which contains oil. The rock is not only the source of the oil but also the reservoir. Shale reservoirs tend to have tighter permeability than sand or carbonate tight oil reservoirs and may require a different type of completion technique. An example of a shale oil reservoir with potential to produce would be the Exshaw formation in southern Alberta.

4.7.2 Technologies Used to Recover Tight Oil

The oil that is produced or extracted from tight reservoirs is the same type of oil produced from conventional reservoirs. It is the application of advanced technologies that makes these developments unconventional. Different technologies are used for different plays, but the most common methods used today are horizontal drilling and multi-stage hydraulic fracturing.

4.7.2.1 Horizontal Drilling

The purpose of drilling a horizontal well is to increase the contact between the reservoir and the wellbore. Wells are drilled vertically to a predetermined depth (typically 1000–3000 m below the surface depending on location) above the tight oil reservoir. The well is then *kicked off* (turned) at an increasing angle until it runs parallel within the reservoir. Once horizontal, the well is drilled to a selected length, which can extend up to 3–4 km. This portion of the well is called the *horizontal leg* (Figure 4.5).

4.7.2.2 Hydraulic Fracturing

Tight oil reservoirs require some form of stimulation once the well has been drilled. The most common type of stimulation used by the oil and gas industry is hydraulic fracturing (fracking). This process applies pressure by pumping fluids into the wellbore, which opens existing or creates new fractures or pathways in the reservoir through which the oil can flow to the wellbore. In conventional oil reservoirs, the reservoir permeability is sufficient that hydraulic fracturing may not be needed to achieve economic production rates.

In unconventional oil, the reservoir permeability is typically very low, and additional pathways must be created to enable the flow of hydrocarbons.

To create the fractures, fluids are pumped under pressure from surface into the reservoir, many hundreds to thousands of meters below ground. The type of fracture fluids used varies depending on the reservoir characteristics. Commonly, water is used as a base fluid. Generally, 3–12 additives are added to the water, based on the characteristics of the source water and also of the formation to be fractured. These additives represent 0.5–2% of the total fracturing fluid volume. Their purpose is to reduce friction, prevent microorganism growth/biofouling, and prevent corrosion. As part of the stimulation process, once the fractures have been created, sand or ceramic beads (proppants) are pumped into these small openings to hold open the fractures created. The volume of fracture fluid and proppant used for each hydraulic fracture varies dependent upon the anticipated production rates following the treatment.

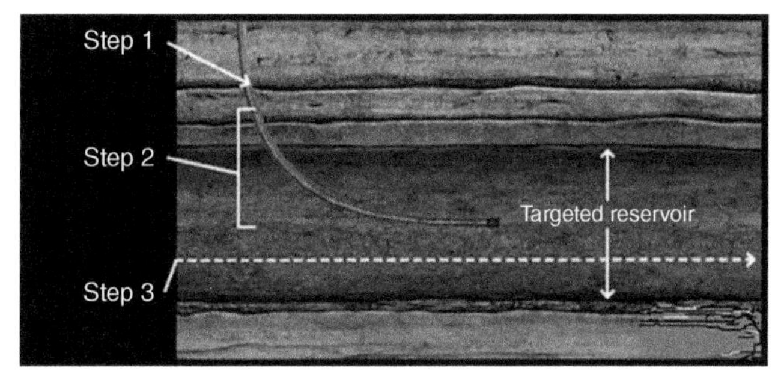

Figure 4.5 Horizontal drilling. Step 1: Drill vertically until the wellbore reaches a point above the targeted reservoir. Step 2: Kick off and begin to drill at an increasing angle until the wellbore runs horizontally through the targeted reservoir. Step 3: Drill horizontally to the desired length.

In tight oil wells, the hydraulic fracturing process typically involves multiple stages along the wellbore. Each stage is isolated using packers or plugs to contain the fracturing fluids and ensure that the fracture grows in the planned direction and distance.

4.7.2.3 Microseismic Events

Magnitude A seismic event may occur from natural or man-made (induced) causes that create sound waves in the earth. It may be caused by an event ranging from a devastating earthquake to something as common as dropping an object from your hand. Microseismic events, as the name suggests, are approximately 1 million times smaller than any tremor that may be felt by a human. Microseismic events associated with hydraulic fracturing are manmade events generated during the process, which create pathways for hydrocarbons to flow to the wellbore. These microseismic events are very small; they release energy roughly equivalent to a gallon of milk falling from a kitchen counter; and their detection, as explained next, requires sensitive and sophisticated equipment. The fractures, or cracks, are generally only wide enough to allow a grain of sand or small ceramic bead to become lodged within them, providing the path for hydrocarbon flow.

Monitoring During fracture stimulation operations, it is important to know where the fractures are being created in the reservoir. Monitoring of the fracturing process in real time can be accomplished by using a variety of techniques. Pressure responses and microseismic monitoring are two such techniques.

Measuring microseismic events that occur as the fracture stimulation takes place provides industry professionals with visual evidence that fractures are being developed both vertically as well as horizontally. Because these microseismic events are measured in real time, immediate adjustments can be made during the operation to ensure that the fractures created stay within the zone that has production potential. The magnitude of seismic events created using hydraulic fracturing techniques is many times smaller than events that can be felt at surface.

Following completion of fracturing operations, the microseismic model can be used to define the limit and reach of each fracture stimulation in the wellbore(s). The horizontal and vertical model is also used to define recoverable resources, areas of insufficient stimulation, and visual assurance that potential groundwater sources are protected.

4.7.3 Initial Production

Once pathways have been created within the tight oil reservoir, allowing the oil to flow to the wellbore, conventional methods are used to produce the well. These can include pump jacks, which lift the oil to the surface; storage tank facilities (commonly referred to as *batteries*); and pipelines and trucks used for transport. Well production is commonly robust in the early stages of production but declines over time.

4.7.3.1 Infill Drilling

In many cases, tight oil development is used to increase the overall recovery of oil from an existing field. Infill wells are located among existing conventional wells. The purpose of these wells is to extract additional oil, which has not been recovered using conventional production technology. In contrast, halo wells are located on the fringes of the existing field and rely upon the utilization of new technology to expand the boundaries of the productive zone or sweet spot within the oil-bearing formation.

4.7.3.2 Wellbore Construction and Groundwater Protection

Proper well construction isolates the wellbore: a critical step taken by the oil and gas industry to protect potential groundwater sources that may be encountered during the drilling process. Typically, three different sets of steel casings are individually cemented into the wellbore to provide barriers that isolate wellbore fluids from rock intervals.

After each string of casing is installed in the well, cement is pumped down the center of the casing (surface, intermediate, or production) and circulates back to the surface in the space outside of the casing. This space is commonly referred to as the *annulus*. After each of these steps is completed, the cement is allowed to set prior to the continuation of drilling; and in some places, a cement bond geophysical log is run to determine the integrity of the cement that surrounds the casing. This extra measure is taken to ensure that the wellbore is adequately cemented and capable of withstanding the pressures associated with hydraulic fracturing. Prior to stimulation, the well is pressure tested to ensure the integrity of the casing system that has been installed in the ground.

4.7.3.3 Minimizing Footprint

Economical production of oil from low-permeability reservoirs has been made commercially viable through the application of technologies such as horizontal drilling and multi-stage hydraulic fracturing. Companies are striving to reduce environmental impacts and to minimize costs associated with tight oil development. The application of multiple wells from a single pad has been recognized as an opportunity to achieve both of these objectives. While the size of a multi-well pad is slightly bigger than a regular oil and gas lease, the cumulative footprint for a tight oil field development is much smaller than it would be using vertical wells. Fewer access roads are required, and the concentration of facilities and pipelines within the pad minimizes surface disturbance.

During the drilling and completion of the well, there is a significant space requirement for the equipment used. Once the well has been completed and commercial production initiated, the lease site requirements are typically reduced by as much as 50%. The space that had been required for drilling and completion activities is reclaimed to the condition found prior to industry activity.

4.7.4 Environmental Impacts of Natural Gas

The extraction and use of shale gas can affect the environment through the leaking of extraction chemicals and waste into water supplies, the leaking of greenhouse gases during extraction, and pollution caused by the improper processing of natural gas. A challenge to preventing pollution is that shale gas extraction varies widely in this regard, even between different wells in the same project; processes that reduce pollution sufficiently in one extraction may not be enough in another.

In 2017, methane accounted for about 10.2 % of all U.S. greenhouse gas emissions from human activities. Human activities emitting methane include leaks from natural gas systems and the raising of livestock. Methane is also emitted by natural sources such as natural wetlands. In addition, natural processes in soil and chemical reactions in atmosphere help remove CH_4 from atmosphere. Methane's lifetime in the atmosphere is much shorter than that of CO_2, but methane is more efficient at trapping radiation than CO_2. The comparative impact of CH_4 is more than 25 times greater than CO_2 over a 100-year period.

Globally, 50–60% of total CH_4 emissions come from human activities. Natural gas and petroleum systems are the largest source of CH_4 emissions in the United States.

4.7.4.1 Water and Air Quality, Methane, and Other Important Greenhouse Gases

Chemicals are added to the water to facilitate the underground fracturing process that releases natural gas. Fracturing fluid is primarily water and approximately 0.5% chemical additives (friction reducer, agents countering rust, agents killing microorganisms). Since (depending on the size of the area) millions of liters of water are used, this means hundreds of thousands of liters of chemicals are often injected into the subsurface. About 50–70% of the injected volume of contaminated water is recovered and stored in aboveground ponds to await removal by tanker. The remaining volume remains in the subsurface; hydraulic fracturing opponents fear that it can lead to contamination of groundwater aquifers, though the industry deems this "highly unlikely." However, the waste water from such operations often leads to foul odors and heavy metals contaminating the local water supply above-ground.

Besides using water and chemicals, however, it is also possible to frack shale gas with only liquefied propane gas. This reduces the environmental degradation considerably. GasFrac, in Alberta, Canada, invented the method.

4.7.4.2 Earthquakes

Hydraulic fracturing routinely produces microseismic events too small to be detected except by sensitive instruments. As mentioned earlier, these microseismic events are often used to map the horizontal and vertical extent of fracturing. However, as of late 2012, there have been three instances of hydraulic fracturing through induced seismicity, triggering quakes large enough to be felt by people: one each in the United States, Canada, and England. All were too small to cause damage.

4.7.5 Conclusion

Shale gas is natural gas that is found trapped within shale formations. It has become an increasingly important source of natural gas in the United States since the start of this century, and interest has spread to potential gas shales in the rest of the world. The EIA predicts that by 2035, 43% of the United States' natural gas supply will come from shale gas. Although the shale gas potential of many nations has been significant, as of 2013, only the United States, Canada, and China produce shale gas in commercial quantities, and only the United States and Canada have significant shale gas production. The US administration believes that increased shale gas development will help reduce greenhouse gas emissions; however, some studies have alleged that it causes the release of more greenhouse gases than conventional natural gas.

Oil recovered from shale and other low-permeability formations is known as tight oil, often in shale or light sandstone. Economic production from tight oil formations requires the same hydraulic fracturing and often uses the same horizontal well technology used in the production of shale gas. It should not be confused with oil shale, which is shale rich in kerogen.

Natural gas production from fractured shales raises tremendous problems. As in the case of tight sandstone reservoirs, horizontal drilling and hydraulic fracturing are the most promising techniques to improve the productivity of these reservoirs. The composition of natural gas reservoirs varies considerably.

4.8 Heavy Oils, Shale, and Tar Sand

Whereas in present-day refining operations light crudes are preferred, increasingly, heavy petroleum sources also must be produced to satisfy ever-increasing needs. These range from commercially usable *heavy oils* (California, Venezuela, etc.) to huge petroleum reserves locked up in *shale* or *tar sand* formations. These more unconventional hydrocarbon accumulations exceed the quantity of oil present in all the rest of the oil deposits in the world taken together. The largest is located in Alberta, Canada, in the form of enormous tar sand and carbonate rock deposits containing some 2.5–6 trillion barrels of extremely heavy oil, called *bitumen*. It is followed by the heavy oil accumulations in the Orinoco Valley, Venezuela, and in Siberia. Another vast commercially significant reservoir is the oil shale deposits of the northwestern United States located in Wyoming, Utah, and Colorado. The practical use of these potentially vast reserves will depend on finding economical ways to extract the oil (by thermal retorting or other processes) for further processing.

The quality of petroleum varies, and, according to specific gravity and viscosity, we talk about light, medium, heavy, and extra-heavy crude oils. Light oils of low specific gravity and viscosity are more valuable than heavy oils with higher specific gravity and viscosity. In general, light oils are richer in saturated hydrocarbons, especially straight-chain alkanes, than are heavy oils, and contain ≤75% straight-chain alkanes and ≤95% total hydrocarbons. Extra-heavy oils, the bitumens, have a high viscosity and thus may by semisolids with high levels of heteroatoms (nitrogen, oxygen, and sulfur) and a correspondingly reduced hydrocarbon content, on the order of 30–40%.

Heavy oils and especially bitumens contain high concentrations of resins (30–40%) and asphaltenes (≤20%). Most heavy oils and bitumens are thought to be derivatives of lighter, conventional crude oils that have lost part or all of their *n*-alkane content along with some of their low-molecular-weight cyclic hydrocarbons through processes taking place in the oil reservoirs. Heavy oils are also abundant in heteroatom (N, O, S)-containing molecules, organometallics, and colloidally dispersed clays and clays organics. The prominent metals associated with petroleum are nickel, vanadium (mainly in the form of vanadyl, VO^{2+} ions), and iron. The former two are (in part) bound to porphyrins from metalloporphyrins. Table 4.7 compares the compositions of typical light and heavy oils.

Processing heavy oils and bitumens represents challenges for presently used refinery processes, as heavy oils and bitumens can poison the metal catalysts used in the current refineries. The use of superacid catalysts, which are less sensitive to these feeds, is one possible solution to this problem.

Further Reading

Bradbury, J. and Obeiter, M. (2013). 5 reasons why it's still important to reduce fugitive methane emissions. World Resources Institute.
The Economist (2012). America's bounty: gas works (14 July).
Bennet, L., Le Calvez, J., Sarver, D.R. et al. (2005). The source for hydraulic fracture characterization. *Oilfield Review* 17: 42–57.
BP (2019). BP energy outlook. http://bp.com/energyoutlook#PPstats.
Brino, A. and Nearing, B. (2013). Shale gas fracking without water and chemicals. Daily Yonder.
Cuderman, J.F. (1982). Multiple fracturing experiments – propellant and borehole considerations. In: Proceedings of the SPE Unconventional Gas Recovery Symposium, 16–18 May, Pittsburgh, Pennsylvania, 535–546. Society of Petroleum Engineers.
Gold, T. (1985). The origin of natural gas and petroleum and the prognosis for future supplies. *Ann. Rev. Energy* 10: 53–77.
Gold, T. and Soter, S. (1980). The deep-earth gas hypothesis. *Sci. Am.* 242 (34): 154–160.
Delahaye, C. and Grenon, M. (1983). Conventional and unconventional world natural gas resources. Paper presented at the Fifth IIASA Conference on Energy Resources, Laxenburg, Austria.
Griswold, E. (2011). The fracturing of Pennsylvania. The New York Times (17 November). http://www.nytimes.com/2011/11/20/magazine/fracking-amwell-township.html.
International Energy Agency (IEA) (2012). Golden rules for a golden age of gas. IEA world energy outlook special report on unconventional gas.
Jarvie, D. (2008). Worldwide shale resource plays. Presentation at a NAPE Forum (26 August).
Kijk magazine (2/2012).
Kuuskraa, V.A. and Meyers, R.F. (1980). Review of world resources of unconventional gas. Paper presented at the Fifth IIASA Conference on Energy Resources, Laxenburg, Austria.
Milam, K. (2011). *Name the Gas Industry Birthplace*. Fredonia, NY: American Association of Petroleum Geologists.
Miller, R., Loder, A., and Polson, J. (2012). Americans gaining energy independence with U.S. as top producer. Bloomberg.
Netherlof, M.H. (1988). The scope for natural gas supplies from unconventional sources. *Ann. Rev. Energy* 13: 95–117.
Rojey, A., Jaffret, C., Cornot-Gandolphe, S., and Durand, B. (1997). *Natural Gas Production Processing Transport*, 429. Technip Editions.
Salehi, I.A. (1992). Unconventional gas development. Horizontal drilling for onshore gas faces many challenges. *Oil Gas J.* 90 (40): 61.
Wikipedia (2013). Shale gas. http://en.wikipedia.org/wiki/Shalegas.
Stevens, P. (2012). *The Shale Gas Revolution: Developments and Changes*. Chatham House.
Takach, N.E., Baker, C., and Kemp, M.K. (1987). Stability of natural gas in the deep subsurface: thermodynamic calculation of equilibrium compositions. *AAPG Bull.* 71 (3): 322–333.
U.S. Energy Information Admininstration (2013). Technically recoverable shale oil and shale gas resources: an assessment of 137 shale formations in 41 countries outside the United States. EIA.
The Associated Press (2013). EPA lowered estimates of methane leaks during natural gas production. The Houston Chronicle.

The Breakthrough Institute (2011). Interview with Dan Steward, former Mitchell Energy vice president.

BP Group (2013). Total energy production. *BP Mag.* 2013 (2): 21.

U.S. Environmental Protection Agency (2013). U.S. greenhouse gas inventory report. EPA.

BP Group (2012). Unconventional gas and hydraulic fracturing. BP Statistical Review.

Canadian Society for Unconventional Resources (n.d.). Understanding hydraulic fracturing. CSUR.

Canadian Society for Unconventional Resources (2013). Understanding tight oil. CSUR.

U.S. Department of Energy (2009). Modern shale gas development in the United States: a primer. DOE, 17.

U.S. Energy Information Administration (2011). World shale gas resources. EIA.

U.S. Energy Information Administration (2013). North America leads the world production of shale Gas. EIA.

U.S. Geological Survey (2013). How is hydraulic fracturing related to earthquakes and tremors? USGS.

Geology.com (2013). What is shale gas? http://www.geology.com/energy/shale-gas.

Smith, K. and Ozimek, A. (2013). Will natural gas stay cheap enough to replace coal and lower us carbon emissions. Forbes.

Trullo, A.H. (2016). Ethane supplier to the World, *CEN.* 94 (44): 28–29.

5

Links with Natural Gas, Crude Oil, and Petroleum Refineries

CHAPTER MENU

5.1 Links with Natural Gas

5.1.1 Introduction

Natural gas is a mixture of alkanic hydrocarbons and certain impurities. In natural gas, methane (CH_4, bp: −159 °C), ethane (C_2H_6, bp: −89 °C), and propane (C_3H_8, bp: −42 °C), are the principal constituents. Butane (C_4H_{10}), pentane (C_5H_{12}), hexane (C_6H_{14}), heptane (C_7H_{16}), and octane (C_8H_{18}) may also be present. In addition to these main components, carbon dioxide, hydrogen sulfide, helium, nitrogen, and water vapor also occur. Aromatic compounds (i.e. benzene) and helium or argon have also been known to occur in natural gas.

The hydrocarbon content of natural gas is usually obtained indirectly by measurement of either the heating value or the specific gravity. However, it must be remembered that the

Petrochemistry: Petrochemical Processing, Hydrocarbon Technology, and Green Engineering,
First Edition. Martin Bajus.

composition of natural gas can vary widely; but because natural gas is a multicomponent system, neither property may be changed significantly. The water content of natural gas is usually expressed by use of the dew point temperature (−80 °C) and pressure.

Hydrogen sulfide and carbon dioxide are the acid gases that are associated with natural gas. In addition to emitting a foul odor at low concentrations, hydrogen sulfide is poisonous and, at concentrations above 600 ppm, can be fatal, having toxicity comparable to hydrogen cyanide. Hydrogen sulfide is corrosive to metals associated with gas transporting, processing, and handling systems (although it is less corrosive to stainless steel), and may lead to premature failure of such systems.

Natural gas liquids (NGLs) are hydrocarbon constituents having a higher molecular weight than methane. They are not true liquids insofar as they are not usually in a liquid from at ambient temperature and pressure. NGLs are ethane, the constituents of liquefied petroleum gas (LPG), and natural gasoline. The constituents of LPG are propane (C_3H_8), butanes (C_4H_{10}), and/or mixtures thereof; small amounts of ethane and pentane may also be present as impurities. On the other hand, natural gasoline (like refinery gasoline) consists mostly of pentane (C_5H_{12}) and higher-molecular-weight hydrocarbons. The term *natural gasoline* has also, on occasion, been applied to mixtures of LPG, pentanes, and higher-molecular-weight hydrocarbons.

5.1.2 Processing

The processes that have been developed to accomplish gas purification vary from a simple once-through wash operation to complex multistep recycle systems. In many cases, process complexities arise because of the need for recovery of the materials used to remove contaminants or even recovery of the contaminants in the original, or altered, form.

Gas-purification processes fall into three categories: (i) removal of gaseous impurities, (ii) removal of particulate impurities, and (iii) ultra-fine cleaning, where the extra expense is only justified by the nature of the subsequent operations or the need to produce a very pure gas stream. In addition, there are many variables in gas treating, and several factors need to be considered: (i) the types and concentrations of contaminants in the gas; (ii) the degree of contaminant removal desired; (iii) the selectivity of acid gas removal required; (iv) the temperature, pressure, volume, and composition of the gas to be processed; (v) the carbon dioxide to hydrogen sulfide ratio in the gas; and (vi) the desirability of sulfur recovery due to process economics or environmental issues.

Process selectivity indicates the preference with which the process will remove one acid gas component relative to (or in preference to) another. For example, some processes remove both hydrogen sulfide and carbon dioxide, whilst other processes are designed to remove hydrogen sulfide only. It is important to consider the process selectivity for, say, hydrogen sulfide removal compared to carbon dioxide removal, which will ensure minimal concentrations of these components in the product. Thus the need to consider the carbon dioxide to hydrogen sulfide ratio change in the natural gas.

5.1.3 Water Removal

Water, in the liquid phase, will cause corrosion or erosion problems in pipelines and equipment, particularly when carbon dioxide and hydrogen sulfide are present in the gas. Thus, there is the need for water removal from gas streams.

The simplest method of water removal (refrigeration or cryogenic separation) is to cool the natural gas to a temperature that is equal to or (more preferentially) below dew point. However, in the majority of cases, cooling alone is insufficient and, for the most part, impractical for use in field operations. Other, more convenient, water removal options use hygroscopic liquids (e.g. di- or triethylene glycol) and solid adsorbents (e.g. alumina, silica gel, and molecular sieves).

5.1.4 Acid Gas Removal: Environmentally Friendly Solvents

The removal of acid gases from natural gas streams can be generally classified into two categories: (i) chemical absorption processes and (ii) physical absorption processes. Several such processes fit into these categories (Table 5.1).

Treatment of natural gas to remove the acid gas constituents (hydrogen sulfide and carbon dioxide) is most often accomplished by contact of the natural gas with an alkaline solution. The most commonly used treating solutions are aqueous solutions of the ethanolamines or alkali carbonates.

5.1.5 Fractionation

Fractionation processes are very similar to processes classed as *liquids removal*, which are used to remove the more significant product stream first or to remove any unwanted light ends from the heavier liquid products.

In general practice, the first unit is for ethane removal (a de-ethanizer) followed by a depropanizer, then a debutanizer, and, finally, a butane fractionator. Each unit can operate

Table 5.1 Processes for acid gas removal.

Chemical absorption (chemical solvent processes)	Physical absorption (physical solvent processes)
Alkanolamines	
MEA	Selexol
SNPA: DEA (DEA)	Rectisol
UCAP (TEA)	Sulfinol[a]
Selectamine (MDEA)	
Econamine (DGA)	
ADIP (DIPA)	
Alkaline salt solutions	
Hot potassium carbonate	
Catacarb	
Benfield	
Giammarco–Vetrocoke	
Nonregenerable	
Caustic	

a) A combined physical/chemical solvent process.

at a successively lower pressure, thereby allowing the different gas streams to flow from column to column by virtue of the pressure gradient and without necessarily using pumps.

5.1.6 Turboexpander Process

In recent years, ethane has become increasingly desirable as a petrochemical feedstock, which has resulted in the construction of many plants that recover ethane from natural gas at from −73 to −93 °C. Combinations of external refrigeration and liquid flash-expansion refrigeration with gas-turbo expansion cycles are employed to attain the low temperatures desired for high ethane recovery.

Dry inlet gas that has been dehydrated by molecular sieves or alumina beds to less than 0.1 ppm water is split into two streams by a three-way control valve. Approximately 60% of the inlet gas is cooled by heat exchange with low-pressure residue gas from the demethanizer and by external refrigeration. The remainder of the inlet gas is cooled by heat exchange with the demethanizer bottoms product, the reboiler, and the side heater. A significant amount of low-level refrigeration from the demethanizer liquids and cold residue gas stream is recovered in the inlet gas stream.

The two portions of the feed stream recombine and flow into the high-pressure separator where the liquid is separated from the vapor and is fed into an intermediate section of the demethanizer with liquid-level control. The decrease in pressure across the level-control valve causes some of the liquid to flash, which results in a decrease in the stream temperature. The pressure of the vapor stream is decreased by way of a turboexpander to recover power and, thus, achieve more cooling than would be possible by Joule–Thompson expansion. The outlet of the turboexpander then is fed into the top of the demethanizer where the separation of liquid and vapor occurs. The vapor is passed as cold residue to the heat exchanger, and the liquid is distributed to the demethanizer top tray as reflux.

Essentially, all of the methane is removed in the demethanizer overhead gas product. High recovery of ethane and heavier components as demethanizer bottom products is commonplace. The work that is generated by expanding the gas in the turboexpander is utilized to compress the residue gas from the demethanizer after it is warmed by heat exchange with the inlet gas. Recompression and delivery to a natural-gas pipeline is performed downstream of the plant. Propane recovery of c. 99% can be expected when ethane recoveries are in excess of 65%.

Recoveries of 90–95% ethane have been achieved with expander processes. The liquid product from the demethanizer may contain 50 liquid vol.% ethane and usually is delivered by a central fractionation facility for separation into LPG products, chemical feedstocks, and gasoline-blending stock.

5.1.7 Solvent Recovery

Solvents may be recovered by use of an adsorbent process in which two carbon-filled adsorbers are employed in sequence and in which each adsorber in turn is successively steamed out, dried, and cooled. In this way, the carbon can be used continuously and almost indefinitely. When non-miscible solvents are used, the recovery plant may be simplified by drying and cooling with solvent-laden air.

Figure 5.1 Raw materials and primary petrochemical.

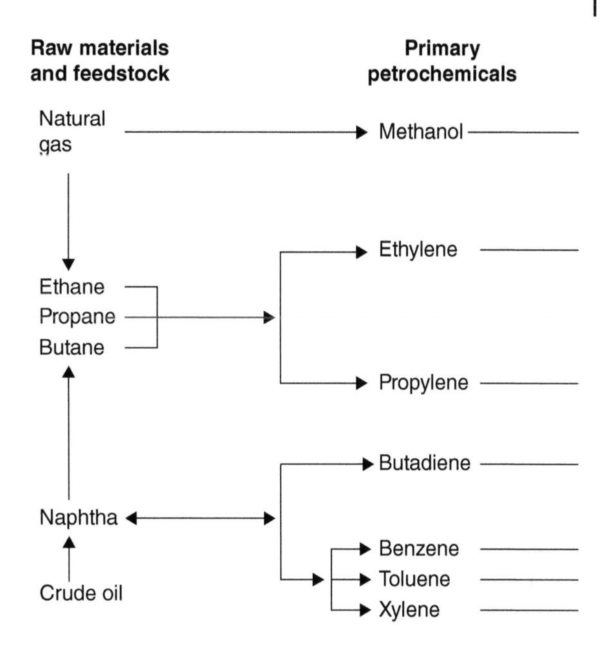

5.1.8 Chemicals From Natural Gas

Natural gas can be used as a source of hydrocarbons (ethane, propane, etc.), which are important chemical intermediates for a wide variety of chemicals (Figure 5.1). Natural gas is also a source of synthesis gas, which also leads to a wide variety of chemicals (Figure 12.1).

Although the emphasis is usually placed on materials that can be prepared directly from natural gas (methane), there are other options for the production of chemical intermediates and chemicals from natural gas by indirect routes (i.e. where other compounds are prepared from natural gas and are then used as further sources of petrochemical products). The preparation of chemicals and chemical intermediates from natural gas is not restricted to those described here, but they should be regarded as some of the building blocks of chemical technology.

5.2 LPG as an Ethylene Feedstock

Although most naphtha crackers can co-crack some LPG, additional investment is required, not only from a technological point, but from the logistical side. In West European steam-crackers, there is considerable technological built-in flexibility for LPG cracking. However, this flexibility is constrained due to insufficient infrastructure such as storage, pipeline distribution, etc.

5.3 Heavy Condensates

Material included in this category is unfractionated liquid hydrocarbons recovered with natural gas. It resembles very light crudes at high naphtha yields (typically 60% and more), containing kerosene and gas–oil fractions and possibly heavier bottoms. The largest heavy condensate usage is currently in refineries and splitting operations. However, selected streams are also dedicated for ethylene production. To be suitable for ethylene operations, i.e. for direct steam-cracking, heavy condensates must match several important quality parameters, including: boiling range (heavy tail, the tendency to form coke), alkanic vs. naphthenic, aromatics content (this determines the attractiveness of condensates to an olefin producer), and contaminants (sulfur, mercury, and other heavy metals). In Western Europe, most heavy condensates used for ethylene take place in a selected number of naphtha crackers with built-in gas oil flexibility; these units were modified to crack heavy condensates. These independent ethylene producers (i.e. those without integration with a nearby refinery that could provide the proper middle distillate cut) wanted access to an alternative to naphtha in international markets and preferred to use liquid feedstocks.

About 80% of West European ethylene production (Table 6.1) has been derived from liquid feedstocks in recent years. With the identified trends of diversification from naphtha, this share is expected to decline, due to growing usage of LPG. Figure 7.6 illustrates recent and forecast developments of West European propylene production. As already noted, steam-cracking operations are the most important propylene source, due to the high share of liquid feedstocks used. However, since propylene's demand is growing faster than that for ethylene, increasing amounts of this product have been sourced from dedicated operations, mostly from refinery gases and, more recently, from propane dehydrogenation.

Europe has about 25.6 million tons per annum (MTPA) of cracking capacity, with North-Central Europe accounting for 18.6 MTPA of this total. The rest of Europe makes up the balance. 900 000 tons of 2018 production was lost, with 470 000 tons of this total the result of scheduled work. Planned and unplanned interruptions in production were lower in 2018 than 2017, lengthening the ethylene market throughout the year. In 2019, according to market sources, capacity will not drop meaningfully compared with the recent past.

Developments in the regional and global natural gas industry may offer some opportunities for West European steam-cracker operators. With natural gas demand growing at a fast pace, more associated NGLs will be produced. In particular, these conditions may increase availability of LPG in international markets and expand heavy-condensate recovery. However, several potential barriers may impede full exploitation of this situation. For LPG, these center on the unpredictability of developments in the logistical infrastructure. When considering heavy condensates, pricing may still be an issue, and potential barriers include uncertainties regarding the quality of future streams and their suitability for direct steam-cracking. However, what are the opportunities and challenges for the West European refining industry? Probably the best way to discuss this point is from a schematic approach.

5.4 Links with Crude Oil

To provide an idea of how petroleum is used as a source of chemicals, we must describe what happens in a petroleum refinery. Consider a simple refinery in which crude oil – a sticky, viscous liquid with an unpleasant odor – is separated by distillation into various fractions. The first most volatile fraction consists of methane and higher alkanes through C_4 and is similar to natural gas. These are dissolved in the petroleum. The methane and ethane can be separated from the higher alkanes, primarily propane and butanes. The methane–ethane mixture is called *lean gas*, from which the ethane can be separated if required. The C_3/C_4 mixture (LPG) may be used as a petrochemical feedstock (Figure 5.1) or a fuel. Butane is also separated for use in gasoline and, to a lesser extent, as a raw material for chemicals.

The first process step a crude oil undergoes after its production is distillation. A first operation on the crude, desalting (washing by water and caustic), extracts salts (NaCl, KCl, and $MgCl_2$ that is converted to NaCl by the caustic), reduces acid corrosion, and minimizes fouling and deposits. Next the crude is distilled into well-defined fractions according to their end uses.

The main distillation products are:

- Refinery gases
- LPGs (propane/butane)
- Gasolines (light/heavy)
- Kerosenes, lamp oils, jet fuels
- Diesel oils and domestic heating oils
- Heavy industrial fuel-oils

Vacuum distillation of the atmospheric residue complements primary distillation, enabling recovery of heavy distillate cuts from atmospheric residue that will undergo further conversion or will serve as lube oil bases. The vacuum residue containing most of the crude contaminants (metals, salts, sediments, sulfur, nitrogen, asphaltenes, Conradson carbon, etc.) is used in asphalt manufacture, for heavy fuel oil, or to feed other conversion processes. Figure 5.2 presents the part of the refining diagram that includes the atmospheric and reduced pressure distillations.

5.4.1 Naphtha

The second fraction comprises a combination of light naphtha, or straight-run gasoline, and heavy naphtha, and is of particular importance to the chemical industry. The term *naphtha* is not well defined, but the material steam cracked (Figure 5.2 and Table 5.2) for chemicals generally distills in a range between 70 and 200 °C and contains C_5—C_9 hydrocarbons. Naphtha contains aliphatic as well as cycloaliphatic materials such as cyclohexane, methylcyclohexane, and dimethylcyclohexane. Smaller amounts of C_9^+ compounds such as polymethylated cycloalkanes and polynuclear compounds such as methyldecahydronaphthalene may also be present. Like the lower alkanes, naphtha may be steam cracked to low-molecular-weight olefins. Its conversion by a process known as *catalytic reforming* into benzene, toluene, and xylenes (BTX) (Section 5.5.2) is, in the United States, its main chemical use. Catalytic reforming is also a source of aromatics worldwide, although in Western Europe benzene and toluene are mainly derived from pyrolysis gasoline, an aromatics fraction that results from steam cracking of naphtha or gas oil.

Light naphtha was at one time used directly as gasoline; hence its alternate name, *straight-run gasoline*. It contains a large proportion of straight-chain hydrocarbons (n-alkanes), and these resist oxidation much more than branched-chain hydrocarbons (isoalkanes), some of which contain tertiary carbon atoms. Consequently, straight-run

Table 5.2 Refinery innovation from 1862 to the present.

Year	Process name	Purpose	By-products
1862	Atmospheric distillation	Produce kerosene	Naphtha, cracked residuum
1870	Vacuum distillation	Lubricants	Asphalt, residua
1913	Thermal cracking	Increase gasoline yield	Residua, fuel oil
1916	Sweetening	Reduce sulfur	Sulfur
1930	Thermal reforming	Improve octane number	Residua
1932	Hydrogenation	Remove sulfur	Sulfur
1932	Coking	Produce gasoline	Coke
1933	Solvent extraction	Improve lubricant viscosity index	Aromatics
1935	Solvent dewaxing	Improve pour point	Wax
1935	Catalytic polymerization	Improve octane number	Petrochemical feedstocks
1937	Catalytic cracking	Higher octane gasoline	Petrochemical feedstocks
1939	Visbreaking	Reduce viscosity	Increased distillate yield
1940	Alkylation	Increase octane number	High-octane aviation fuel
1940	Isomerization	Produce alkylation feedstock	Naphtha
1942	Fluid catalytic cracking	Increase gasoline yield	Petrochemical feedstocks
1950	Deasphalting	Increase cracker feedstock	Asphalt
1952	Catalytic reforming	Convert low-quality naphtha	Aromatics
1954	Hydrodesulfurization	Remove sulfur	Sulfur
1956	Inhibitor sweetening	Remove mercaptans	Disulfides and sulfur
1957	Catalytic isomerization	Convert to high-octane products	Alkylation feedstocks
1960	Hydrocracking	Improve quality and reduce sulfur	Alkylation feedstocks
1974	Catalytic dewaxing	Improve pour point	Wax
1975	Residual hydrocracking	Increase gasoline yield	Cracked residua
1980s	Heavy oil processing	Increase yield of fuels	Gas oil, coke
Post-1980s	– Selective oligomerization (SO)	1-alkenes (α-olefins)	Higher polymers
	– Light olefin etherification (LOE)	MTBE, ETBE, MTBEG	Oligomers, ethers dealkylation
	– Very low pressure catalytic reforming with continuous catalyst (VLPCR) with continuous catalyst regeneration (CCR)	High-quality gasoline	Aromatics, ethylene, propylene

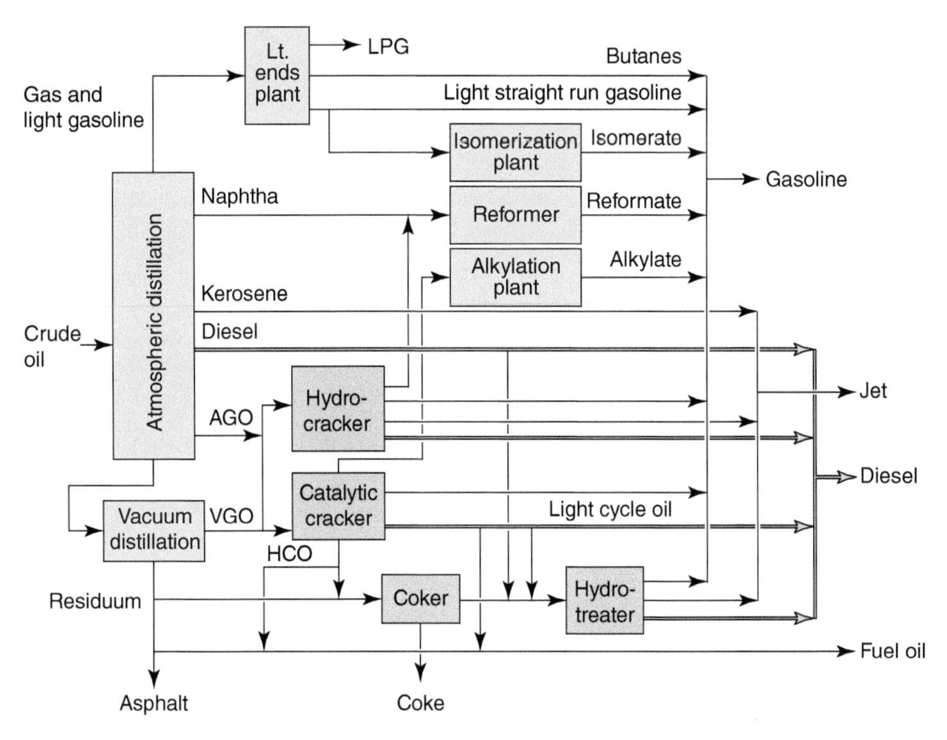

Figure 5.2 Schematic of a modern refinery.

gasoline has poor ignition characteristics and a low octane number of about 60. It is of little use in gasoline for modern high-compression-ratio automobile engines, and its properties are even worse if it is unleaded. Isomerization of its components to branched compounds increases its octane number and thus its utility. Chemically, however, its significance is like that of naphtha, for it can be cracked to low-molecular-weight olefins. It does not perform well in catalytic reforming, giving large amounts of cracked products and small amounts of benzene. Benzene is no longer a welcome constituent of gasoline because of its toxicity and relatively low octane number.

As noted earlier, international practice is different. The United States has preferred to crack ethane and propane from natural gas, while the rest of the world has cracked naphtha. The shortage of natural gas in the United States projected in the 1970s led to an increased interest in liquid feedstock cracking. The preferred liquid feedstock is naphtha, which has traditionally been preempted in the United States for gasoline manufacture. Accordingly, when a natural gas shortage loomed, gas oil steam cracking was developed. This was not considered when natural gas was plentiful because cracking of gas oil to olefins is accompanied by tar and coke formation. The prediction of a shortage, however, motivated techniques for ameliorating this latter problem, and it is now possible to crack gas oil as well as naphtha. The industry has been loath to do so because it is usually more economical to crack ethane and propane. They are easier to handle and provide fewer coproducts and less coke.

In the 1980s and 1990s, the switch to liquid feedstocks lost momentum for three reasons. First, US natural gas production was maintained. Although reserves are being depleted – the reserves/production ratio was down to 8.8 years in 1993 – production was down only 5% from the 1973 peak. Second, natural gas discoveries in Canada meant that cheap natural gas could be imported and was just under 10% of consumption. Third, Saudi Arabia decided to base its chemical business on the cracking of ethane only, making LPG available on the world market.

Because of availability of gas from the North Sea, the percentage of gas cracked in Western Europe doubled by 1995, but the predominant feedstock in Europe continues to be liquids. LPG cannot be used as readily in Western Europe as in the United States because most European crackers are not flexible, accommodating only liquids. This lack of flexibility is in part due to the lack of the expensive infrastructure required to handle gas. Most crackers in the United States are flexible. If the price of propylene or butadiene goes up, more liquids can be cracked; indeed, the optimum ratio of feedstocks is determined hourly by linear programming. The situation in Japan approximates that in Western Europe except that even less gaseous feed is cracked, and what is cracked consists almost entirely of butane. The liquid feed is entirely naphtha, and it is not expected that gas oil will be cracked in the foreseeable future.

5.4.2 Middle Distillates

Kerosene is a fuel for tractors, jet aircraft, and domestic heating and has applications as a solvent. Gas oil is further refined into diesel fuel and light fuel oil of low viscosity for domestic use. Its use as feed for cracking units for olefin production has already been mentioned. Both the kerosene and gas oil fractions may be catalytically cracked to gasoline range materials. Actually, the term *gas oil* is applied to two types of material, both useful for catalytic cracking. One is so-called *atmospheric gas oil*, which, as its name indicates, is produced by atmospheric pressure distillation. The other is *vacuum gas oil* (VGO), which results from the vacuum distillation of so-called *residual oil*. It has a much higher boiling range of 430–530 °C.

Residual oil boils above 350 °C. It contains the less volatile hydrocarbons together with asphalts and other tars. Most of this is sold cheaply as a high-viscosity heavy fuel oil (*bunker oil*), which must be burned with the aid of special atomizers. It is used chiefly on ships and in industrial furnaces. A proportion of the residual oil is vacuum distilled at 7 kPa to give, in addition to gas oil as mentioned, fuel oil (bp < 350 °C), wax distillate (350–560 °C), and cylinder stock (>560 °C). The cylinder stock is separated into asphalts and hydrocarbon oil by solvent extraction with liquid propane in which asphalts are insoluble. The oil is blended with the wax distillate, and the blend is mixed with toluene and methyl ethyl ketone and cooled to −5 °C to precipitate *slack wax*, which is filtered off. The dewaxed oils are purified by countercurrent extraction with such solvents as furfural, which remove heavy aromatics and other undesirable constituents. The oils are then decolorized with fuller's earth or bauxite and blended to give lubricants.

Part of the vacuum distillate and the slack wax can be further purified to give paraffin and microcrystalline waxes used for candles and the impregnation of paper. The petroleum industry is constantly trying to find methods by which the less valuable higher fractions from petroleum distillation can be turned into gasoline or petrochemicals.

5.4.3 Heavy Condensates Recovery

An alternative opportunity for a refiner to expand petrochemical feedstocks production is by increasing heavy condensates recovery. Due to their characteristics, heavy condensates are processed in conventional refineries or in dedicated operations to yield naphtha and, to lesser extent, middle distillates. In conventional refineries, there is a limit to the amount of condensates that can be processed. Because condensates have a much lighter specific gravity than crude oil, less efficient topping operations and lower utilization of downstream (DS) conversion units occurs. Processing heavy condensates in a dedicated condensate splitter (or even in spare distillation capacity, when available) would avoid these problems, thus offering several advantages. Sharing of infrastructures with the base-load operation certainly lowers investment and is cost-effective. The product slate from the splitter may compensate for/optimize that of the conventional refinery, particularly in terms of middle distillate quality.

With some approximation, petrochemical feedstocks that refineries can release on the merchant markets are restricted to naphtha and, to a much lesser extent, LPG. Due to the growing regional naphtha deficit, local refinery operations will become important providers. If refiners do not provide the necessary raw materials, then an independent petrochemical producer without the necessary source in merchant markets may have to "back-integrate" into feedstock production, with heavy condensate processing in a dedicated splitter. This option provides direct access to merchant raw materials and allows optimum usage of both the light- and heavy-naphtha cuts. For a petrochemical producer, the associated investment is relatively high. Also, the condensates also yield middle distillates and heavies. These co-products must be marketed separately.

This processing scheme requires that the condensate is a good precursor for both alkanic (pro-ethylene) and naphthenic (pro-aromatic) feedstock. Undoubtedly, a refiner will be in a better position than a petrochemical producer to exploit the non-naphtha content of heavy condensates. This may promote the processing of heavy condensates in dedicated facilities at Western European refineries, to release naphtha on the regional petrochemical market. While this development may also characterize integrated refinery/ethylene operations, caution should be noted. Under traditional price differentials, the investment for a grassroots plant that basically provides for an incremental hydroskimming operation may be difficult to justify, since the economics are extremely sensitive to the price differential between naphtha and heavy condensates.

Middle distillates used for ethylene production can vary from distillates produced from a dedicated hydro-cracking operation (also called *unconverted oils*), to atmospheric gas–oil (but this requires access to highly alkanic crudes) and to vacuum gas–oil (in specifically designed steam-crackers). Due to dedicated investment from both the refining and olefins industries, this feedstock in Western European steam-crackers really belongs to dedicated, rigid operations that aim at optimizing the use and value of a relatively heavy feed. Ethane and middle distillates have typically accounted for 15% of regional ethylene production. Little change is expected in the short- to medium-term.

Normal alkanes are used in the petrochemical industry for the production of plasticizers, linear alkyl benzene sulfonates, detergent alcohols, and ethoxylates. Typical carbon numbers are C_6—C_{10} for plasticizers, C_{10}—C_{14} for linear alkylbenzenes, and C_{13}—C_{22}^+ (usually heavier than C_{16}) for detergent alcohols. Current world demand is over two million tons. They are obtained from kerosene and lighter fractions by separation of the normal paraffins from the branched and cyclic paraffins by adsorption using molecular sieves. The main technology used is the UOP Molex process.

This process uses UOP's Sorbex separation technology, which employs zeolites. Isothermal liquid-phase operation facilitates the processing of heavy and broad-range feedstocks. Vapor-phase operations, in addition to having considerable heating and cooling requirements, require large variations of temperature and/or pressure through the adsorption–desorption cycle in order to make an effective separation. Vapor-phase operations also tend to leave a certain residual level of coke on the adsorbent, which must then be regenerated on a cyclic basis. Operation in the liquid phase, on the other hand, allows for uninterrupted continuous operation over many years without regeneration.

The Molex process is a simulated moving-bed adsorption process with all process streams in the liquid phase and at constant temperature within the adsorbent bed. A separate liquid of different boiling point is used to displace the feed components from the pores of the adsorbent. Two liquid streams emerge from the bed: an extract and a raffinate stream, both diluted with adsorbent. The desorbent is removed from both product streams by fractionation and is recycled back to the system. The simulated moving-bed effect is obtained by moving the feed, desorbent, and product streams past a fixed adsorbent by means of a specially designed rotary switching valve. Thus the adsorbent remains fixed while liquid streams flow down through the bed. A shift in the positions of the liquid feed and withdrawal, in the direction of fluid flow through the bed, simulates the movement of the solid in the opposite direction.

The economics of the process are favored by close integration with the refinery, with the refinery taking back the branched chain and cyclic alkanes at a similar value to the feed kerosene/naphtha. In the lighter range, these have higher octane values and superior gasoline-alkylation characteristics.

5.5 Links with Petroleum Refineries

Refineries produce a number of streams, which can be used directly as petrochemical feedstocks. The nature and quantity depend on crude type, refinery complexity, and other operating variables (Figure 5.2 and Table 5.2). In many cases, these streams are retained within the refinery for fuel, gasoline, or other refinery uses. However, they can also support competitively sized petrochemical operations, which are often closely integrated with the refinery and thereby add value to and/or optimize the operation of the refinery. The interaction of refinery and petrochemical operations is complex (Figure 5.2), with the petrochemical business in turn benefiting from close integration with the refinery by being able to return surplus gasoline and fuel streams.

5.5.1 Fluid Catalytic Cracking

The most important sources of petrochemical feedstocks within the refinery are the crude distillation unit (CDU), the fluid catalytic cracker (FCC), and the reformer (Figure 5.2

and Table 5.2). Thus the FCCs are an important source of propylene and butylenes, and also of small amounts of ethylene. Delayed cokers produce ethylene, propylene, and butylenes, although in relatively small quantities compared to FCC operations. Reformate from catalytic reformers is the major source of aromatics including benzene, toluene, and mixed xylenes. There is increasing interest in extracting benzene from reformate in order to reduce the benzene content of motor gasoline. Light naphtha reforming to benzene using the Aromax process (IFP/Chevron) provides an alternative use for low-octane-value naphtha, although this is not currently available for license.

Naphtha and other hydrocarbon streams from the CDU provide the feedstock for production of ethylene via steam cracking. Kerosene provides the feedstock for *n*-alkanes used for linear alkyl benzene (a detergent raw material). The polygasoline unit may provide feedstock for oligomer production.

Methane, obtainable from petroleum as natural gas or as a product from various conversion (cracking) processes, is an important source of raw materials for aliphatic petrochemicals. In addition, ethane (also available from natural gas and cracking processes) is an important source of ethylene which, in turn, provides more valuable routes to petrochemical products. Ethylene is consumed in larger amounts to produce aliphatic petrochemicals than any other hydrocarbon, but it is by no means the only source of aliphatic petrochemicals.

As mentioned earlier, in the 1980s and 1990s, the switch to liquid feedstocks lost momentum for three reasons. First, US natural gas production was maintained. Although reserves are being depleted – the reserves/production ratio was down to 8.8 years in 1993 – production was down only 5% from the 1973 peak. Second, natural gas discoveries in Canada meant that cheap natural gas could be imported and was just under 10% of consumption. Third, Saudi Arabia decided to base its chemical business on the cracking of ethane only, making LPG available on the world market.

Propane and butane are also important sources of aliphatic hydrocarbons. Propane is usually converted to propylene by thermal cracking, although some propylene is also available from refinery gas streams. The various butylenes are more commonly obtained from refinery gas streams. Butane cracking to butylene is known, but it is more complex than ethane or propane cracking, and product distributions are not always favorable. The production of gasoline and other liquid fuels consumes large amounts of butane.

By contrast, the remaining feedstock categories in Figure 5.2, LPG and heavy condensates, are more representative of really flexible operations. The majority of LPG and heavy condensates fed to West European steam crackers are sourced on international markets, basically substituting naphtha. However, LPG pricing is very seasonal, making this feedstock attractive only during summer months. Opportunistic usage occurs in the summer. Thus, the volume of LPG cracking in Western Europe is extremely sensitive to the naphtha/LPG price relationship and can vary considerably.

The naphtha produced in the fluid catalytic cracking unit (Figure 5.3) has a relatively high octane number but is less chemically stable compared to other gasoline components due to presence of olefins, which are also responsible for the formation of deposits in storage tanks, fuel lines, and injectors. The hydrocarbon gases from the fluid catalytic cracking unit are an important source of propylene and butylenes as well as iso-butane, which are essential feedstocks for the alkylation process (which produces high-octane gasoline

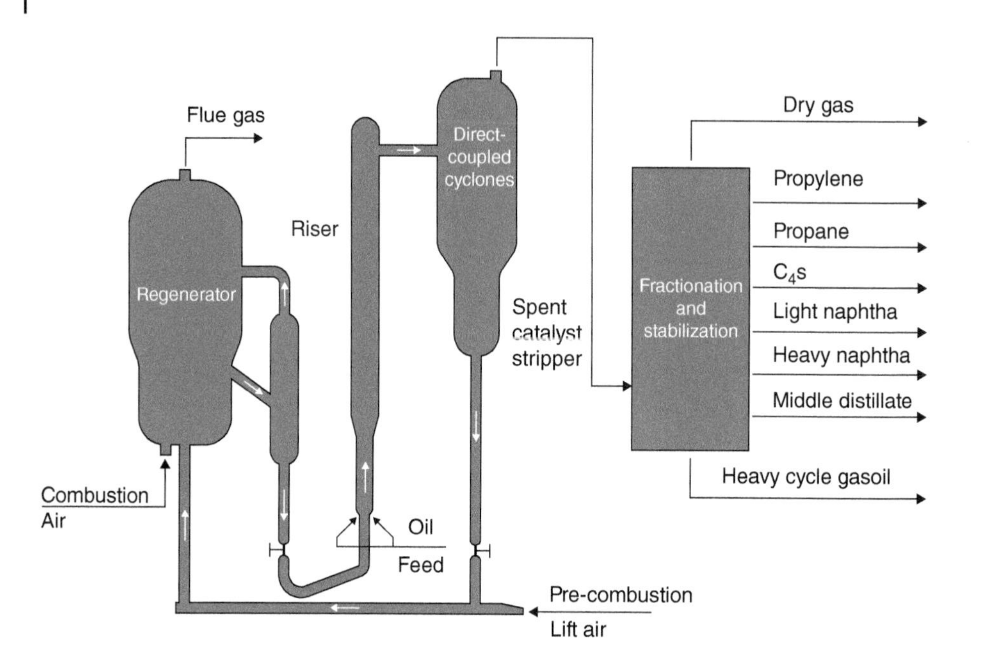

Figure 5.3 Process fluid catalytic cracking (FCC) flow diagram.

components) (Table 5.3). Refinery streams are very often used for production of petrochemicals (Table 5.4).

The Indmax FCC (I-FCC) catalyst is a unique, proprietary, multicomponent, multifractional catalyst formulation that promotes selective component cracking to provide very high conversion and yield of light olefins. The I-FCC process operates in a dynamic heat balance with hot regenerated catalyst supplying the net heat demand required by reaction system. Finely sized solid catalyst continuously circulates in a closed loop between reaction systems and the catalyst regeneration system. The feed and catalyst are in close contact in the riser reactor at the proper ratio and temperature to achieve the desired level conversion. The reaction products are disengaged from the spent catalyst using a novel riser/reaction termination device. The catalyst passes through a highly efficient, patented, spent catalyst stripper where any hydrocarbon product vapors entrained with the catalyst are removed and recovered. The regeneration system restores the catalytic activity of the coke-laden spent catalyst by combustion with air. It also provides heat of reaction and heat of feed vaporization by returning hot, freshly regenerated catalyst back to the reaction system.

Hot regenerated catalyst flows to the base reaction system riser where it comes in contact with feed supplied through the feed injector. Vaporized feed and catalyst travel up the riser where catalytic reactions occur. The reacted vapor is rapidly disengaged from the spent catalyst in closely coupled riser cyclones and routed directly to product fractionation, minimizing the time for non-selective cracking. Reactor vapors are quenched and fractionated in the product recovery system, which yields dry gas, LPG, naphtha, and middle distillate products.

The spent catalyst separated by the riser cyclones is degassed of most of the reaction vapor while flowing via diplegs into a catalyst stripper. In the stripper, hydrocarbons are

Table 5.3 Deep catalytic cracking (DCC) vs. fluid catalytic cracking (FCC) yield (wt. on feed).

	DCC[a]	FCC (maximum propylene)
C_2	11.9	3.5
C_3—C_4	42.2	17.6
C_5^+ naphtha	26.6	54.8
Light cycle oil	6.6	10.2
Decanted oil	6.1	9.3
Coke	6.0	4.3
Loss	0.6	0.3
Total	100.0	100.0
Olefins		
Ethylene	6.1	0.8
Propylene	21.0	4.9
Isobutylene	5.1	1.9
Total butylene	14.3	8.1

a) Using paraffinic vacuum gas oil (VGO).

Table 5.4 Refinery streams used for petrochemicals.

Refinery stream	Petrochemicals	Alternative use for refinery stream
	Base petrochemicals	
Fluid catalytic cracking (FCC) offgas	Ethylene	Fuel gas
FCC olefins	Propylene, butylenes	Alkylation/polygasoline
Reformate	Benzene, toluene, xylenes (BTX)	Gasoline blending
	Petrochemical derivatives	
Naphtha	Olefins	Gasoline
Gasoil	Olefins	Diesel, jet kerosene, heating fuel
Liquefied petroleum gas (LPG)	Olefins	
FCC ethylene	Ethylbenzene	Fuel gas
FCC propylene	Cumene	Alkylation/polygasoline
	Isopropanol	
	Oligomers	
FCC butylenes	Methyl ethyl ketone	Alkylation/polygasoline
	MTBE	
Kerosene	*n*-Paraffins	Refinery product
FCC light cycle oil	Naphthalene	Diesel blendstock after hydrotreating

effectively removed from the catalyst by efficient contacting with steam. The spent catalyst is transported from the stripper into the regenerator. The hydrogen-rich portion of the coke deposits reacts with the lift air at lower combustion temperature relative to the regenerator dense-bed temperature, which reduces the catalyst hydrothermal deactivator. Regeneration flue gases are first routed through cyclones to minimize catalyst losses and then sent to energy recovery and environmental treatment before being ejected from the stack. Hot regenerated catalyst overflows into an external catalyst hopper where it is aerated to the proper density before flowing back to the base of the riser.

5.5.2 Catalytic Reforming

Today, most refineries are equipped with catalytic reforming units. There are around 600 refineries in operation in the world. Catalytic reforming converts low-octane gasoline into high-octane gasoline (reformate). Catalytic reforming produces reformate with an octane number on the order of 90–95. Catalytic reforming is conducted in the presence of hydrogen over hydrogenation-dehydrogenation catalysts, which can be supported on alumina or silica alumina. All current catalysts are derived from platinum on chlorinated alumina, as introduced by UOP in 1949. The first bimetallic catalysts were introduced in the late 1960s. They consist of platinum associated with another metal (iridium, rhenium, tin, or germanium). While maintaining an acceptable run duration, they allow activity to be boosted at identical operation conditions (Ir), operating pressure to be lowered with the same run duration (Re), or low-pressure yields to be improved (Sn and especially Ge, seldom used today). The use of Pt/Re is now most common in semi-regenerative (SR) processes and Pt/Sn in moving beds. Sulfur is a poison reforming catalyst, which requires that virtually all sulfur must be removed from heavy naphtha by a hydrotreating process prior to reforming.

Several different types of chemical reactions occur in the reforming reactors: alkanes are isomerized to branched chains and to a lesser extent to naphthenes, and naphthenes are converted to aromatics. The addition of a hydrogenation–dehydrogenation catalyst to the system yields a dual-function catalyst complex. Hydrogen reactions – hydrogenation, dehydrogenation, dehydrocyclization, and hydrocracking – take place on one catalyst, and cracking, isomerization, and olefin polymerization take place on the acid catalyst sites.

The required reactions are indicated in Table 5.5 and illustrated by hydrocarbons with eight carbon atoms. There is hydrogenation of aromatics into cyclohexanic naphthenes, then isomerization of cyclohexanic into cyclopentanic naphthenes and dehydrogenation into aromatics. Dehydrogenation of cyclohexanic naphthenes is highly influenced by the hydrogen partial pressure. Low hydrogen pressures and high temperatures promote aromatics production.

Dehydrogenation is a primary chemical reaction in catalytic reforming, and hydrogen gas is consequently produced in large quantities. The hydrogen is recycled through the reactors where the reforming takes place to provide the atmosphere necessary for the chemical reactions and also prevents carbon from being deposited on the catalyst, thus extending its operating life. An excess of hydrogen above whatever is consumed in the process is produced, and as a result, catalytic reforming processes are unique in that they are the only petroleum refinery processes to produce hydrogen as a by-product.

Table 5.5 Hydroisomerization of aromatics C_8.

H-D HYDROGENATION – DEHYDROGENATION SITE; **A** – ACID SITE

Table 5.6 Operating conditions for present-day processes.

	Catalyst	P MPa	H_2/HC (mol/mol)	Space velocity ($m^3/m^3/h$)	RON
SR fixed bed	Monometallic	>2.5	>7	1–2	90–92
	Bimetallic	1–2	4–6	2–2.5	91–98
Cyclic fixed bed	Bimetallic	1.5–2.0	4	2	96–98
Moving bed			100–102		
Continuous Regeneration	Bimetallic	0.3–1.0	2	2–3	>104 for aromatic production

Table 5.7 Typical yields on a Middle Eastern feed: RON = 98 P = 1.5 MPa.

Products	wt.%/feed
H_2	2.5
CH_4	1.7
C_2H_6	3.1
C_3H_8	4.2
$(i + n)\ C_4H_{10}$	6.0
C_5^+	82.5

The thermodynamics of the desired reactions determines operating conditions: high temperature, around 500 °C, and hydrogen pressure as low as possible. Since the reaction produces hydrogen, the minimum pressure is determined by the desired aromatics conversion.

The commercial processes available for use can be broadly classified as moving-bed, fluid-bed, and fixed-bed types. The continuous catalyst regeneration (CCR) configuration is the most complex configuration and enables the catalyst to be continuously removed for regeneration and replaced after regeneration. The benefit of the more complex configurations are that operating severity may be increased as a result of higher catalyst activity, but this does come at an increased capital cost for the process.

Operating conditions are summarized in Table 5.6 for operating pressure, H_2/HC ratio, space velocity, and octane number (severity). SR reforming units, based on monometallic catalysts, are characterized by high operating pressure, high H_2/HC, low space velocity, and rather low RON.

Catalytic reforming produces C_5^+ gasoline, hydrogen, and also a small amount of methane, ethane, propane, and butanes. In the last few decades, the yield in target products, C_5^+ and H_2, has risen gradually, with pressures reduced to 1.0 MPa or less and improved catalysts. Table 5.7 gives the average product distribution from alkanic feed on a bimetallic catalyst at 1.5 MPa and RON = 98.

In countries that consume large amounts of LPG, there are reforming units working at high pressure (4.0 MPa) on more acid catalysts (presence of fluorine or zeolites). Naturally,

Table 5.8 Maximum aromatics run, Middle Eastern type feed.

Feed (distillation range) (°C)	60–103	70–150
Products		wt.%/feed
C_5^+	66.5	72.5
Benzene	20.4	5.7
Toluene	27.5	25.7
Xylene	3.1	25.7
C_9^+	0.7	7.7
Total aromatics	51.7	64.7

this type of production occurs at the expense of C_5^+ and especially of hydrogen yield. Hydrogen yield may be lower than 0.7 % wt.

5.5.2.1 Maximum Aromatic Production

As mentioned earlier, reforming yields octane numbers by producing aromatics and by extending severity to RON = 102 to 104, so aromatics are concentrated. Accordingly, catalytic reforming, along with steam cracking of naphthas, is today the major source of benzene and xylenes (Table 5.8).

Certain aromatic hydrocarbons may be a specific production objective – benzene, for example – and more economical feeds than naphtha may be needed, such as light gasoline or even LPG. In this case, it is possible to use new processes such as Chevron's Aromax, UOP/BP's Cyclar, or IFP's Aroforming.

Aromax is based on a new generation of catalyst with platinum laid on barium and potassium-exchanged L zeolite. It is far superior to conventional reforming with preferably linear C_6 and C_7 alkanic feeds. It operates in low pressure reforming conditions. Cyclar and Aroforming use catalysts of the gallium type laid down on MFI zeolite along with CCR. They are suited to feeds and are characterized by considerable hydrogen and aromatics production, as well as methane and ethane. These processes are expected to penetrate the market very slowly.

5.5.2.2 Aromatics Complex

The UOP Parex process is an innovative adsorptive separation method for the recovery of *para*-xylene from mixed xylenes. The term *mixed xylenes* refers to a mixture of C_8 aromatic isomers that includes ethylbenzene, *p*-xylene, *m*-xylene, and *o*-xylene. These isomers boil so closely together that separating them by conventional distillation is not practical. The Parex process provides an efficient means of recovering *p*-xylene by using a solid zeolitic adsorbent that is selective for *p*-xylene. Unlike conventional chromatography, the Parex process simulates the countercurrent flow of a liquid feed over a solid bed of adsorbent. Feed and products enter and leave the adsorbent bed continuously at nearly constant compositions. This technique is sometimes referred to as *simulated-moving-bed* (SMB) separation.

In a modern aromatics complex (Figure 5.4), the Parex unit is located downstream of the xylene column and is integrated with a UOP Isomar unit. The bed to the xylene column consists of the C_8^+ aromatics products from the CCR Platforming unit (Figure 5.4) together

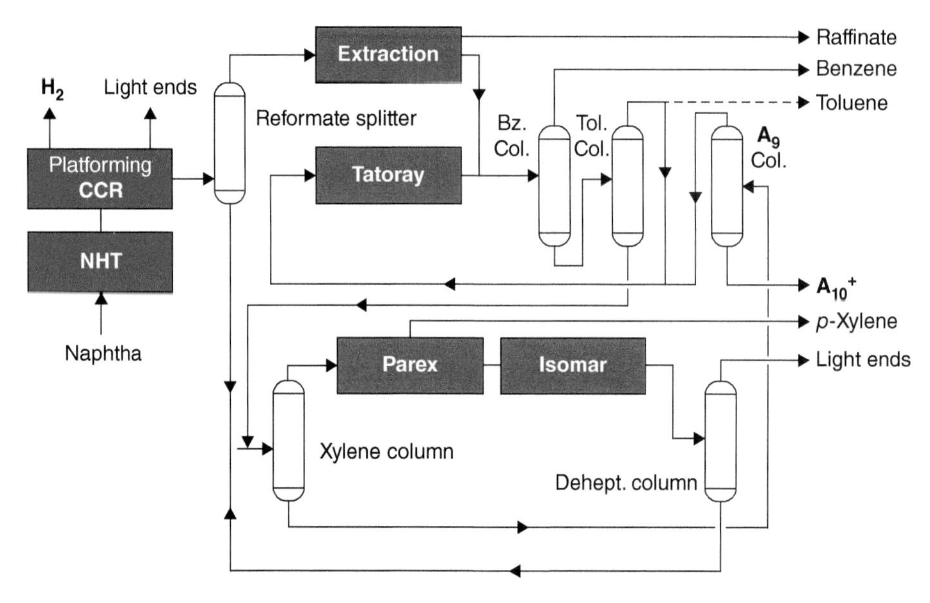

Figure 5.4 UOP aromatics complex.

with the xylenes produced in the Tatoray unit. The C_8 fraction from the overhead of the xylene column is fed to the Parex unit, where the other C_8 aromatic isomers are converted to additional *p*-xylene and recycled to the xylene column.

UOP Parex units are designed to recover more than 97 % wt. of the *p*-xylene from the feed in a single pass at product purity of 99.9 % wt. or better. The Parex design is energy-efficient, mechanically simple, and highly reliable. On-stream factors for Parex units typically exceed 95%.

Before the introduction of the Parex process, p-xylene was produced exclusively by fractional crystallization. In crystallization, the mixed-xylenes feed is refrigerated to −75 °C, at which point the *p*-xylene isomer precipitates as a crystalline solid. The solid is then separated from the mother liquor by centrifugation or filtration. Final purification is achieved by washing the *p*-xylene crystals with either toluene or a portion of the *p*-xylene product.

The principal advantage of the Parex adsorptive separation process is increase in the percentage of the *p*-xylene in the feed per pass. Crystallizers must contend with a eutectic composition limit that restricts *p*-xylene to about 65% per pass.

Most of the mixed xylenes used for *p*-xylene production are produced from petroleum naphtha by catalytic reforming. Modern UOP CCR Platforming units operate at such high severity that the $C_8{}^+$ fraction of the reformate contains virtually no nonaromatic impurities. As a result, these C_8 aromatics can be fed directly to the xylene recovery section of the complex.

6

Hydrocarbon Technology, Trends, and Outlook in Petrochemistry

CHAPTER MENU

6.1 Definition

Petrochemistry is a branch of chemical technology based on hydrocarbon raw materials, all derived from petroleum and natural gas. Petroleum is perhaps the most important substance consumed in modern society. Natural gas is increasingly being recognized as hydrocarbon raw material, the preferred chemical feedstock for the twenty-first century (Chapter 4). The two major factors that have driven petrochemical technology development in recent years have been identified as economic pressure and environmental concerns. These factors impinge directly on the activities of the petrochemical industry, most clearly in the field of petrochemical processing. The petrochemical industry is made up of very large, global, integrated companies that can produce finished or semi-finished goods from hydrocarbons. In petrochemical products, the major impact is felt via the consumer, both the direct consumer who buys petrochemicals and the intermediary consumer, whose product provides the use. In fact, the term *petrochemical* may not exist as such, and petrochemicals might just be intermediates inside a more complex production module. Petrochemicals, in some form, will probably always exist and be marketed globally. There will always be a need for some of the petrochemicals used in the manufacture of products that are relatively small and that do not lend themselves to integrated, mass manufacture (Figure 6.1).

A relatively small portion (~10%) of oil and natural gas is used as raw material to produce chemical products essential to our everyday life, ranging from plastics to textiles to pharmaceuticals and so on. They provide not only raw materials for the ubiquitous plastics and other products, but also fuel for energy, industry, heating, and transportation. The petrochemical industry grew rapidly in the 1950 and 1960s and today provides well over 90% by tonnage of all organic chemicals. Petroleum and natural gas products are the basic materials

Petrochemistry: Petrochemical Processing, Hydrocarbon Technology, and Green Engineering,
First Edition. Martin Bajus.

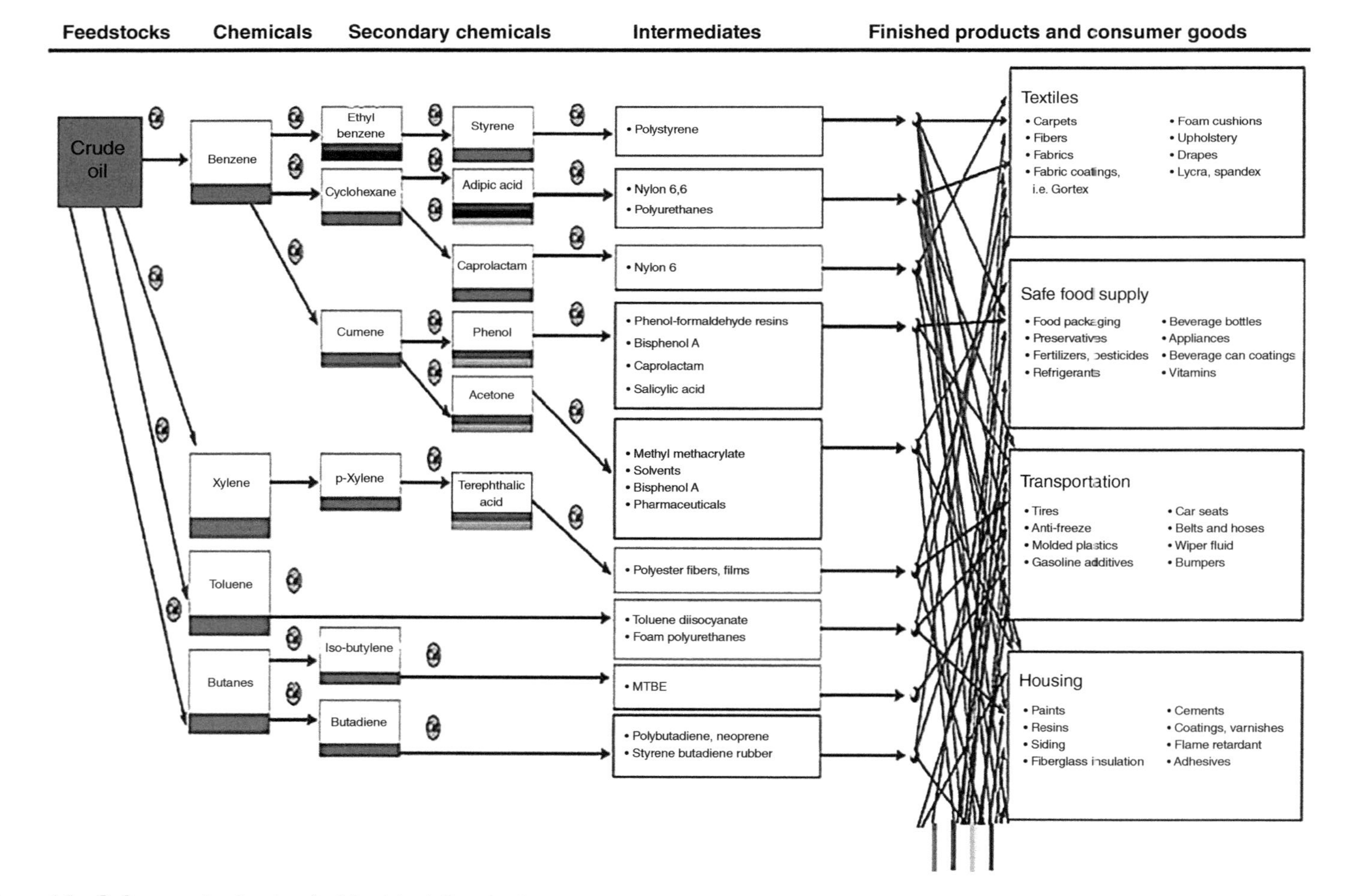

Figure 6.1 Refinery and petrochemical feedstock flowchart.

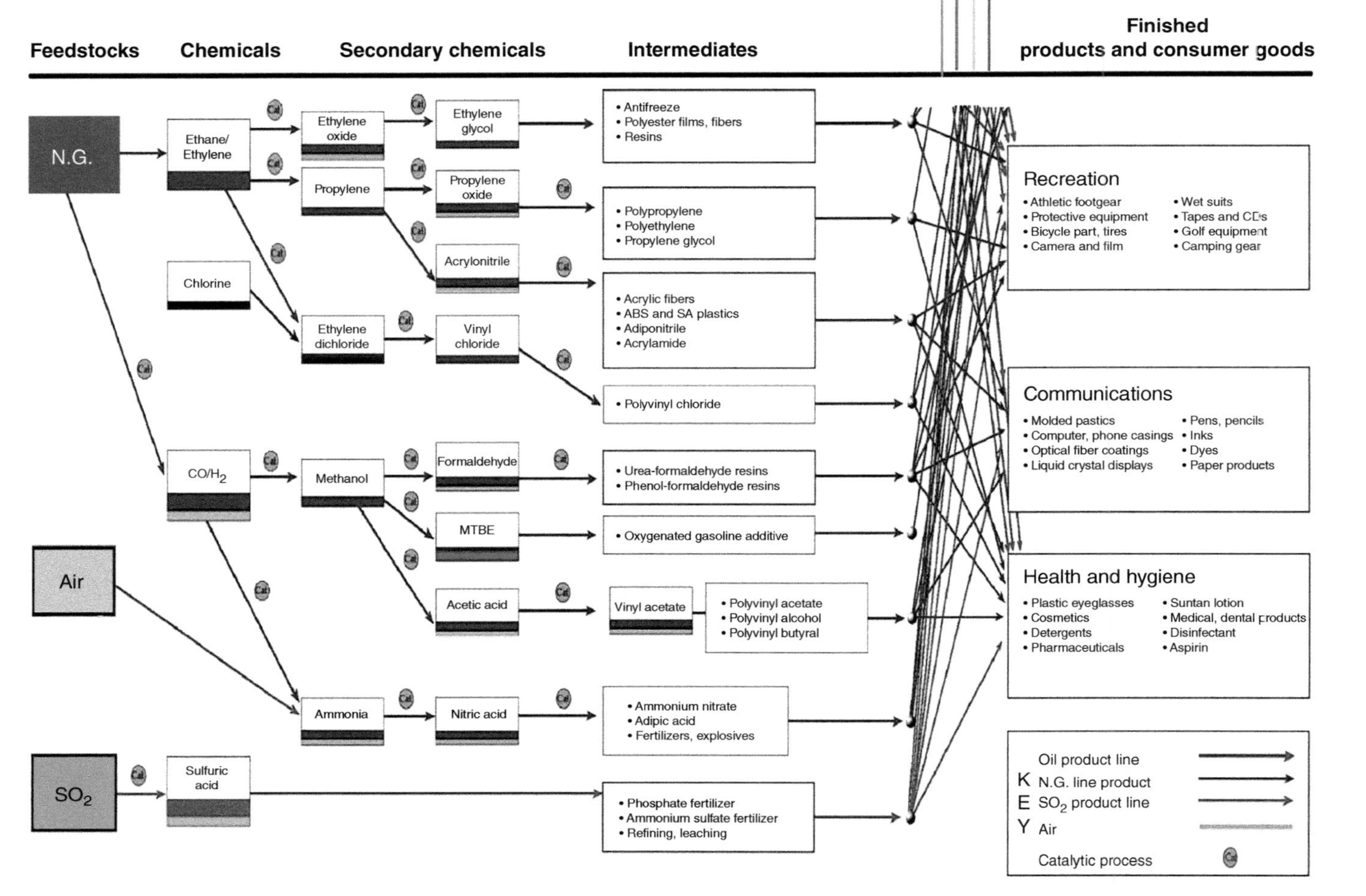

Figure 6.1 (*Continued*)

used for the manufacture of synthetic fibers for clothing and in plastics, paints, fertilizers, insecticides, soaps, and synthetic rubber. The uses of petroleum and natural gas as sources of raw materials in manufacturing are central to the functioning of modern industry.

On the basis of chemical structure, petrochemicals are categorized into the following categories (the pillars or building blocks of petrochemistry) of petrochemical products:

- *A pillar – alkenes:* These include ethene, propene, and butenes, which are important sources of industrial chemicals and plastics. Butadiene is used in making synthetic rubber.
- *B pillar BTX aromatics:* These include benzene, toluene, and xylenes, which have a variety of uses. Benzene is a raw material for dyes and synthetic detergents, and benzene and toluene for isocynates, while xylenes are used in making plastics and synthetic fibers.
- *C pillar – $CO + H_2$ synthesis gas (C_1-technologies)*: A mixture of carbon monoxide and hydrogen is sent to a Fischer–Tropsch reactor to product gasoline-range and diesel-range hydrocarbons as well as methanol, dimethyl ether, and methyl tert-butyl ether (MTBE).
- *D* pillar – *diversification of petrochemicals:* These include petrochemicals containing oxygen, halogens, nitrogen, and sulfur; *derived processes and products* such as polymers, agrochemicals, surfactants, dyes, textile chemicals, and related products; fuels, lubricants, and additives on a petrochemical basis; and application of petrochemicals in other chemical technologies.

Ethene, propene, and butenes, the major parts of alkenes, are the basic sources for preparing several industrial chemicals and plastic products, whereas butadiene is used to prepare synthetic rubber.

Benzene, toluene, and xylenes are major components of aromatic chemicals. These benzene, toluene, xylene (BTX) aromatic petrochemicals are used in manufacturing of secondary products like synthetic detergents, polyurethanes, plastic, and synthetic fibers. Synthesis gas comprises carbon monoxide and hydrogen, which are basically used to produce methanol and ammonia, which are further used to produce other chemical and synthetic substances.

Many new processes and products have been introduced, and large economics of scale have proven possible. The prices of chemicals and polymers have dropped so that they can compete with traditional materials. Cheerfully colored plastic housewares, highly functional packaging, and easy-care garments of synthetic fibers are no longer exciting new technology but have become an accepted and routine part of modern life.

While Germany is probably credited as starting the chemical industry, the United States with its vast availability of natural gas and oil is credited with founding the petrochemical industry (Table 6.1). An international survey of ethylene from steam crackers in 2014 is shown in Table 6.2. As of 2014, production had increased by 2.6 million tons per annum (MTPA) from 2012, which was nearly as much as capacities increased by the same period in 2011. At the end of 2010, global capacities set a record, exceeding capacity added in 2009 and 2008.

China and the United States lead the world in ethylene capacity additions. Ethylene capacity is poised to see considerable growth over the next nine years, potentially increasing from 146 MTPA in 2014 through 184.1 MTPA in 2017 to 279.4 MTPA in 2026. Around 140 planned and announced plants are slated to come online in the next nine years, with the top

10 countries planning capacity additions as follows: China (number of plants 36), United States (17), Iran (17), Russia (11), Indonesia (7), South Korea (7), India (12), Qatar (2), Egypt (2), and Brunei (1). Saudi Arabian Oil Co, ExxonMobil, and National Petrochemical Co. are the top three companies by planned capacity additions globally over the next nine years.

Table 6.3 shows the rankings of the 10 largest ethylene production complexes in the world in 2013. The order for 2013 changed from 2012, with the entry of one complex leading to the removal of another: ExxonMobil Chemical Company's Jurong Island complex in Singapore entered the list to displace NOVA Chemicals' Joffre, Atlanta, complex as the world's largest, pushing EQUATE Petrochemical Company's Shuaiba, Kuwait, complex into eleventh place.

Table 6.4 ranks ethylene production capacity by region; it shows changes for individual countries between 2013 and 2014. Table 6.5 lists the top owners of ethylene capacities, based on assets owned entirely by one company.

By the 1970s, growth was leveling off. The first and second oil shocks increased the price of crude oil and hence of its downstream products. Economies of scale ran out. The petrochemical industry had matured. As its technology became better known, developing countries started their own petrochemical industries, competing with developed countries and depressing profitability. Further, the impact of the petrochemical industry on the environment became evident. The downside is course not absent from the picture: oil and politics, oil and the environment. However, for a long time yet to come, oil will have a definite influence in the world and perhaps even more so in the developing countries. As such, some changes are already making themselves felt, and others can be glimpsed in the distance. The petrochemical industry will have to respond to increasingly stringent quality requirements, particularly with respect to environmental concerns, and has already made good progress in this area.

In the 1980s and early 1990s, new products were no longer the name of the game, in part because the 1960 and 1970s had provided an arsenal of them to attack new applications. Also, the petrochemical industry became subject to strict government monitoring. Extensive testing was required before a new compound could be introduced.

Rather than developing bigger, better plants to manufacture novel chemicals, the petrochemical industry became concerned with lessening pollution, improving processes, and developing specialty chemical formulations and niche products that could be sold at higher profit margins. Research and development became highly process oriented, in part to combat maturity and gain an edge over competition with money-saving technology.

The petrochemical industry has played a part in the major events of the past 70 years. It has kept up with the great discoveries and thereby made economic development possible.

Table 6.1 Feedstocks for ethylene: United States vs. Western Europe.

Years	1982		1985		1995–2000		2016		
Feedstock (% of total)	US	Western Europe	US	Western Europe	US	Western Europe	US	Western Europe	World
Ethane	80	–	79	–	70	–	65	10	36
LPG	–	10	–	14	–	20	25	22	14
Naphtha/gas oil	20	90	21	86	30	80	10	68	46

Table 6.2 National ethylene capacity.

Country	Jan. 1, 2014	Jan. 1, 2013	Change, tpy
Algeria	133.000	133 000	–
Argentina	838 500	838 500	–
Australia	502 000	502 000	–
Austria	500 000	500 000	–
Azerbaijan	330 000	330 000	–
Belarus	193 000	193 000	–
Belgium	2 460 000	2 460 000	–
Brazil	3 500 000	3 500 000	–
Bulgaria	400 000	400 000	–
Canada	5 530 794	5 530 794	–
Chile	45 000	45 000	–
China	13 778 000	13 778 000	–
China, Taiwan	4 006 000	4 006 000	–
Colombia	100 000	100 000	–
Croatia	90 000	90 000	–
Czech Republic	544 000	544 000	–
Egypt	330 000	330 000	–
Finland	330 000	330 000	–
France	3 373 000	3 373 000	–
Germany	5 757 265	5 743 000	14 265
Greece	20 000	20 000	–
Hungary	660 000	660 000	–
India	3 315 000	3 315 000	–
Indonesia	600 000	600 000	–
Iran	4 734 000	4 734 000	–
Israel	200 000	200 000	–
Italy	2 170 000	2 170 000	–
Japan	6 935 000	6 935 000	–
Kazakhstan	130 000	130 000	–
Kuwait	1 650 000	1 650 000	–
Libya	350 000	350 000	–
Malaysia	1 723 000	1 723 000	–
Mexico	1 384 000	1 384 000	–
Netherlands	3 965 000	3 965 000	–
Nigeria	300 000	300 000	–
North Korea	60 000	60 000	–
Norway	550 000	550 000	–
Poland	700 000	700 000	–
Portugal	330 000	330 000	–

Table 6.2 (Continued)

Country	Jan. 1, 2014	Jan. 1, 2013	Change, tpy
Qatar	2 520 000	2 520 000	–
Romania	844 000	844 000	–
Russia	3 490 000	3 490 000	–
Saudi Arabia	13 155 000	13 155 000	–
Serbia and Montenegro	200 000	200 000	–
Singapore	5 380 000	2 780 000	2 600 000
Slovakia	220 000	220 000	–
South Africa	585 000	585 000	–
South Korea	5 630 000	5 630 000	–
Spain	1 430 000	1 430 000	–
Sweden	625 000	625 000	–
Switzerland	33 000	33 000	–
Thailand	3 172 000	3 172 000	–
Turkey	520 000	520 000	–
Ukraine	630 000	630 000	–
UAE	2 050 000	2 050 000	–
UK	2 855 000	2 855 000	–
US	28 121 132	28 121 132	–
Uzbekistan	140 000	140 000	–
Venezuela	1 900 000	1 900 000	–
Total	**146 016 691**	**143 402 426**	**2 614 265**

Table 6.3 Top 10 ethylene complexes (2013).

Company	Location	Capacity, tpy
1. ExxonMobil Chemical Co.	Jurong Island, Singapore	3 500 000
2. Formosa Petrochemical Corp.	Mailiao, Taiwan, China	2 935 000
3. NOVA Chemicals Corp.	Joffre, Alberta, Canada.	2 811 792
4. Arabian Petrochemical Co.	Jubail, Saudi Arabia	2 250 000
5. ExxonMobil Chemical Co.	Baytown, Texas, USA	2 197 000
6. Chevron Phillips Chemical Co.	Sweeny, Texas, USA	1 865 000
7. Dow Chemical Co.	Terneuzen, Netherlands	1 800 000
8. INEOS Olefins & Polymers	Chocolate Bayou, Texas, USA	1 752 000
9. Equistar Chemicals LP	Channelview, Texas, USA	1 750 000
10. Yanbu National Petrochemicals Co.	Yanbu, Saudi Arabia	1 705 000

Table 6.4 Regional capacity breakdown.

	Ethylene capacity, tpy			
	Jan. 1, 2014	Jan. 1, 2013	Change tpy, %	
Asia-Pacific	45 701 000	43 101 000	2 600 000	6,03
Eastern Europe	7 971 000	7 971 000	–	–
Middle East, Africa	26 007 000	26 007 000	–	–
North America	35 035 926	35 035 926	–	–
South Amcrica	6 383 500	6 383 500	–	–
Western Europe	24 918 265	24 904 000	14 265	0,06
Total capacity	**146 016 691**	**143 402 426**	**2 614 265**	**1,82**

Table 6.5 Top 10 ethylene producers.[a]

	Company	Sites[b]	Of wholly owned complexes	Of partially owned complexes
			Capacity, tpy	
1	Saudi Basic Industries Corp.	15	13 392 245	10 273 759
2	ExxonMobil Chemical Co.	21	15 115 000	8 550 550
3	Dow Chemical Co.	21	13 044 841	10 529 421
4	Royal Dutch Shell PLC	13	9 358 385	5 946 693
5	Sinopec	13	7 895 000	7 275 000
6	Total S.A.	11	5 933 000	3 471 750
7	Chevron Phillips Chemical Co.	8	5 607 000	5 352 000
8	LyondellBasell	8	5 200 000	5 200 000
9	National Petrochemical Co.	7	4 734 000	4 734 000
10	INEOS Olefins & Polymers	6	4 656 000	4 286 000

a) As of Jan. 1, 2014.
b) Wholly owned plus partially owned.

Petrochemicals are manufactured from an abundant raw material with a low production cost that is easy to transport and store. Meeting the most varied requirements, they are present in our daily lives and have often become synonymous with comfort and quality of life.

6.2 Petrochemistry and Its Products

The major sources, petroleum and natural gas, provide basic chemicals or chemical groups from which most petrochemicals are made (Table 3.1 and Figure 3.1). These building blocks were listed in Section 6.1.

Whereas acetylene was very important 65 years ago, its significance has steadily decreased as a result of newer chemistry based on ethylene and propylene. The chemistry of methane, a relatively unreactive molecule, nonetheless is the source of synthesis gas. Its small amount of chemistry is based on alkanes other than methane. In Western Europe, petrochemical demand remains strong, in contrast to stagnant oil requirements. Under these conditions, refiners have new opportunities not only as manufacturers but also as petrochemical feedstock suppliers. Tighter gasoline specifications are rejecting olefins and aromatics from local gasoline pools. Thus, refiners can recover propylene and xylenes from processing streams for sale or feedstock purposes. These two petrochemicals have very strong consumption demand.

The global recession of 2008–2009 led to a drop in petrochemical consumption, but consumption recovered quickly. Global ethylene capacity is expected to see considerable growth of 27% from 207.58 MTPA in 2019 to 264.13 MTPA in 2023, led by Asia and North America, with petrochemical consumption driven by growth in economic activity and population.

Since over half of all petrochemicals manufactured end up in polymers, we are devoted to polymerization processes and polymer properties. We have also expanded the scope of new processes to include many apparently less important reactions that are significant because they rise the more profitable specialty chemicals. So-called *fine chemicals* such as ingredients for pharmaceutical and pesticides, dyestuffs, and food additives make a larger contribution to the chemical industry's added value than they do to its shipments. They tend to be high-priced products with specialized markets, and their manufacture is less capital-intensive and more labor-intensive than the manufacture of run-of-the mill general chemicals. Their importance to petrochemistry is best represented by the value-added figure, which, for example, emphasizes the importance of the pharmaceutical sector (Figure 6.1).

The relative amounts of usable fractions obtainable from a crude oil do not coincide with commercial needs. Also, the qualities of the fractions obtained directly by distillation of crude oil seldom meet required specifications. Hydrocarbon feeds of the refining operation are further converted or upgraded to needed products. Major hydrocarbon refining and conversion processes include cracking, dehydrogenation (reforming), alkylation, isomerization, addition, substitution, oxidation-oxygenation, hydrogenation, metathesis, oligomerization, and polymerization.

The chemistry of the high-temperature conversion and transformation processes of hydrocarbons is based on homolytic processes. Thermal cracking, oxidation, hydrogenation-dehydrogenation, cyclization, and so on proceed through free radicals (Section 1.2.8.2).

Low-molecular-weight olefins, such as ethylene and propylene, are very rare or absent in hydrocarbon sources (Sections 9.2 and 9.3). Demand for these olefins requires their preparation from readily available petroleum sources. Cracking is the process in which higher-molecular-weight hydrocarbons, olefinic or alifatic, are converted to more useful low-molecular-weight materials through carbon-carbon bond fission. Cracking is affected by one of three general methods: thermal cracking, catalytic cracking, or hydrocracking. Each process has its own characteristics concerning operating conditions and product composition. Because of the different chemistry of cracking processes, their products have different compositions.

The major industrial source of ethylene and propylene is pyrolysis (steam cracking) of hydrocarbons. Ethane, butane, and naphtha are the most useful feeds in steam cracking. Because of their ready availability, natural-gas liquids are used in the manufacture of ethylene and propylene in the United States. In Europe and Japan the main feedstock is naphtha. Dehydrogenation of saturated compounds also takes places during refining processes and is practiced in petrochemical synthesis of olefins and dienes. Several industrial processes have been developed for olefin production through catalytic dehydrogenation of C_4-alkanes. The current high and increasing demand for oxygenates, especially MTBE, in new gasoline formulations calls for a substantial increase in the capacity of isobutylene production. The higher olefin processes are commonly used to manufacture detergent-range (C_{10}–C_{14}) olefins via dehydrogenation of corresponding alkanes. All commercial processes use the catalytic dehydrogenation of ethylbenzene for manufacture of styrene (Sections 11.1 and 11.2).

Demand for olefins will account for the largest part of regional petrochemical demand and is expected to show high growth. Its share represented about 75% of total petrochemical demand in 2010, up from 70% in 1990. These trends are supported by strong polyolefin growth, which makes up more than half of the consumption of ethylene and propylene. In 2015, over 320 million tons of polymers, excluding fibers, were manufactured across the globe. Polyolefins are dominant for a few reasons: (i) they can be produced using relatively inexpensive natural gas; (ii) they are the lightest synthetic polymers produced at large scale; and (iii) they resist damage by water, air, cleaning solvents, and so on. These commodity petrochemicals represent the bulk of olefin consumption. The other important point affecting olefin requirements is the faster growth of propylene demand relative to ethylene.

Further decline in EU petrochemical production is expected. Petrochemical production in Western Europe has a competitive disadvantage due to high energy and labor costs. Imports of US ethane began in 2015 and will grow to two to three MTPA. In Eastern Europe, demand growth is strong and feedstock prices are low. Feedstock prices give Russia a cost advantage and will drive growth. Naphtha steam cracking will provide the majority of petrochemical growth in the long term. Diversification of feedstocks and petrochemicals will also drive Middle Eastern growth.

Further Reading

Amghizar, I., Vandewalle, L.A., Van Geem, K.M. et al. (2017). New trends in olefin production. *Engineering* 3 (2): 171–178.

Brelsford, R. (2014). International survey of ethylene from steam crackers. *Oil and Gas Journal* 98 (14): 99–117.

7

Pillar A of Petrochemistry

Production of Lower Alkenes

CHAPTER MENU

7.1 Steam Cracking (Pyrolysis)

Alkenes are major building blocks for petrochemicals. Because of their reactivity and versatility, there has been tremendous growth in the demand for olefins – especially light olefins like ethylene, propylene, butenes, butadiene, etc. Olefins are finding wide application in the manufacture of polymers, chemical intermediates, and synthetic rubber.

Petrochemistry: Petrochemical Processing, Hydrocarbon Technology, and Green Engineering,
First Edition. Martin Bajus.

Ethylene itself is basic building block for large number of petrochemicals and is referred to as the "king of chemicals." Some ethylene manufacturing companies and their capacity are given in Table 6.3. Global ethylene capacity on January 1, 2014, net additions and closings, was more than 146 million tons per annum (MTPA), increased to 184.1 MTPA in 2017, and is expected to reach 279.4 MTPA in 2026 (Section 6.1). Global ethylene capacity is given in Table 6.4.

Steam-cracking (SC) operations are the most important alkene source. Ethylene is the main product; however, propylene is an important co-product and a function of the feedstock slate. The greater the share of liquid and heavier feedstocks (i.e. naphtha, middle distillates, and heavy condensates) used in steam crackers, versus gaseous feedstocks (i.e. ethane and liquefied petroleum gas [LPG]), the larger the amount of co-products manufactured. Naphtha is the most important raw material for the local SC industry and is used in integrated and merchant feedstock-based operations. In 1995, around 64% of West European ethylene production was derived from naphtha, vs. 67% in 1990. The diversification from naphtha feedstock is expected to continue. However, naphtha is expected to remain dominant.

The steam cracker remains the fundamental unit and is the heart of any petrochemical complex and mother plant and produces a large number of products and byproducts such as alkenes – ethene, propene, butadiene, butane and butenes, isoprene, etc. – and pyrolysis gasoline. The choice of the feedstock for alkene production depends on the availability of raw materials and the range of downstream products. Naphtha has made up about 50–55% of ethylene feedstock sources since 1992. Although basic SC technology remains the same for naphtha, gas oil, and natural gas, different configuration of SC plants are available from various process licensors.

Olefins are considered to be key components of chemical technology. Ethylene and propylene are the most important olefins. Production rates are expected to increase as a result of an increasing global population combined with rising living standards. Light olefins are the mainstay of modern life, as many different derivatives used in our daily lives are produced from these building blocks. Traditionally, olefin production depends mainly on natural gas processing products or crude oil fractions. The current leading technology for olefin production is SC. In this process, hydrocarbons that primarily originate from fossil resources are cracked at elevated temperatures in tubular reactors suspended in a gas-fired furnace. In recent decades, this process has been highly optimized and its capacities have been increased, resulting in a well-established technology whose economics can hardly be challenged. Declining crude oil reserves and increasing social awareness of the human impact on the environment have had very little impact on the petrochemical industry. Investments in alternative processes and feedstocks are still to come; the lack of economic viability of such processes in an uncertain commodity market threatens their large-scale implementation. However, some exceptional cases have showed economic viability as a result of limited supply, or have benefited from favorable policies. A continuing search for alternative – and preferably also more sustainable – processes and feedstocks will eventually be required in order to fulfill the future demand for commodity chemicals. Potential alternative feedstocks are coal, natural gas, biomass, waste streams, and their derivatives.

7.1.1 Reaction in Steam Cracking

SC yields a large variety of products, ranging from hydrogen to fuel oil. The product distribution depends on the feedstock and on processing conditions. These conditions are determined by thermodynamic and kinetic factors (Sections 1.2.4–1.2.5).

7.1.2 Thermodynamics

In general, lower alkenes, especially ethene, propene, and butadiene, are the desired products of SC. From a thermodynamic view, the reaction temperature should be high for sufficient conversion. The forward reaction is also favored by a partial pressure of the alkanes, because for every molecule converted, two molecules are formed. A process under vacuum would be desirable in this respect. In practice, it is more convenient to apply dilution with steam, which has essentially the same effect.

7.1.3 Mechanism

The reactions involved in thermal cracking of hydrocarbons are quite complex and involve many radical steps (Section 1.2.8.2). The thermal cracking reaction proceeds via a free-radical mechanism. Two types of reactions are involved: (i) primary cracking, where the initial formation of alkane and olefin takes place; and (ii) secondary cracking, where light products rich in olefins are formed. The total cracking reactions can be grouped as follows:

- Initiation reaction
- Propagation reaction
- Addition reaction
- Isomerization reaction
- Termination reaction
- Molecular cyclization reaction

7.1.4 Kinetics

First-order kinetics implies the rate of reaction of alkanes. The reactivity of alkanes increases with chain length. The rate constants for cracking of a number of alkanes increase as a function of temperature (Section 1.2.6.4).

7.2 Industrial Process

From the previous discussion, a number of requirements concerning the SC process can be derived:

- Considerable heat input at a high temperature level
- Limitation of hydrocarbon partial pressure
- Very short residence times (<1 second; 0.1–0.3 seconds)
- Rapid quench of reaction product to preserve composition

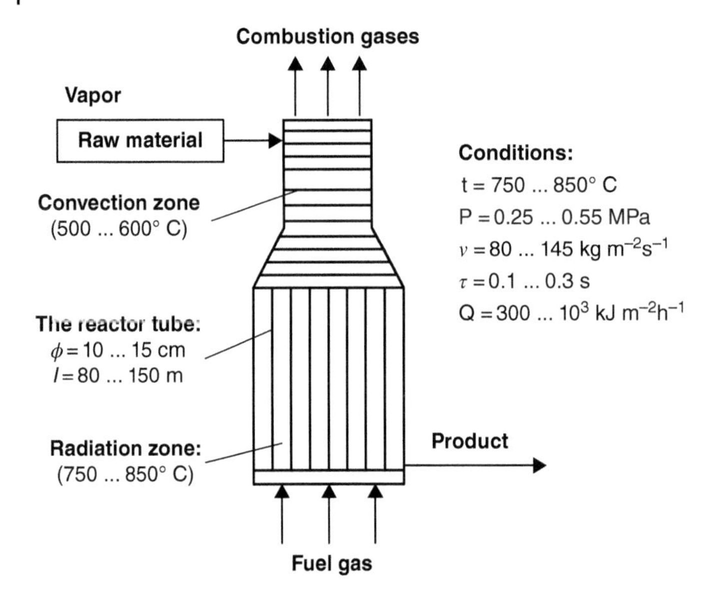

Figure 7.1 Schematic representation of a tubular reactor.

Figure 7.1 shows a simplified flow scheme of a steam cracker for cracking ethane, gas condensate, and naphtha.

The main operating variables in the pyrolysis of hydrocarbons are composition of feedstock, reaction temperature, residence time, hydrocarbon partial pressure, and severity.

7.2.1 Composition of Feedstock

Naphthas are mixture of alkanes, cycloalkanes, and aromatic hydrocarbons, depending on the type of oil from which the naphtha was derived. The group properties of these components greatly influence the yield pattern of the pyrolysis products. A full naphtha boiling range of approximately 20–200 °C would contain compound, with from 4 to 12 carbon atoms. The short naphtha boiling point range is from 100 to 140 °C, and the long-chain naphtha boiling point lies around 200–220 °C.

SC naphtha yields a wide variety of products, ranging from hydrogen to highly aromatic heavy liquid fractions. The thermal stability of hydrocarbons increases in the following order: alkanes, naphthenes, and aromatics. The yield of ethylene as well as that of propylene is higher if the naphtha feedstock is rich in alkanes. The effect of feedstock on the yield of various gases is given in Table 7.1. It may be seen that relative production of ethylene decreases as the feedstock becomes heavier. The percentage of pyrolysis gasoline C_5–200°C cut increases. Simultaneously, butadiene yield varies slightly with feedstock in the treatment of liquid petroleum fractions.

7.2.2 Pyrolysis Temperature and Residence Time

The effluent exit temperature is generally considered a significant indicator of the operation of a furnace. As the furnace exit temperature rises, the yield also rises, while the yields

Table 7.1 Ethylene yield from various hydrocarbon feedstocks.

Ultimate yields from steam-cracking various feedstocks						
	Feedstock					
	Ethane	Propane	Butane	Naphtha	Gas oil	Saudi NGL
Yield, wt.%						
$H_2 + CH_4$	13	28	24	26	18	23
Ethylene	80	45	37	30	25	50
Propylene	2.4	15	18	13	14	12
Butadiene	1.4	2	2	4.5	5	2.5
Mixed butenes	1.6	1	6.4	8	6	3.5
C_5^+	1.6	9	12.6	18.5	32	9

of propylene and pyrolysis gasoline (C_5–200°C) decrease. With respect to ethylene yield, each furnace exit temperature corresponds to an optimum. The highest ethylene yields are achieved by operating at high severity: around 850 °C with residence time ranging from 0.2 to 0.4 seconds. However, operating at high temperature results in high coke formation.

7.2.3 Partial Pressure of Hydrocarbon and Steam-to-Naphtha Ratio

Pyrolysis reactions producing light olefins are more advanced at lower pressure. Decreasing the partial pressure of hydrocarbons by dilution with steam reduces the overall reaction rate but also help to enhance the selectivity of pyrolysis substantially in favor of the light olefins desired. Other roles of steam during pyrolysis are (i) to increase the temperature of the feedstock, (ii) to reduce the quantity of heat to be furnished per linear meter of tube in the reaction section, and (iii) to partially remove coke deposits in furnace tubes. The ethylene yield decreases as the partial pressure of hydrocarbons increases. The effect of H_2O/naphtha on ethylene yield is given for economic reasons a value of 0.5–0.64 of steam per ton of naphtha, which is generally adopted as the upper limit.

7.2.4 Severity and Selectivity

The term *severity* is often used to describe the depth of cracking or extent of conversion. The definition varies with different manufacturers and may differ accordingly to the type of hydrocarbon treated. Applying the first-order dissociation of *n*-pentane, the definition of *severity* is this:

$$\mathrm{KSF} = \ln \frac{1}{1 - x}$$

KSF = kinetic severity function
x = conversion of *n*-pentane

In the case of SC ethane and propane, it is convenient to express the severity of the operating conditions in terms of feed conversion. At very high severities, the methane and

Table 7.2 Yield of products vs. kinetic severity function (KSF).

% conversion	KSF	
82	1.7	Lowest commercial severity; max. propylene
90	2.3	Max. butadiene
93	2.7	Max. combined olefins
98	3.9	Highest commercial severity; max. ethylene

ethylene yield level off, while those of propylene and C_4 cut reach a peak and then decline. The ratio of ethylene and propylene yield increases with severity, which favors the formation of ethylene. The relative production of C5+ cut passes through a minimum and at very high severity tends to increase. Modern ethylene plants are normally designed for near-maximum cracking severity because of economic considerations.

Applying this definition, the severities listed in Table 7.2 may be developed. Typical product maxima are listed for reference.

7.2.5 Furnace Run Length

Furnace run length can be calculated from the equation:

$$\text{Run length} = \frac{\text{Tmd} - Tmc}{\Delta T/\text{day}}$$

T_{md} = maximum allowable tube skin temperature
$\Delta T/\text{day}$ = average rise in tube skin temperature per day
T_{mc} = maximum metal skin temperature in a clean, uncoked condition

For any feedstock, the heater section run length depends on the pyrolysis coil selectivity, cracking severity, and transfer-line exchanger design. Run length varies between 21 and 60 days for gas-based furnaces and 21–40 days for liquid-feed-based furnaces.

7.3 Ethylene Furnace Design

Pyrolysis furnace design during the recent decades has made significant development. Prior to 1960, ethylene pyrolysis furnaces were box types with horizontal radiant tubes. A typical two-cell cracking furnace (two furnaces sharing one stack) has a capacity of 50 000 to over 100 000 t/a of ethene. With the present plant sizes of about 500 000–800 000 t/a, a plant based on naphtha typically consists of 10 naphtha crackers and one or two ethane crackers, in which recycled ethane is processed.

High-thermal-efficiency furnace design can contribute greatly to minimum overall plant utility costs. Higher efficiency can be achieved by (i) upgrading pyrolysis furnace capacity, (ii) increasing cracking severity, (iii) improving ethylene selectivity, (iv) improving thermal efficiency, (v) reducing downtime for decoking, and (vi) reducing maintenance costs. This can be achieved by radiant coils with shorter residence time and lower pressure drop, combustion air preheating, and short residence time. Small-diameter coils coupled with

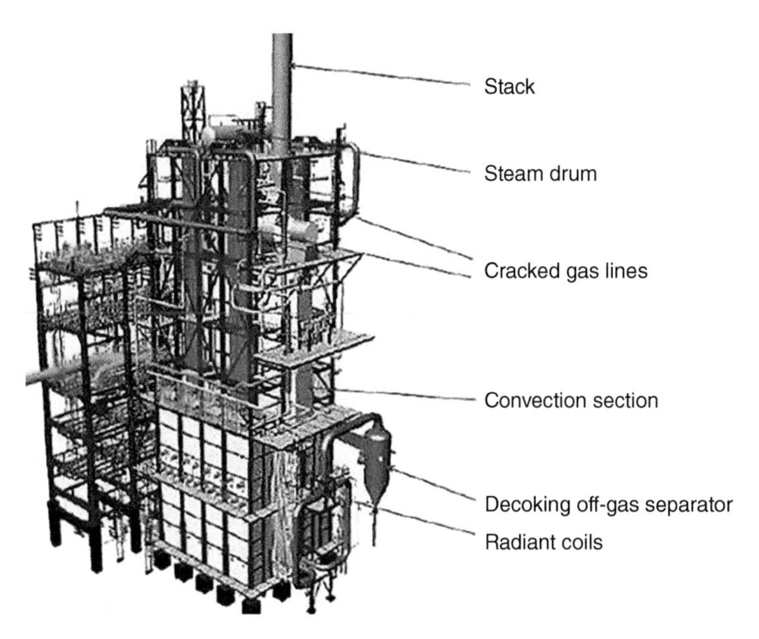

Figure 7.2 Cracking furnace setup.

increased-dilution steam, with the use of booster compressors to reduce furnace outlet pressure, can increase efficiency ethylene selectivity. Radiant coils with a short residence time and low hydrocarbon partial pressure give higher ethylene selectivity. Figure 7.2 shows a commercial steam cracker (Linde), which indicates the size of the furnace.

7.3.1 Heat Exchanger

The cracked gases usually leave the reactor at a temperature exceeding 1070 K. They should be instantaneously cooled to prevent consecutive reactions. Quenching can be direct or indirect or a combination of both. Direct quenching involves injection of a liquid spray, usually water or oil, and cooling can be very fast. However, indirect cooling proceeds through the use of *transfer-line exchanger (TLE)*. It allows the generation of valuable high-pressure steam; this generated steam is advantageous when used in the refrigeration compressor and the turbines driving the cracked gas (see Chapter 3).

In designing TLEs, the following points are of concern:

- Minimum residence time in the section between furnace outlet and TLE.
- Low pressure drop.
- High heat recovery. The outlet temperature for liquid cracking should not be too low (condensation of heavy components).
- Acceptable run times (fouling occurs by coke deposition and of heavy components).

Usually a TLE is used, followed by direct quenching. In ethane cracking, where only small amounts of fuel oil are produced, cooling is mainly accomplished in the TLE. During SC of heavy naphtha or gas oil, in which large amounts of fuel oil are produced, the heat-exchange area of the TLE is reduced and most of the cooling is performed by direct quench. In the extreme case of vacuum gas oils, the TLEs are totally eliminated and supplanted by direct quench.

7.4 Coke Formation During Pyrolysis and Decoking Measures

Pyrolysis of any hydrocarbon feedstock is always accompanied by coke formation, which deposits on the walls of the tubular reactor. Under typical operating conditions, the coke formation in naphtha pyrolysis is about 0.01 wt. % of the feed. The coke deposits on the walls of the reactor reduce the overall heat transfer coefficient and increase the pressure drop across the reactor. This results in a gradual decrease of both the reactor tube metal temperature and the pressure drop across the reactor, necessitating periodic shutdowns. The coke formation inside the tube depends upon (i) characteristics of the feedstock and coking precursor, (ii) hydrocarbon partial pressure, (iii) thermal condition of the coil, and (iv) mass velocity, which controls the dynamics of the gas film close to the wall. Controlling the coking rate permits increasing the severity of the furnace to increase the conversion rate; and reducing the cycle rate and unloading downstream limiting equipment, which increases throughput. However, frequent decoking operations result in loss of production, affect the coil life, and increase fuel and utility costs.

The run length between two successive decoking operations varies according to the installation and the type of feedstock, but can be estimated at a few weeks on average. In ideal conditions, a furnace operating on naphtha can run for 65 days without decoking. However, the run length is always shorter due to the inevitable fouling of the quench boiler. In practice, the run length is as long as 90 days on ethane feedstock, 65 days on naphtha, and 40 with gas oil.

Various approaches for coke mitigation are based on reduced coke production in the coils and increased rates of coke removal or removal of coke precursors during pyrolysis. By using improved metallurgy and innovative coating systems, ethylene producers seek to improve unit reliability, increase carburization resistance, extend processing run length, reduce downtime and plant shutdowns, increase yields and throughput, and extend furnace tube life. Mechanical decoking and steam air decoking are the methods used for decoking: a mechanical decoking process takes four to seven times longer than steam/air decoking. The principal methods of decoking with steam/air are spalling and burning. The coke is burned in the presence of steam and air at a temperature from 600–800 °C. In industrial practice, this means at least one furnace is always undergoing decoking.

Decoking of the cracking coils takes place by gasification in an air/steam mixture or coating reactor coils; the TLE is also partly decoked. The procedure for complete decoking of the TLE is usually shutting down the furnace, disconnecting the TLE from the coil, and removing the coke mechanically or hydraulically.

7.4.1 Catalytic Gasification of Coke During Production

It is very attractive to avoid coke formation during pyrolysis reactions. In principle, this would be possible if a catalytic coating could be applied that reacted the coke away by steam gasification. Five years ago, it was claimed that a certain ceramic coating would perform this highly desired catalytic function. It would mean significant cost savings, in particular due to reduced downtime.

In tubular flow reactors, the amount of coke formed at a given temperature, pressure, and surface is independent of the amount of starting feeds or residence time. The quality of hydrocarbon feedstocks and the internal surface are decisive:

$$r_c = k_o C^o = k_o$$

r_c = coking rate
k_o = rate constant
C = reactant concentration

While the size of the inner surface of the reactor during pyrolysis does not change, qualitative changes take place on the wall of the reactor and its surroundings as a consequence of deposition of pyrolytic carbon in the form of coke. The quality of the reactor's wall material, especially a reactor with a small diameter, does not influence the formation of coke to any considerable measure. The rate of coke formation on a stainless steel surface is an order of magnitude higher than the rate of coking on copper, titanium, and non-metal surface (Table 7.3).

A higher rate of coke formation on a metal surface is related to the catalytic activity of iron, chromium, and nickel, which are components of steel. Carboneous sediments on copper, titanium, and non-metal surface have their origin in reactions taking place in the gaseous phase without any catalytic effect on the surface. A layer of silica or enamel has an inhibiting effect on coke formation. Silica was coated on steel from a solution of tetraethyl silicate, and after decomposition at 800 °C, a layer 6–9 μm thick had been deposited. The enamel was formed from a suspension of enamel coating by decomposition at 820 °C. The silica not only abolished the catalytic effect of the material of the reactor wall, but also formed a barrier preventing the transfer of carbon into the metal structure (carburization).

For the decoking the pyrolysis reactor, the air/steam method is used. A certain part of the steam reacts with pyrolytic carbon and forms carbon monoxide and carbon dioxide. This reaction is slower than burning out with air and requires higher temperatures to reach a marked increase in rate. It seems that burning with steam is catalyzed by iron and nickel,

Table 7.3 Rate of coke formation on different surfaces.

Reactor material	k_o, s^{-1}
Stainless steel	0.6923
Nickel	0.0396
Silica deposited on steel	0.0234
Quartz (SiO_2)	0.0207
Titanium	0.0179
Alundum (α-Al_2O_3)	0.0147
Enamel	0.0140
Copper	0.0056

which diffuse into the layer of deposited coke from the material of the reactor. Sulfur acts as a poison of these catalysts and retards of the gasification of coke. Another part of the steam reacts with the inner surface of the reactor according to reaction (Eq. (7.1)):

$$(\text{METAL})_{\text{surface}} + H_2O \rightleftarrows (\text{METAL OXIDES})_{\text{surface}} + H_2 \tag{7.1}$$

Metal oxides = NiO, Fe_3O_4, Fe_2O_3

Metal oxides are also formed in reaction (Eq. (7.2)) by reaction of air oxygen:

$$(\text{METAL})_{\text{surface}} + O_2 \rightleftarrows (\text{METAL OXIDES})_{\text{surface}} \tag{7.2}$$

Metal oxides = Fe_2O_3, Fe_3O_4, Cr_2O_3

Metal oxides formed by reactions (Eqs. (7.1) and (7.2)) in air/steam decoking markedly favor the course of secondary reactions, as a result of which coke is formed in the initial operational phase of the pyrolysis.

During the pyrolysis of hydrocarbon feedstocks, the presence of sulfur and its compounds in the reaction system makes itself felt in several ways in the course of specific reactions in the gaseous phase and also on the heterogeneous surface of the reactor. What is most required is its positive influence on the rate of decomposition, improvement of the selectivity of conversion, and retardation of secondary reactions leading to pyrolytic carbon, which in the form of coke is deposited in the pyrolysis system. Irreversible reactions also occur between the wall reactor, hydrocarbons, and sulfur compounds. In short-duration use of the reactor (a few months), no substantial changes of the reactor steel occur. In long-duration use (a few years), interactions occur between the metal surface of the reactor and pyrolytic carbon when metal carbides are formed (carburization). If sulfur compounds are present in the reaction system, metal sulfides are formed. After prolonged use of the reactor, the quantity of carbon increases eight times and that of sulfur six times.

In the presence of hydrogen sulfide, a reaction takes place with the inner surface. Under the conditions of pyrolysis, part of the hydrogen sulfide can undergo fission according to reaction (Eq. (7.3)):

$$H_2S \rightleftarrows {}^{\cdot}SH + H^{\cdot} \tag{7.3}$$

In the presence of hydrogen radicals, the decomposition of hydrogen sulfide can take place with the formation of thiyl radicals according to reaction (Eq. (7.4)):

$$H_2S + H^{\cdot} \rightleftarrows HS^{\cdot} + H_2 \tag{7.4}$$

Under specific conditions (temperature of 2700–3800 K and concentrations of hydrogen sulfide in the reaction zone of 25–200 ppm), a course of reactions is assumed where there is a release of elemental sulfur in atomic form and of S_2 with paramagnetic properties. The HS· radicals or S_1 and S_2 can take part in reactions in the gaseous phase or in reactions with the metal surface of the reactor where metal sulfides are formed according to reaction (Eq. (7.5)):

$$(\text{METAL})_{\text{surface}} + HS^{\cdot}(S_1, S_2) \rightleftarrows (\text{METAL SULFIDES})_{\text{surface}} + H^{\cdot} \tag{7.5}$$

With the formation of metal sulfides, the quality of the reactor wall significantly changes, which can be seen especially in the formation of carbonaceous sediments.

7.4.2 Sulfur Addition to Ethane Feedstocks

Sulfur (sulfur compounds) is the most common additive and pretreatment chemical for the reduction of radiant coil coking. Sulfur also has the benefit of reducing the formation of CO both at the start of run (SOR) and to a lesser extent during steady-state operation (Lummus 2013).

Most naphtha and heavier feedstocks contain sufficient sulfur so that the addition of sulfur during the run is not required and has no impact on the run length or reduction in CO produced. However, with gas feedstocks (C_4 and lighter that typically contain no sulfur) and desulfurized naphthas (reformate), the addition of sulfur during the furnace run normally has a positive effect on decreasing CO production and extending the furnace run length. This serves as evidence of the presence of H_2S in the coking process. However, the use of H_2S is not recommended due to handling problems that can arise. Complex sulfur compounds that decompose at cracking temperatures to yield H_2S and other sulfur compounds are preferred.

Numerous complex sulfur compounds are available for use in ethylene plants. These include:

- Diethyl sulfide (DES)
- Dimethyl disulfide (DMDS)
- Dimethyl sulfide (DMS)
- Ethyl mercaptan (EM)
- Tertiary butyl polysulfide (TBPS)
- Tertiary nonyl polysulfide (TNPS)

The choice of sulfur source is usually based on availability, price, and environmental/safety considerations (odor, toxicity, high flash point, etc.).

Based on contained sulfur, DMS, DMDS, and DES are generally the cheapest, although this varies from one geographic area to another. Each end user should determine the most economical source of sulfur that meets plant environmental/safety requirements. The other consideration is to make sure all the sulfur added is converted, since unconverted heavy compounds can enter the gasoline product. DMS requires somewhat higher temperatures to fully convert than either DMDS or EM.

Pre-sulfiding may only be necessary for plants that have front-end acetylene converters. CO acts as a temporary poison to the catalyst. During steady-state operation, Lummus typically recommends between 50 and 100 wppm of contained sulfur and avoiding sulfur rates of 200 wppm and higher, as these levels can lead to sulfur attack and damage to the radiant coils.

7.5 Product Processing

The requirement of steam will depend upon the type of feedstock; lighter hydrocarbons require less steam as compared to heavier feedstocks. The relative cost of SC based on feedstock varies from 1.00 for ethane to 1.84 for vacuum gas oil. The steam requirement in steam crackers is various from 0.2–0.4 for ethane to 0.8–1.0 for gas oil (feed kg steam/kg of hydrocarbon).

Modern ethylene plants incorporate the following major process steps: cracking, compression, and separation of the cracked gas by low-temperature fractionation. The nature of the feedstock and the level of pyrolysis severity largely determine the operating conditions in the cracking and quenching section. Various steps involved in the pyrolysis of naphtha and separation of the products are discussed next. In the case of gas cracking, separation of ethane and propane from natural gas is involved. A flow diagram for pyrolysis of naphtha is given in Figure 7.3.

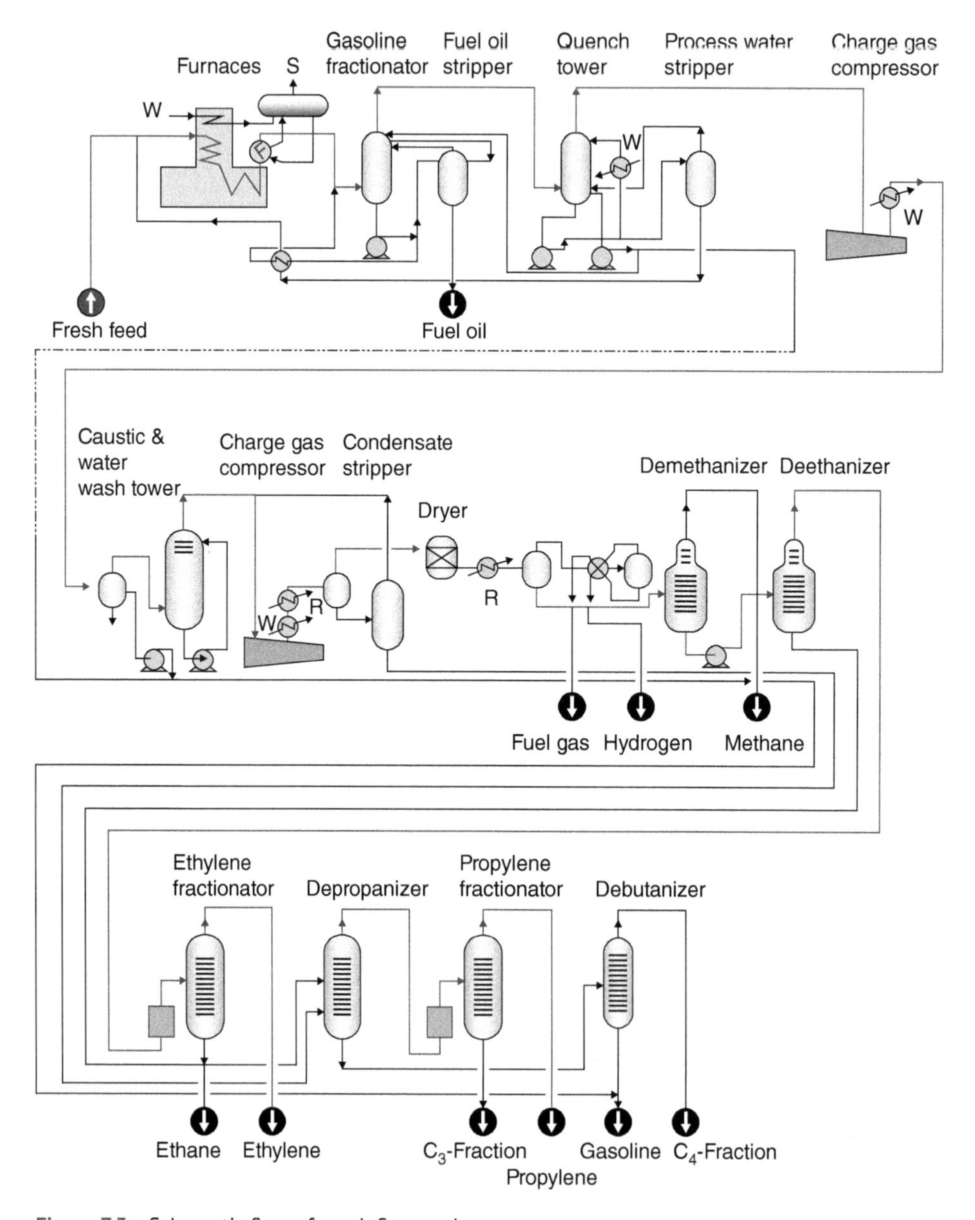

Figure 7.3 Schematic flow of an olefins cracker.

7.5.1 Hot Section

The hot section consists of a convection zone and a radiant zone. In the convection zone, the hydrocarbon feedstock is preheated and mixed with steam and heated to a high temperature. In the convection zone, a rapid rise in temperature takes place, and the pyrolysis reaction takes place. The addition of dilution steam enhances the ethylene yield and reduces the coking tendency in the furnace coils. The production of the pyrolysis reaction consists of a wide range of saturated and unsaturated hydrocarbons.

7.5.2 Quench Section

To avoid subsequent reaction, the effluents are fixed in their kinetic development by a sudden quench, first by indirect quench with water to 400–450 °C in a transfer line exchanger or quench boiler. This is a large heat exchanger that is a bundle of metal tubes through which the gases pass and around which water is circulated under pressure. The hot water produced is used to generate steam for use in the plant. In the next step, the quench is done by heavy products of pyrolysis.

Separation of cracked gases (Figures 7.3 and 7.4):

- Steam
- Compressors
- Caustic oxidation
- Demethanizer
- Deethanizer and C_2 hydrogenation
- Depropanizer and C_3 hydrogenation
- Final products: ethylene, propylene, etc.

7.6 Typical Naphtha Cracker Plant

7.6.1 Hot Section

Convection zone	Feedstock is pyrolyzed and the effluents are conditioned.
Radiation zone	The products formed are separated and purified.
Quench	To avoid subsequent reaction, the effluents are fixed in their development with a sudden quench.
I Indirect	Indirect quench by water to 400–500 °C, generating high-pressure steam.
II Direct	Direct quench by heavy residue byproduct of pyrolysis.
Primary fraction-atator column	Separation of light products of pyrolysis as the top and bottom as pyrolysis products.
Compression	Compression of light products.
Caustic scrubbing and drying	Scrubbing with caustic followed by molecular sieve adsorption to remove sulfur compounds, mercaptan, etc.

7.6.2 Cold Section

After compression, caustic scrubbing, and drying, the light effluents enter the cold section of the unit, which performs the following:

- Hydrogen separation to various concentrations
- Ethylene separation (99.9%)
- Propylene separation (95%)
- A C_4 cut containing 25–50% butadiene
- Complementary fraction of pyrolysis gasoline that is rich in aromatic hydrocarbons.

The complexity of the separation section of a cracker increases markedly as the feed changes from ethane.

Demethanizer	Methane is condensed at the top, around − 100 °C, pressure 32 Pa.
Deethanizer	The C_2 cut is separated (ethane and ethylene). Acetylene is eliminated by selective hydrogenation. Catalyst: palladium or nickel, 40–80 °C, 3 kPa.
Separation of ethylene	Ethylene is fractionated, and unreacted ethane is recycled.
Depropanizer	The C_3^+ cut from the bottom of the deethanizer is fractionated. The C_3 cut from the top of depropanizer is selectively hydrogenated to remove methyl acetylene and propadiene. Propylene content of the cut is 95%. Separation takes place in the supplementary column for more pure propylene.
Removal of propane from propylene	Separation in the supplementary column for more pure propylene.
Debutanizer	The C_4 stream is separated from the C_5^+ stream

7.7 Gas-Feed Cracker Process Design

Thanks to technological advancements and refinements in fracking (Section 4.5), production of natural gas (methane) has increased from 3030 billion cubic meters in 2008 to 3870 billion cubic meters in 2018. This makes shale gas a game changer and an interesting cost-competitive feedstock. The large availability of shale gas has reinforced the interest in routes for valorizing methane in the form of olefins and higher hydrocarbons, either directly or indirectly. These methane-conversion processes could be an excellent way to valorize large amounts of methane from stranded gas rather than from shale gas.

Although the growth of shale gas has triggered increasing research into methane-upgrading processes, methane from shale gas is relatively expensive due to stringent specifications on maximum ethane content. This has caused the price of ethane to drop significantly, falling even below its calorific value. The abundance of cheap ethane created by shale gas exploitation has enabled cheap low-olefin production via SC and has had a profound impact on the local US olefin market. Many newly built crackers are ethane based, and many existing liquid crackers are retrofitted to lighter gaseous feeds, as such feeds offer an economic advantage when accessible. New ethane steam crackers are expected to come online soon and to add 1×10^7 t of ethylene capacity by 2020 in the United States. Globally, ethylene producers will experience a substantial capacity expansion in the next five years, with an annual increase from 1.76×10^8 t to 2.18×10^8 t and with investments of nearly $45 billion USD in upcoming projects. These large investments suggest that SC of hydrocarbons will remain the main pathway in the production of light olefins. Ethane cracking is highly selective toward ethylene; hence, few or no other olefins are coproduced. Therefore, the shift toward lighter feedstocks has resulted in increasing interest in developing on-purpose production routes toward propylene and higher olefins.

Ethylene and propylene production will continue to proceed mainly by the SC of hydrocarbons. The SC of hydrocarbons is both one of the most important and one of the most energy-intensive processes in the petrochemical industry. It is the leading technology for light olefin production, and is a well-established technology. In this process, a hydrocarbon feedstock is mixed with steam and cracked at elevated temperatures in a tubular reactor (Figure 7.4). In order to attain these elevated temperatures, and hence initiate the thermochemical conversion, the reactors are suspended in a gas-fired furnace. The feedstock ranges from light alkanes such as ethane and propane to complex mixtures such as naphthas and gas oils. Since the first commercial steam cracker, the capacity of this process has grown; up to 1.5×10^6 t of ethylene is now produced by this process annually, taking full advantage of the economy of scale.

A steam cracker unit can be roughly divided into two distinct parts: a hot section and a cold section. In the latter, the reactor effluent is separated into different products, usually at high purity as required for polymer production.

Ethane crackers have a less-intensive separation train than liquid crackers, and hence a lower capital investment cost. The choice of the feedstock is mainly based on availability and profitability. An ethane cracker is preferred when the feedstock is available in sufficient amounts; otherwise, a liquid cracker is more beneficial, as the transportation of gases is relatively costly. The global steam cracker portfolio is dominated by naphtha crackers. As a result of shale gas developments in the United States, ethane cracking has become more attractive due to its low cost. The abundance of ethane has widened the ethane cracking margins. Therefore, there is a strong push to minimize liquid feedstocks and, preferably, even to fully replace them with ethane. However naphtha crackers produce the largest fraction of the total ethylene production capacity. Increasing ethane cracking has led to a decline in the production volumes of some co-products such as propylene, as shown in Figure 7.5.

Liquid crackers have the opportunity to diversify these kinds of by-products and hence sustain their margins. In addition, revamping SC facilities from liquid to ethane cracking is not always straightforward. Steam crackers are not standalone units but are highly integrated with downstream production units and the downstream demand for a range of cracker products. Furthermore, the separation trains for liquid and gaseous feedstocks differ considerably. Completely switching the feed leads to a different reactor effluent, whose effect puts pressure on different downstream units such as, for example, the methane refrigerant cycle, the performances of the columns, and the cracked-gas compressor. Hence, the opportunity to take advantage of low ethane prices is highly dependent on the downstream logistics of the plant.

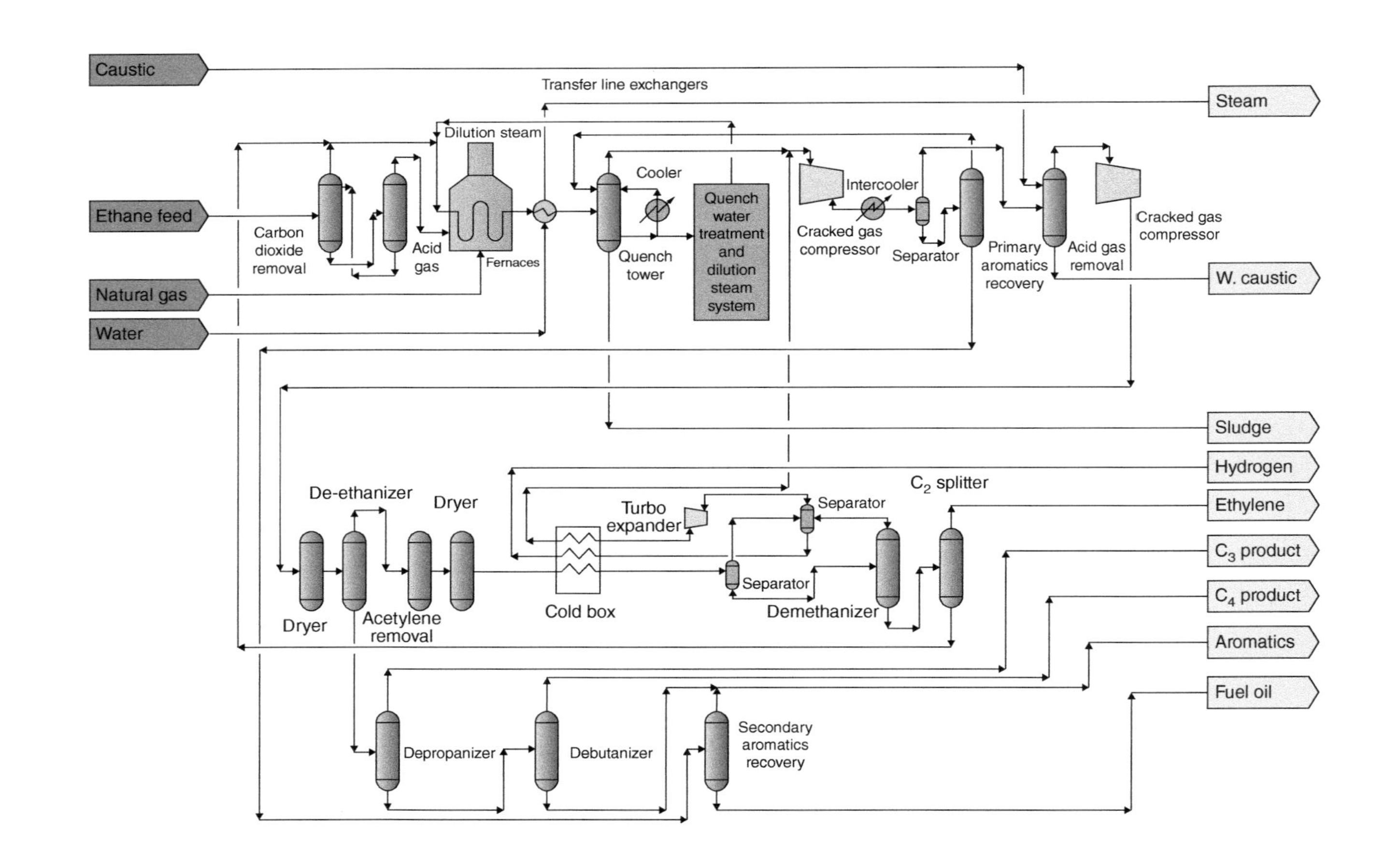

Figure 7.4 Ethane/propane cracking flow scheme.

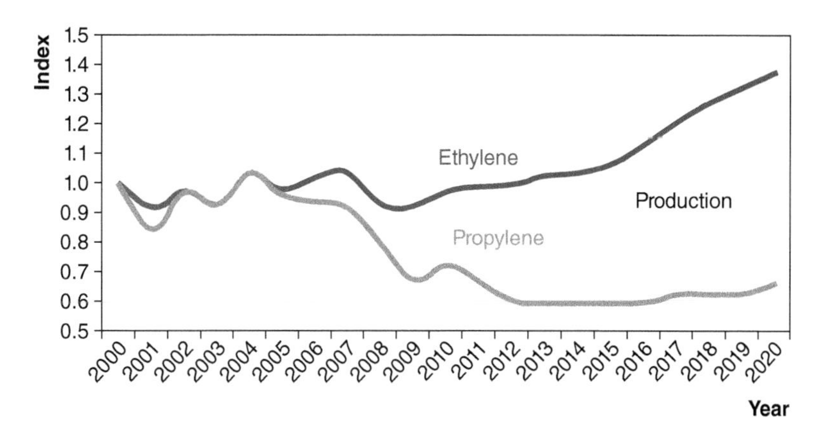

Figure 7.5 North American olefin production from steam crackers in 2010.

On the other hand, the large availability of crude oil has increased interest in using it directly as feedstock for the production of chemicals. This interest led to the first world-scale facility that feeds crude oil into a steam cracker, with a production capacity of 1×10^6 t of ethylene. The driving force is production cost savings. In this way, refinery costs are omitted, and the cost savings have been estimated to be as high as $200 USD/t of ethylene. The steam cracker includes a flash pot between the convective and radiant sections of the furnaces. This modification allows the removal of the heavier component, prior to sending it to the radiant coils.

7.8 Trends in Technological Development of Steam Crackers for Production of Ethylene

From the late 1960s through the 1970s, the petrochemical industry built a generation of new steam crackers with an ethylene capacity of several million tons capacity with residence time 0.3–0.6 seconds, thermal efficiency below 90%, central waste-heat recovery systems, and nitrogen oxide (NO_x) emissions of 75–100 ppm. Present-day olefin plants have a production capacity of more than 1 000 000 tons of ethylene per year with modern cracking furnaces using twin-cell designs. They have a short residence time of 0.1–0.3 seconds and radiant coils with smaller diameters to increase yields. The higher selectivity of modern coils reduces specific energy consumption. The modern olefin plants have better ethylene selectivity, and they improve health, safety, and environment effects by incorporating current emission and safety standards.

Major energy improvements from revamping or replacing existing furnace sections can be achieved by:

- Increased thermal efficiency.
- Higher radiant efficiency and less excess air with new burner technology and better instrumentation.

- Reduced heat losses due to fewer and bigger furnace units, or new refractories with coiled materials.
- Higher yields with new radiant coils, reducing specific energy demand.
- Higher availability by application of new and highly reliable technology, reducing losses due to unplanned shutdowns.

Performance of the SC furnace can be upgraded by: (i) increasing furnace capacity, (ii) increasing cracking severity, (iii) improving ethylene selectivity, (iv) improving thermal efficiency, and (v) reducing downtime for decoking and reducing maintenance.

The other aspect of the SC furnace is the relatively large spectrum of raw materials that are processed in Western European steam crackers. It is important to add a few considerations about these feedstocks. Two ethylene feedstock categories are typically part of highly integrated operations: ethane (including some light gases from refinery operations) and middle distillates. Not only are these operations characterized by physical integration, but in most cases, they are used as base-load raw material, rather than on an "opportunistic" basis. Ethane-based ethylene production depends on locally available natural gas resources, since this is not traded in international markets. The most important Western European operations of this kind rely on North Sea light natural gas liquids (NGLs) that are piped to dedicated facilities, i.e. with limited flexibility for alternatives.

Also, using light refinery gases for ethylene production requires a dedicated and integrated operation with a refinery. This is also true of the second category: middle distillates. In Western Europe, middle distillates-based ethylene production typically takes place in naphtha crackers, with the built-in flexibility for cracking gas-oil type material. However, middle distillates used in this operation are tailor-made from a nearby refinery and supplied on a base-load mode. Table 7.1 illustrates the yield of ethylene from various feedstocks.

As can be seen, the yield of ethylene varies according to the feedstock. The co-products of ethylene production are in many cases as important as the ethylene itself. Both propylene and butadiene are important chemical feedstocks that are in short supply. Butadiene is widely used in car tire production and for the production of nylon polymers. Propylene is used both in polypropylene and for the production of other chemical products such as methacrylates, phenol, and propylene oxide.

Other routes to propylene are gaining capacity, and a new type of fluid catalytic cracking (FCC) has been developed that is configured for the production of petrochemicals rather than fuels and that can produce 40–50% C_3 and C_4 fractions. Increasing amounts of petrochemicals are now obtained from refineries. For those interested, Figure 7.3 shows the process flow for a naphtha olefins cracker.

A large proportion of liquid feedstocks in Western European steam crackers are also influencing the aromatics supply structure, via pyrolysis gasoline production. Regional SC operations typically provide about 47–49% of primary aromatics, with an almost equal share derived from reformate, and a minor amount from coal-derived feedstocks. However, on a product-by-product basis, this situation is different. Figure 11.2 shows recent and forecast developments in Western European gross aromatics production, i.e. including the conversion of toluene to benzene and xylenes via hydro-dealkylation and disproportionation.

SC operations are the most important source for benzene, accounting for 72% of the total. In contrast, the majority of toluene and xylenes originate from the refining industry through reforming. Due to the faster growth of xylenes over benzene, increasing volumes of aromatics will be required from refineries. However, this will continue for xylenes production, with most of the incremental benzene derived from increased pyrolysis gasoline availability from SC. With regional SC profitability highly dependent on co-product revenues, ethylene operators will search for added value through the aromatics-derivative chain.

Despite declining importance, gasoline operations will remain the most important outlet for naphtha, representing almost 60% of naphtha consumption in 2010, down from 65% in 1995. The vast majority of regional naphtha is derived from refinery operations, which are a function of oil demand. Anticipated Western European energy requirements point to relatively moderate growth for oil. With total naphtha requirements growing at 0.9% annually but showing increases of 1.9% for olefins and aromatics production, future incremental availability of naphtha may be constrained. Consequently, dedicated investment may be required.

In Western Europe, this aspect assumes a particular relevance. This methodology outlines potential shortcomings in the existing system and, therefore, it identifies actions that should be undertaken. While this approach is common to all oil product balances from post-translation modification (PTM) analysis, an additional factor must be considered for naphtha.

Since all future ethylene feedstocks other than naphtha are based on what is known today, the projected naphtha requirements as steam-cracker feedstocks are calculated by difference. This clearly amplifies the potential for large naphtha deficits, particularly in the long term on a regional and global basis. Consequently, new investment is needed not only from the supply side but also from the consuming industry, to reduce dependence on naphtha. Since the local refining industry is geared toward gasoline production, the local naphtha balance has typically shown a large deficit, as opposed to a growing gasoline surplus. Also, olefin producers are a major factor in naphtha import requirements. Developments toward feedstock diversification in local SC operations have recently limited naphtha import requirements by 10–13 million tons. However, in the absence of dedicated investments, increased petrochemical demand may result in a disproportionate increase in the regional naphtha deficit.

Considering that Western European gasoline demand is expected to stagnate, if not decrease, in the long term, a first correction may come from more naphtha diverted to petrochemicals at the expense of gasoline production. However, due to the possible deficit, and considering that worldwide incremental naphtha availability may be constrained, a further correction may come from feedstock diversification, i.e. involving the demand side. Given the relative rigidity of naphtha feedstock requirements that characterizes the aromatics industry, these developments will mostly involve SC operations.

7.8.1 Direct Involvement in Petrochemical Production

Developments in the supply–demand balance of petrochemical products indicate areas of opportunity for Western European refiners in propylene and xylenes production (Figures 7.6, 9.2, and 11.2). Further incentive is provided by the current and projected gasoline surplus. Because both propylene and xylenes are produced at the expense of gasoline from cat-cracking and reforming operations, increased release of these two

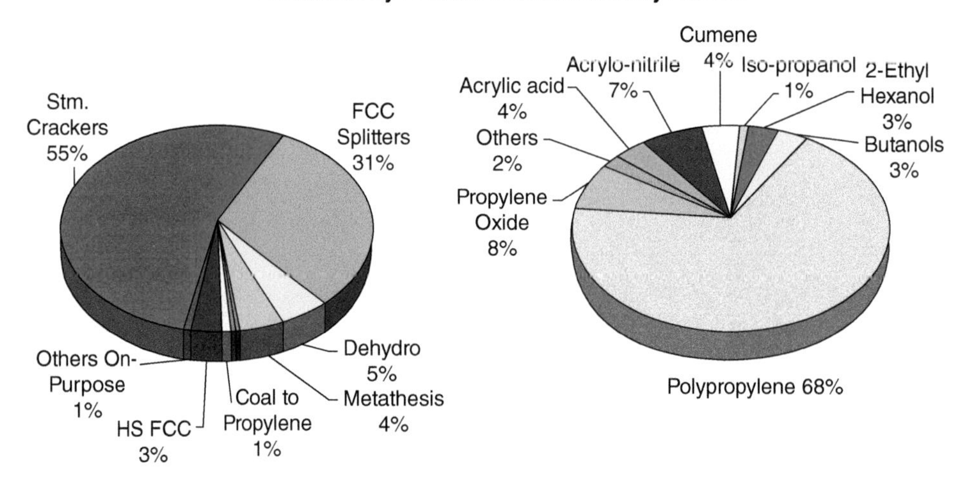

Figure 7.6 Propylene production and demand profile 2013. Production: 84 million metric tons.

petrochemical products may look attractive, particularly when available capital is short. Further support may come from considering future gasoline quality requirements, where more aromatics and olefins must be rejected from the local gasoline pool. This trend will involve local gasoline markets, but with mandated tighter quantities for gasoline on a global basis, export material will have to gradually comply in the long term. For benzene, however, competition with steam-cracker operators is likely to make large releases of this product an unattractive option. The most dynamic area appears to be increased production and recovery of propylene from cat-cracking operations. Several licensors are evaluating new technology that maximizes light olefin yields.

7.8.2 Integrating SC Operations

As already discussed, integration with SC operations may offer attractive opportunities to refineries (Tables 5.3 and 5.4). In addition to providing feedstocks, refiners would have access to several products from the steam-cracker – i.e. hydrogen, gasoline component returns, and others – that contribute to optimizing global operation. In existing refinery/ethylene integrated operations, there may be some incremental feedstock supply, particularly from using cat-cracking derived off-gases. However, considerable investment is required to provide raw material for a grassroots steam-cracker. A new world-scale steam cracker with an ethylene capacity of 700 thousand tons per year, if based on liquid feed, would require more than 2 MTPA of feedstock. If the steam cracker is based on naphtha, in addition to the investment for new topping capacity, a refiner would have to considerably increase crude runs to produce sufficient naphtha. Consequently, about 80% of the incremental crude would be distilled into other products, i.e. a factor not necessarily rewarding. Alternatively, the refiner would have to invest heavily in conversion capacities to increase the naphtha output. This latter solution is unlikely, since the investment would be high and could be considered only as part of an overall refinery expansion plan.

Another option is to provide middle-distillates feedstock. Although this operation may by attractive under particular conditions, it has limitations. Considering the growing regional middle-distillates deficit caused by tighter requirements from the transportation sector, some competition could arise for the same hydro-cracked material, which could go to ethylene and/or to diesel. However, this event is extremely site specific.

Aromatic compounds are the most widely used and of the most important classes of petrochemicals (Figure 11.2). They are excellent solvents and constitute an important component of synthetic rubbers and fibers. Catalytic reforming has become the most important process for the preparation of aromatics. The two major transformations that lead to aromatics are dehydrogenation of cyclohexanes and dehydrocyclization of alkanes. Additionally, isomerization of other cycloalkanes followed by dehydrogenation (dehydroisomerization) also contributes to aromatic formation (Section 5.5.2.1). benzene, toluene, and xylene (BTX) processing, the major source of these important chemicals, is connected to catalytic cracking and catalytic reforming. In the United States, SC is the major source of aromatics. In Europe and Japan, catalytic reforming is the major process for aromatics production. A cracking process, the dealkylation of alkylbenzenes, became an established industrial synthesis to produce aromatics. Several technologies for both catalytic hydrodealkylation and thermal dealkylation are operated industrially. Dealkylation, isomerization, disproportionation, and transalkylation processes are used in the manufacture of benzene and xylene.

Naphtha remains the most important petrochemical feedstock. However, future shortages will force petrochemical manufacturers to seek alternative raw material sources. Olefin operators are considering other sources such as heavy condensates, middle distillates, and ethane and refinery gas. Flexible olefin-cracking operations and integration with refineries will affect future alternative feedstock usage. Increased processing of heavy condensates by dedicated splitters may entice some Western European refineries to participate in the local merchant petrochemical feedstock markets. Continued strong petrochemical demand can open revenue opportunities for refiners.

Figure 11.2 illustrates demand developments for major aromatics. In particular, this chart highlights the importance of benzene. Also, it shows toluene's role as an incremental source for both benzene and xylenes. Similar to ethylene and propylene, benzene is the single important derivative in the aromatic's demand structure. While styrene represents about half of the benzene usage with a minor change during the period analyzed, xylene's demand is primarily for para-xylene, whose importance is growing continuously. Recent and expected evolution of the regional xylene/benzene ratio is shown in Figure 11.2. As in the case of the propylene/ethylene ratio, this figure is expected to grow, although the regional para-xylene market could be strongly influenced by the commissioning of a single new world-scale facility.

Methane and carbon monoxide are the two materials of practical importance in C_1 hydrocarbon technology. According to present industrial practice, natural gas (methane) can be converted to a mixture of carbon monoxide and hydrogen called *synthesis gas* (syngas; Figure 12.1). Hydrocarbons can be produced from syngas through Fischer–Tropsch synthesis. Of the technological modifications, Fischer–Tropsch synthesis in the liquid phase may be used to produce light alkenes under appropriate conditions in a very efficient and economical way. The favorable surface composition of the catalyst suppresses secondary transformation, thus ensuring selective α-olefin formation.

The transformation of syngas to methanol is a process of major industrial importance. The conversion of methanol to alkenes (Figure 8.4) cycloalkanes or a mixture of alkanes and aromatics requires the elimination of oxygen. In general, medium-pore zeolites (ZSM-5 and ZSM-11) are the best catalysts to produce hydrocarbons from methanol. Hydroformylation gained substantial industrial importance in the manufacture of *n*-butyraldehyde and certain alcohols. Hydroformylation is the metal-catalyzed transformation of alkenes with carbon monoxide and hydrogen to form aldehydes.

All these processes exist and are already being implemented. However, since they are due to take on ever more importance; their performance will have to be improved by developing more sophisticated, active, and selective catalysts to allow less demanding operating conditions (pressure, velocity, etc.). A large segment of the modern petrochemical industry is based on catalytic processes. Technology in the petrochemical industry tends to be evolutionary rather than revolutionary. Rarely does a researcher in petrochemistry come up with an entirely new product or a process that will open up a completely new market application for petrochemicals or that will give an entirely new route to the production of a product. Most research and development is oriented toward evolutionary improvement – how can we achieve better yields from an existing process, how can we improve the quality of exploration data, or how can we produce our products to new quality standards that meet ever-higher expectations of regulators and customer?

Technological development in the petrochemical industry also derives predominantly from applied technology rather than pure "blue sky" science. If we look at major technological developments in offshore engineering, the story is the same; materials scientists and computing specialists have developed stronger, lighter materials and more advanced CAD and structural analysis tools, respectively, which the petrochemical industry has successfully applied to the design of safer, cheaper, more reliable structures as its operations move into deeper waters.

Further Reading

Bajus, M. and Veselý, V. (1976). Process of pyrolysis in the present of coking inhibitors with promotion effect to olefins production. Czechoslovak Republic patent 180859.

Bajus, M., Veselý, V., Baxa, J. et al. (1981). *Ind. Eng. Chem. Prod. Res. Dev.* 20: 741.

Bajus, M., Veselý, V., Baxa, J. et al. (1983). *Ind. Eng. Chem. Prod. Res. Dev.* 22: 336.

Lummus Technology INC (2013). Position addition to ethylene cracking feedstock.

Hlinšťák, K., Huba, E., and Kopernický, I. et al. (1981). Olefins production by pyrolysis of hydrocarbons. Czechoslovak Republic patent 214405.

Jukič, A. (2013). Petroleum refining and petrochemical processes; production of olefins: steam cracking of hydrocarbons. www.gasandoil.com.

Lummus Technology (2013). Position on sulfur compounds addition to ethylene feedstocks. www.gaylordchemical.com.

Mlynková, B., Hájeková, E., and Bajus, M. (2010). *Chem. List.*, ISSN 0009-2770 104 (10): 926–933.

Ethylene l Toyo Engineering Corporation (2019). https://www.toyo-eng.com>company>news.

8

Pillar A of Petrochemistry

Other Sources of Lower Alkenes

CHAPTER MENU

8.1 Catalytic Dehydrogenation of Light Alkanes

The most obvious disadvantage of the steam cracking (SC) process is its relatively low selectivity for desired products. A wide range of products is produced with limited flexibility. This is inherent in the non-catalytic nature of the process. In this chapter, alternative feedstocks and processes are presented for the production of the lower alkenes (e.g. for propylene in Figure 8.1).

Dehydrogenation is a chemical reaction that involves the elimination of hydrogen and is the reverse of hydrogenation. Dehydrogenation reactions are of major industrial importance, especially in the production of constituents for the manufacture of high-octane gasoline.

The dehydrogenation of low-boiling alkanes has become of great industrial importance because it presents an alternative method for obtaining alkenes (e.g. butenes, butadiene) from low-cost saturated hydrocarbon feedstocks.

Process chemistry:

$$C_nH_{2n+2} \rightleftarrows C_nH_{2n} + H_2;\ \Delta_{r,T}H^o = +125 \text{ to } 120\ \text{kJ/mol}$$

$$C_nH_{2n} \rightleftarrows C_nH_{2n-2} + H_2;\ \Delta_{r,T}H^o = +110 \text{ to } 125\ \text{kJ/mol}$$

$$C_nH_{2n+2} \rightleftarrows C_nH_{2n-2} + 2H_2;\ \Delta_{r,T}H^o = +235 \text{ to } 255\ \text{kJ/mol}$$

Petrochemistry: Petrochemical Processing, Hydrocarbon Technology, and Green Engineering, First Edition. Martin Bajus.

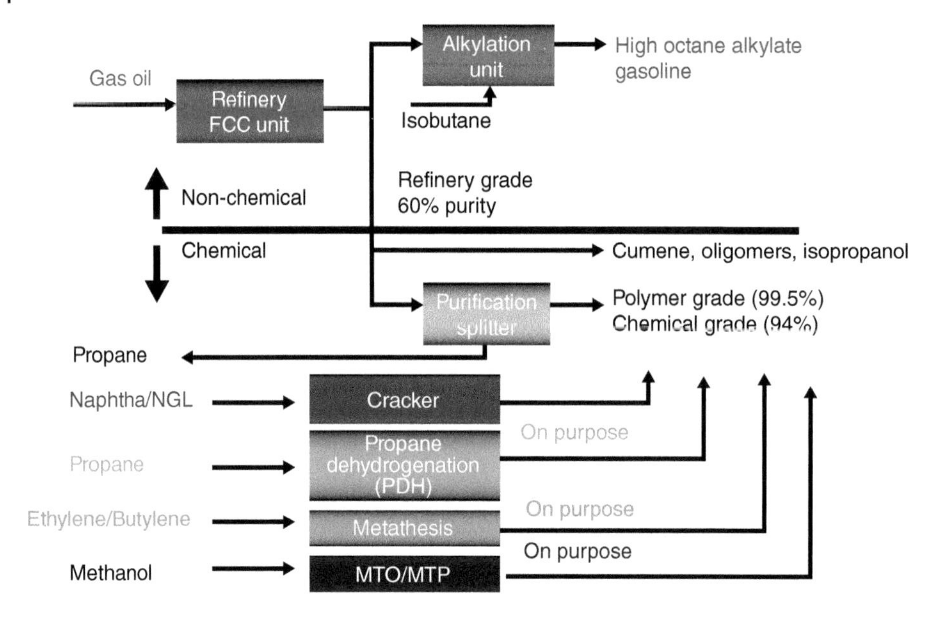

Figure 8.1 Alternative feedstocks and processes for the production of the propylene.

Alkane dehydrogenation is highly endothermic, and high temperatures are required. However, at these high temperatures, secondary reactions such as cracking and coke formation become appreciable. Deposition of coke on catalysts makes regeneration necessary. Furthermore, the thermodynamic equilibrium limits the conversion per pass so that a substantial recycle stream is required. The standard Gibbs energy $\Delta_T G^o$ is negative to at high temperatures, e.g. for:

- *n*-Butane to butenes at $T > 960$ K
- *n*-Butenes to *1,3*-butadiene at $T > 975$ K
- Isopentane to methylbutenes at $T > 873$ K
- Methylbutenes to isoprene at $T > 990$ K

The shift from naphtha toward light feeds that are derived from tight oil, for the production of ethylene in steam crackers, has impacted global propylene and crude C_4 production capacity. Therefore, routes for the on-purpose production of light olefins have received considerable interest. Catalytic dehydrogenation provides the possibility of high selectivity to a single olefin product – much higher than can be expected from SC alone. Given the relatively high abundance of cheap light alkanes from shale gas, this option is worth considering. The amount of propylene produced by dehydrogenation was 5×10^6 t in 2014 and is increasing, as new PDH plants are built worldwide. The profitability of this process is emphasized by the rather low prices of propane, compared with those of propylene. Two patented industrial processes for the dehydrogenation of alkanes are currently in commercial use: Oleflex (UOP, Figure 8.2) and CATOFIN (Lummus Technology, Figure 8.3). Both technologies use an alumina/supported catalyst: Pt-Sn/Al_2O_3 and Cr_2O_3/Al_2O_3 for Oleflex and CATOFIN, respectively. The occurring reaction is highly endothermic and equilibrium-limited, so high temperatures and low pressures are favored.

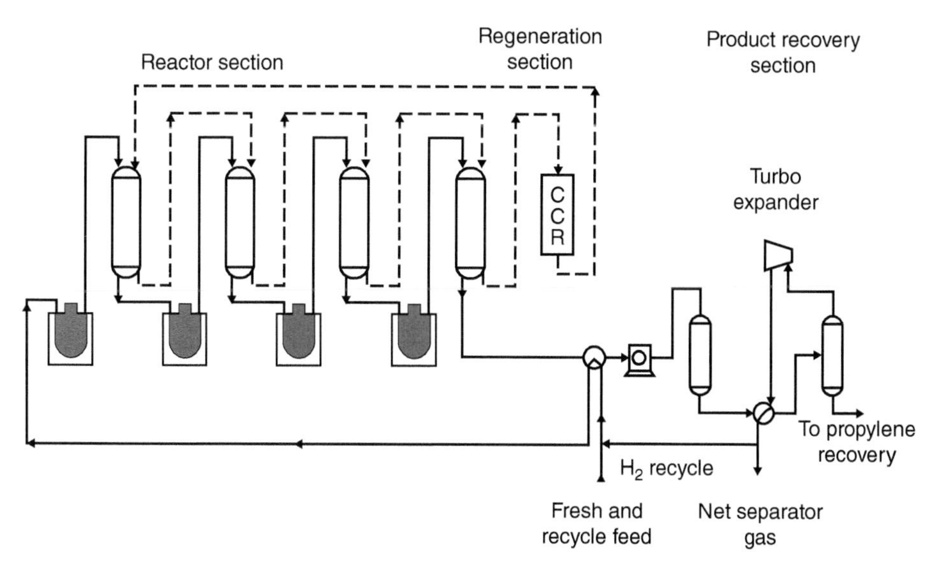

Figure 8.2 Oleflex process: Pt/Al_2O_3 catalyst, adiabatic moving beds, operation under light pressure (0.2 MPa), temperature: 820–890 K, continuous regeneration. Source: Reproduced with the permission of UOP.

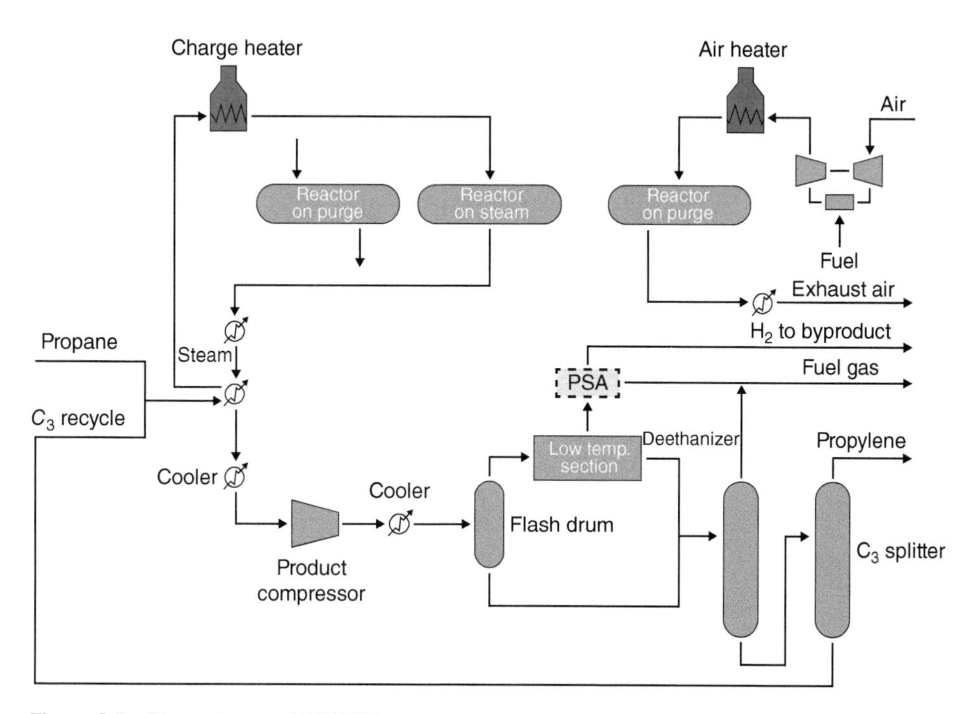

Figure 8.3 Flow scheme of CATOFIN process.

These high temperatures hamper high selectivity and favor side reactions such as coke deposition, introducing the need for a regeneration step; however, this step is not fully reversible at this point, thus affecting the activity of the catalyst. The catalysts and technologies are available, and the process can run continuously for several years.

Although the process has been successfully commercialized, there is room for improvement. Further advances with respect to catalyst design, such as lower noble metal loading and increased stability, are essential. Further improvements are still possible by increasing the energy efficiency of the hydrogenation process and overcoming the biggest drawback: the limited equilibrium conversion and the occurrence of side reactions at relatively high temperature. It is notable that the reaction equilibrium could be shifted toward the product. Different strategies are already applied in industry, such as lowering the partial pressure and the selective in situ combustion of hydrogen. The latter is of great interest, as the combustion heat promotes the dehydrogenation step. A third strategy is to remove hydrogen from the reacting system by using membranes; nevertheless, thermal stability remains an issue.

Figure 8.3 is for the production of propylene. For isobutylene production, the deethanizer and C_3 splitter are replaced by a depropanizer. A deoiler is also included to reject a small amount of C_4s and heavier material. For amylene production, an extractive distillation system with a quench oil tower is used instead of the deoiler/splitter combination.

The cost of the alumina-based proprietary catalyst is low, because it uses nonprecious metal. A comparison of individual processes of propane and isobutane dehydrogenation is in Table 8.1.

Table 8.1 Performance of propane and isobutane dehydrogenation processes.

Process (licensor)	Reactor	Catalyst life (a)	T (K)/P(kPa)	Conversion (%)	Selectivity (%)
Oleflex (UOP)	Adiabatic moving beds in series with intermediate heating	Pt/Al_2O_3	820–890	P : 25	P : 89–91
		1–3	20	IB: 35	IB: 91–93
CATOFIN (ABB Lummus crest)	Parallel adiabatic fixed beds with swing reactor	Cr/Al_2O_3	860–920	P : 48–65	P : 82–87
		1–2	5	IB: 60–65	IB: 93
STAR (Phillips	Tubular	Pt-Sn/Zn-	750–890	P : 30–40	P : 80–90
Petroleum)	reactors in	Al_2O_3, 1–2	50	IB: 40–55	IB: 92–98
	furnace				
FBD-4	Fluidized bed	Cr/Al_2O_3	820–870	P : 40	P : 89
(Snamprogetti)			13	IB: 50	IB: 91

P = propane dehydrogenation, IB = isobutane dehydrogenation.

8.2 Methanol to Alkenes

Methanol to alkenes (olefins) (MTO) is one of the technologies that can produce basic petrochemicals. Methanol is produced mainly catalytically via syngas, which is a valuable gas mixture of hydrogen and carbon monoxide. Methanol is used in large volumes for the production of a wide variety of commodity chemicals. Syngas can be obtained from different carbonaceous resources through the gasification of natural gas, coal, or biomass. Hence, this process presents a suitable alternative in order to produce base chemicals from resources other than crude oil. However, the lowest production costs of syngas and the highest carbon efficiencies are based on methane. Biomass and coal yield hydrogen-deficient syngas, introducing the need for a water gas shift facility, which has a negative impact on overall carbon efficiency and produces immense carbon emissions. MTO was introduced in the late 1970s by ExxonMobil scientists and was later patented by ExxonMobil and UOP/Hydro. The process converts methanol to hydrocarbons over a zeolite containing active acid sites. In the ExxonMobil MTO process, Zeolite Socony Mobil-5 (ZSM-5) is used as a catalyst; in contrast, in the UOP/Hydro process, silico-alumino phosphate (SAPO)-34 is used. Research has been performed on this process since its discovery, leading to significant progress in our understanding of the reaction and of catalyst design; this progress has led to significant improvements in process performance.

An example of the comparison of product gas compositions from a naphtha cracker and the UOP/Norsk Hydro MTO process are shown in Table 8.2. Figure 8.4 illustrates a simple flow diagram for the UOP/Hydro MTO process.

8.2.1 MTO Catalyst

The MTO catalyst is based on SAPO-34, a template-based silicoaluminumphosphate molecular sieve with a chabazite structure and a unique pore size of about 0.38 nm. By taking

Table 8.2 Comparison of product gas compositions from naphtha cracker and MTO.

	Naphtha cracker (vol.%)	**MTO (vol.%)**
Ethene	31.27	57.9
Propene	10.63	26.6
Butenes	2.33	5.8
Light alkanes	32.54	6.25
C_5+	3.8	0.7
Acetylenes and dienes	3.16	0.25
Hydrogen	16.27	1.9
CO and CO_2	*n.a.*	0.6
Total	100.0	100.0

UOP/Norsk Hydro MTO process.

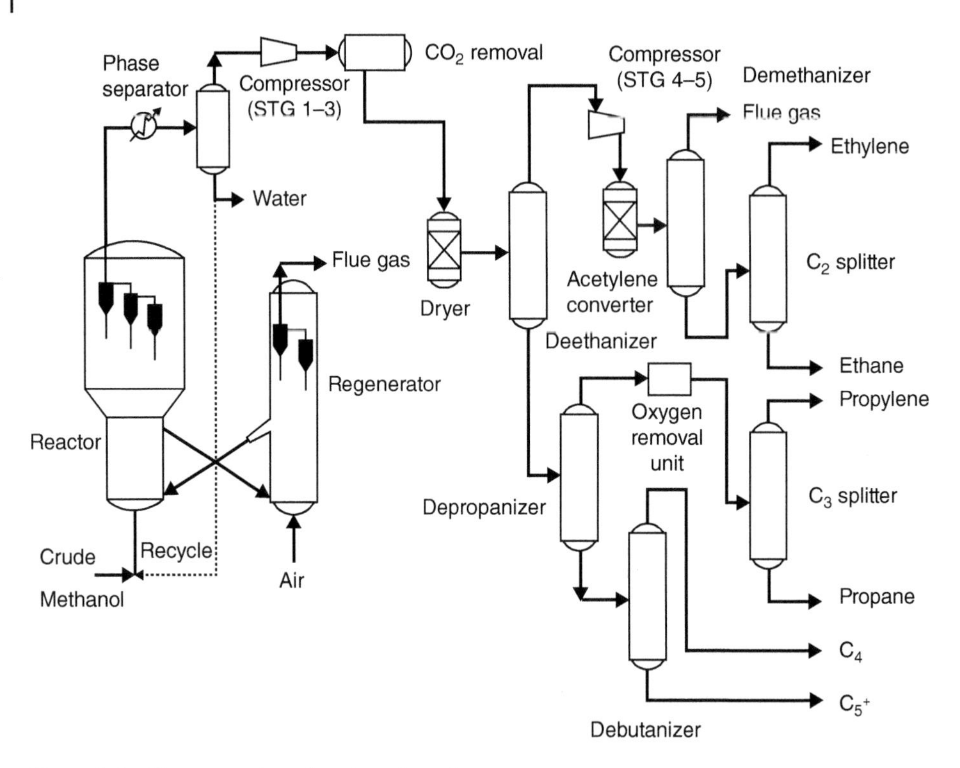

Figure 8.4 The conversion of methanol to alkenes.

advantage of shape selectivity and the lower acidity of the SAPO-34 catalyst, a 75–80% carbon selectivity toward ethylene and propylene has been reported. In addition, the process conditions can be altered to favor propylene formation. The ethylene-to-propylene ratio can be varied between 0.5 and 1.5. Furthermore, by integrating MTO with the olefin-cracking process that is based on a technology developed and demonstrated by Total Petrochemicals and UOP, the light olefin yield can be increased and a higher flexibility to anticipate the olefin market can be offered. The MTO technology has been well developed and demonstrated in China, which has led to the successful construction and commissioning of the world's first CTO plant. China's CTO production capacities have grown from 1.1×10^6 t.a^{-1} in 2010 to the current rate of 1.55×10^7 t.a^{-1}, mainly due to favorable government policies.

CTO requires high capital investment due to its complexity. Furthermore, the hydrogen deficiency of the syngas results in very low carbon efficiency and high water consumption. The CO_2 emissions of these plants are estimated to be between 6 t and 10 t per produced ton of olefin, compared with 1 t for the SC of hydrocarbons.

When efforts toward reducing CO_2 emissions are performed globally, the trend of the large-scale development of coal-based olefin production under favorable governmental policies is highly questionable. In order to improve the MTO process, some efforts are still required to face the main challenge of the process: rapid deactivation of the catalyst. The use of zeolites in catalytic reactions often includes side reactions leading toward coke deposits. This is known to be the major cause of deactivation, affecting both activity and selectivity. A more fundamental understanding of the formation of these carbonaceous deposits and the reaction mechanism is a basis for improving the process, and much research effort has been focused on improving the stability of the catalyst.

8.3 Metathesis

Chemists first recognized the metathesis reaction of olefins in 1955 (the Nobel Prize in Chemistry 2005 for metathesis was awarded to Y. Chauvin, R.H. Grubbs, and R.R. Schrock). The first significant application of metathesis for propylene production was designed by ABB Lummus Global and commissioned by Lyondell Petrochemical in 1985.

8.3.1 Process Chemistry

For the production of propylene from ethylene and butenes, two main equilibrium reactions take place: metathesis and isomerization. Propylene is formed by the metathesis of ethylene and 2-butene, and 1-butene is isomerized to 2-butene as 2-butene is consumed in the metathesis reaction. In addition to the main reaction, side reactions between olefins also occur:

Main reactions:

$$\underset{\text{Ethylene}}{C_2H_4} + \underset{\text{2-Butene}}{2\text{-}C_4H_8} \leftrightarrows \underset{\text{Propylene}}{2C_3H_6} \quad \text{Metathesis}$$

$$\underset{\text{1-Butene}}{1\text{-}C_4H_8} \leftrightarrows \underset{\text{2-Butene}}{2\text{-}C_4H_8} \quad \text{Isomerization}$$

Typical side reactions:

$$\underset{\text{Propylene}}{C_3H_6} + \underset{\text{1-Butene}}{1\text{-}C_4H_8} \leftrightarrows \underset{\text{Ethylene}}{C_2H_4} + \underset{\text{2-Pentene}}{2\text{-}C_5H_{10}}$$

$$\underset{\text{2-Butene}}{2\text{-}C_4H_8} + \underset{\text{1-Butene}}{1\text{-}C_4H_8} \leftrightarrows \underset{\text{Propylene}}{C_3H_6} + \underset{\text{2-Pentene}}{2\text{-}C_5H_{10}}$$

$$2\ \underset{\text{1-Butene}}{1\text{-}C_4H_8} \leftrightarrows \underset{\text{Ethylene}}{C_2H_4} + \underset{\text{3-Hexene}}{3\text{-}C_6H_{12}}$$

The reaction takes places in a fixed-bed reactor over a mixture of WO_3/SiO_2 (metathesis catalyst) and MgO (isomerization catalyst) at >260 °C and 3.0–3.5 MPa (Figure 8.5).

Whether combined with a steam cracker or an FCC unit or as a stand-alone project, the OCT process is a low-cost option for propylene production. It provides product flexibility to increase propylene production and upgrade excess butylenes for higher total product value and improved margins.

Typical steam crackers with liquid feedstocks produce ethylene, propylene, and mixed C_4s as a byproduct. With C_4s and ethylene readily available, OCT provides an excellent vehicle to upgrade C_4s to high-value propylene. When integrated with a steam cracker, OCT optimizes propylene production and lowers energy consumption. OCT typically is applied

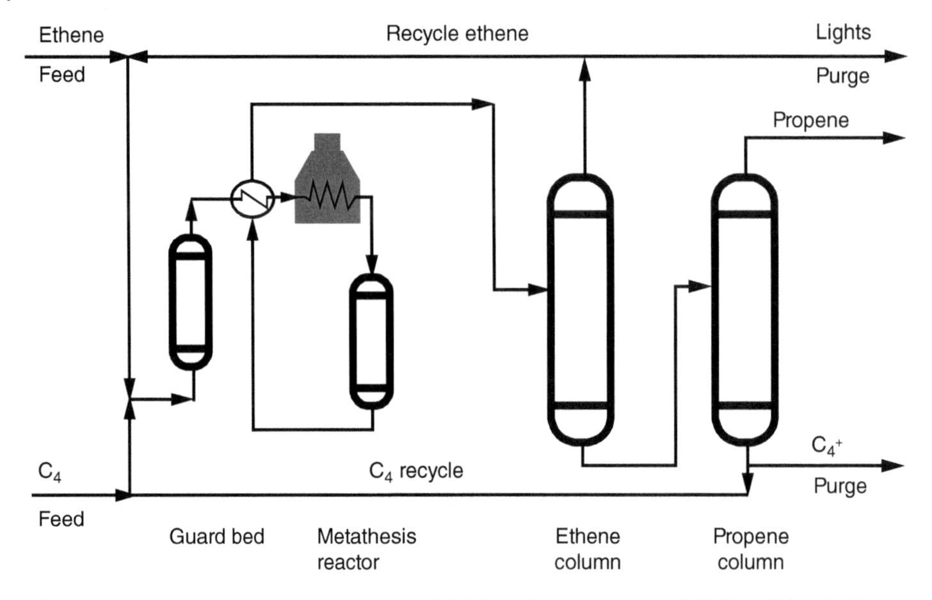

Figure 8.5 Olefins conversion technology OCT flow diagram (reversed Phillips Triolefin Process).

to the ethylene plant material balance to produce a propylene-to-ethylene product (P/E) ratio of greater than 1.0.

8.4 Oxidative Coupling of Methane

The direct conversion of methane (the largest constituent of natural gas and shale gas) into valuable chemicals seems very attractive. The direct conversion of methane into ethene is possible by applying so-called *oxidative coupling*, i.e. the catalytic reaction of methane with oxygen (see the reactions in the oxidative coupling of methane). The reaction produces ethane, which is converted to ethene in situ, while through sequential reactions higher hydrocarbons are also formed in small amounts. The main problem with oxidative methane coupling is the formation of large amounts of CO_2 and CO by oxidation reactions. This not only reduces the selectivity toward ethene, but the high exothermicity also presents formidable heat removal problems.

Desired reactions:

$$2CH_4 + 1/2\,O_2 \rightarrow CH_3CH_3 + H_2O \qquad \Delta H^o_{298} = -177\ \text{kJ/mol}$$

$$CH_3CH_3 + 1/2\,O_2 \rightarrow CH_2 = CH_2 + H_2O \qquad \Delta H^o_{298} = -105\ \text{kJ/mol}$$

Undesired reactions:

$$CH_4, CH_3CH_3, \text{etc.} + O_2 \rightarrow CO_2, CO \qquad \Delta H^o_{298} \ll 0$$

In the past three decades, oxidative methane coupling has received much attention. Since the first publication in 1982, considerable progress has been made in the development of catalysts for the oxidative coupling reaction. However, the best combinations of conversion and selectivity (c. 30 and 80%, respectively) thus far achieved in laboratory fixed-bed reactors are still well below those required for economic feasibility. Accordingly, the emphasis in oxidative coupling research has shifted somewhat to innovative reactor design, such as staged oxygen addition and membrane reactors for combined reaction and ethene removal.

A strong economic interest exists in developing processes that allow methane conversion into more valuable products. MTO and FTS (discussed in Section 12.2.10) can be used to convert methane to higher hydrocarbons, but only in an indirect way; that is, they require the production of syngas as a first step. This initial step to produce syngas represents an inherent inefficiency. On the other hand, the direct activation of methane and its conversion into other useful products remain one of the most challenging topics facing the catalysis community today.

In this regard, OCM is one of the most promising direct routes to convert methane into ethylene and higher hydrocarbons. Ever since the pioneering work of Keller and Bhasin, OCM has attracted both industrial and academic interest. Although the benefits of OCM have been known for over 30 years, the issue of finding a viable catalyst with the necessary performance for the commercialization of the process is still crucial among researchers. Furthermore, the low yields of ethylene and the strong exothermicity of this reaction need to be addressed in an appropriate reactor technology before OCM can be used as an alternative to SC for the production of ethylene and higher olefins. The main challenges for OCM to be economically successful are hence twofold, with catalyst development on the one hand and a novel reactor design on the other.

In earlier research, the selectivity and catalyst stability at elevated temperatures were considered the most attractive features of a potential OCM catalyst. Considerable literature can be found on OCM catalysts that are based on La_2O_3 doped with alkaline earth metals (Sr, Mg, and Ca), on Li/MgO, and on a catalyst generally represented by $Mn/Na_2WO_4/SiO_2$. Recently, however, the research focus shifted to OCM catalysts that enable low-temperature performance. It was found that not only does the metal composition have an influence on catalyst activity, but the particle size and morphology do as well. Specifically, by using nanostructured catalysts (nanofibers, nanowire, and nanorods), methane can be activated at lower temperatures, and better OCM performance can be obtained compared with powder-form catalysts. The development of nanostructured catalysts has played a major role in the evolution of the OCM technology. In 2018, Siluria Technologies, Inc. (Siluria for short) announced the first commercial OCM process, in which a series of so-called *nanowire catalysts* are reported to operate at process temperatures below 600 °C. However, Siluria's patent application reveals that the single-pass C_2 yield does not meet the target of 25% (Figure 8.6).

At the moment, catalytic packed-bed reactors constitute the majority of all laboratory-scale OCM reactors; these are also used in Siluria's demonstration plant. Because of the high exothermicity of the OCM reaction, thermal control of the reactor is an important

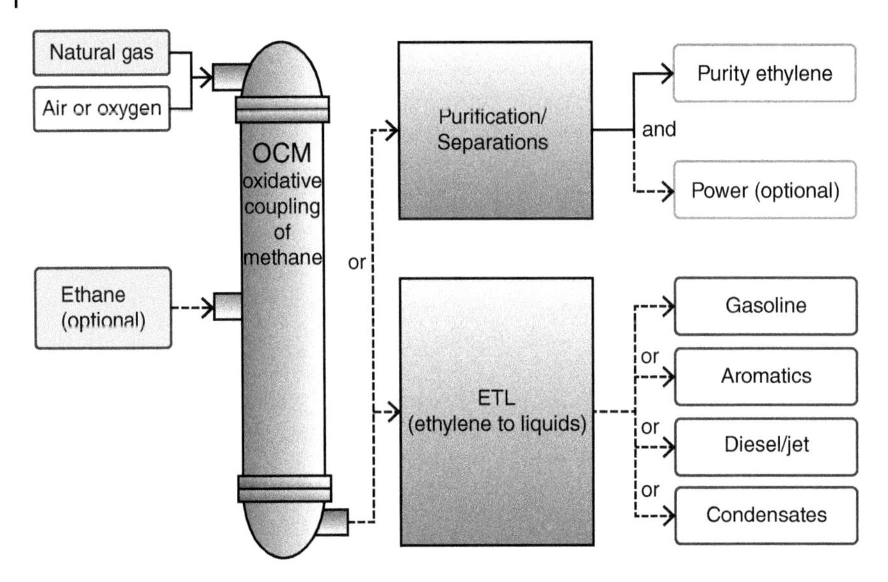

Figure 8.6 Oxidative coupling of methane by the Siluria process.

issue. Most laboratory-scale experimental setups use small-diameter tubes and run with very dilute mixtures and low methane conversions. Of course, this is not practical for an actual large-scale process. According to a review of process intensification for natural gas conversion, thermal control in OCM reactors may follow three approaches: microchannel reactors, membrane reactors, and staged reactors are listed as possible options, while combinations of these are possible in more innovative designs. Different authors designated the fluidized-bed reactor as the best reactor concept for OCM. This is mainly because the beneficial heat-transfer characteristics in fluidized-bed reactors cannot be achieved in any type of packed-bed reactor.

Based on these considerations, it is clear that some effort is still required before OCM can be considered an alternative to SC for the production of ethylene and other olefins. Although the door toward successful commercialization has already been opened by Siluria, further investigation to enhance ethylene yields is still necessary. Catalyst design and the development of novel reactor technologies are important to this purpose.

8.5 Current and Future Developments

This chapter discussed some of the most promising alternative technologies and feedstocks for alkene production. Alternative feedstocks mainly arise from the abundance of cheap propane, ethane, and methane from shale gas and stranded gas. Although biomass and waste streams were also mentioned here, they are believed to be useful as a complement to rather than a complete replacement for fossil resources in the near future. Coal, natural gas, shale condensates, and crude oil will remain dominant in the petrochemical industry. However, the use of coal as feedstock for carbochemicals is rather questionable from an environmental point of view, as it has very low carbon efficiencies and hence huge CO_2 emissions. Therefore, recent developments of coal as a feedstock for the production of carbochemicals are debatable. As a response to low ethane prices and large ethane availability, the SC feedstock portfolio is making a shift to ethane feedstocks. Shipping and transport via pipes of ethane has become advantageous and allows ethylene producers outside of the United States to take advantage of the local low ethane prices. On the other hand, the large availability of crude oil has caused ethylene producers to shift toward this extreme as well. ExxonMobil was the first to commission a large-scale plant that uses crude oil to produce olefins.

Recent investments and the huge capital in current production facilities suggest that SC of hydrocarbons will remain the leading technology for the production of ethylene. The large number of projects coming online will lead to a substantial capacity expansion for ethylene producers. As a result, biomass and waste-stream conversion technologies will have to become even more competitive than before, or else their focus should be on olefin complexes. Nevertheless, it will eventually be necessary to deploy biomass and waste streams, although this shift will need to be pushed by world governmental regulatory bodies, because the contribution of these streams in the coming lustrum will remain marginal. Drop-in feedstocks for the current generation of steam crackers would be a first step. However, the change in the global steam cracker portfolio – that is, the significant lighter feedstocks – creates the need to develop routes for the on-purpose production of propylene and other light olefins. The catalytic dehydrogenation of light alkanes is a well-established process that accomplishes this goal. The main improvements that are possible here are related to energy efficiency and catalyst deactivation.

Furthermore, the abundance of methane has caused increased interest in developing processes to valorize methane to higher hydrocarbons or chemicals. Several of these processes were identified as potential alternatives for the SC process: FTS, MTO, and OCM. Both the FTS and the MTO processes are proven technologies, with some plants already operational worldwide. Despite the maturity of these technologies, both are inherently inefficient due to the syngas production step. In addition, FTS is not selective enough to solely produce light olefins, and it produces a considerable amount of fuel-range hydrocarbons. However, improvements to these processes are still possible, especially with regard to catalyst design. For OCM, major efforts are still required before it can be considered as an alternative to SC. Catalyst design and the development of a reactor technology that is able to deal with the strong exothermicity of this reaction are mandatory.

In the near future, SC will still be the predominant process for the production of olefins, albeit with a mainly lighter feedstock. The alternative technologies discussed in this chapter are promising, provided that the bottlenecks in their process efficiency and CO_2 footprint are resolved.

What than really drives the technological development of the petrochemical industry, an industry whose operational ethos tends to be risk-averse and frequently conservative in applying new technology? The answer lies in three words – economics, safety, and environment – the three principal preoccupations of petrochemical industry management.

Further Reading

Bajus, M. and Back, M.H. (1995). *Applied Catalysis* 128: 61–77, 142/100.

9

Pillar A of Petrochemistry

Petrochemicals from C_2 – C_3 Alkenes

CHAPTER MENU

9.1 Introduction

Petrochemicals are chemical products derived from petroleum, although many of the same chemical compounds are also obtained from other fossil fuels such as coal and natural gas or from renewable sources such as corn, sugar cane, and other types of biomass.

Petrochemical production relies on multiphase processing of oil-associated petroleum gas. Petrochemicals and petroleum products are the second-level products derived from crude oil after several refining processes.

Petrochemicals are used for production of several feedstocks and monomers or monomer precursors. The monomers after the polymerization process create several polymers, which ultimately are used to produce gels, lubricants, elastomers, plastics, and fibers.

Petrochemical intermediates are generally produced by chemical conversion of primary petrochemicals to form more complicated derivative products. These can be made in a variety of ways: directly from primary petrochemicals; through intermediate products that still contain only carbon and hydrogen; and through intermediates that incorporate chlorine, nitrogen, or oxygen in the finished derivate. In some cases, they are finished products; in others, more steps are needed to arrive at the desired composition.

Some typical petrochemical intermediates are (i) vinyl acetate for paint, paper, and textile coatings; (ii) vinyl chloride for polyvinyl chloride (PVC); (iii) ethylene glycol for polyester textile fibers; and (iv) styrene, which is important in rubber and plastic manufacturing.

Of all the processes used, one of the most important is polymerization. It is used in the production of plastics, fibers, and synthetic rubber, the main finished petrochemical derivatives.

Petrochemistry: Petrochemical Processing, Hydrocarbon Technology, and Green Engineering,
First Edition. Martin Bajus.

The production of petrochemicals commences with refining of petroleum and natural gas. The chemical technology is the chemical process by which a variety of chemicals are manufactured. This chemical process technology is subdivided into other categories:

- Chemicals and allied products in which chemicals are manufactured from a variety of feedstocks and are put to further use.
- Rubber and miscellaneous products, which focus on the manufacture of rubber and plastic materials.
- Petroleum refining and related industries.

Thus, the petrochemical industry falls under the subcategory of petroleum and related industries.

For the purpose of this book, there are four general types of petrochemicals: (i) aliphatic compounds, (ii) aromatic compounds, (iii) inorganic compounds, and (iv) synthesis gas (syngas: carbon monoxide and hydrogen). Syngas is used to make ammonia and methanol. Ammonia is used primarily to form ammonium nitrate, a source of fertilizer. Much of the methanol produced is used in making formaldehyde. The rest is used to make polyester fibers, plastics, and silicone rubber.

9.2 Chemicals from Ethylene

Ethylene is the hydrocarbon feedstock used in the greatest volume in petrochemical technology (Figure 9.1). From ethylene, for example, are manufactured ethylene glycol, used in polyester fibers and resins and in antifreezes; ethyl alcohol, a solvent and chemical reagent; polyethylene, used in film and plastics; styrene, used in resins, synthetic rubber, and polyesters; and ethylene dichloride, for vinyl chloride, used in plastics and fibers.

9.3 Chemicals from Propylene

Propylene is also an important source of petrochemicals (Figure 9.2) and is used in making such products as acrylics, rubbing alcohol, epoxy glue, and carpet fibers, paper coatings, and plastic pipes.

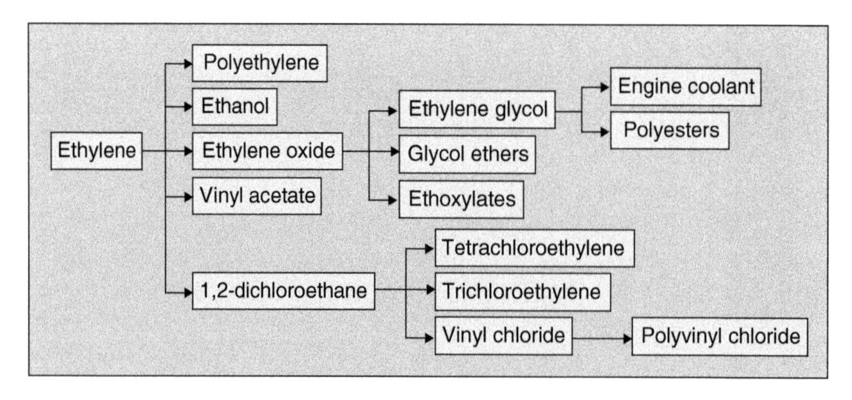

Figure 9.1 Chemicals from ethylene.

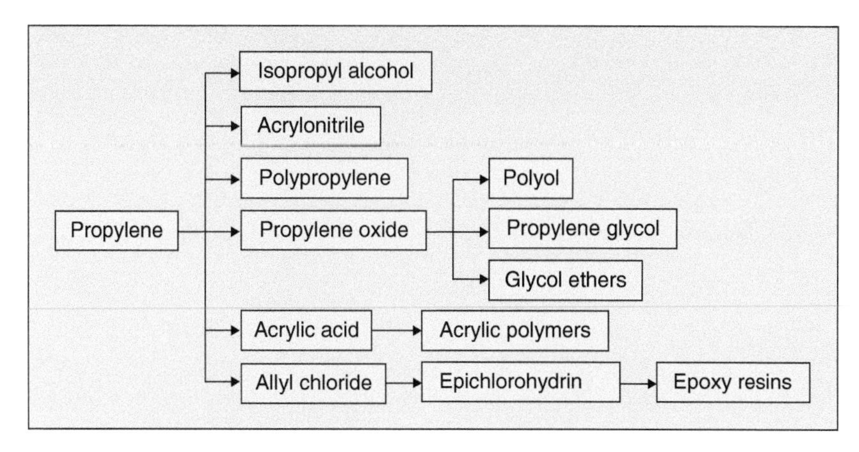

Figure 9.2 Chemicals from propylene.

A highly significant proportion of these basic petrochemicals are converted into plastics, synthetic rubbers, and synthetic fibers. Together these materials are known as *polymers*, because their molecules are high-molecular-weight compounds made up of repeated structural units that have combined chemically. The major products are polyethylene, polyvinyl chloride, and polystyrene, all derived from ethylene; and polypropylene, derived from monomer propylene. Major raw materials for synthetic rubbers include butadiene, ethylene, benzene, and propylene. Among synthetic fibers, the polyesters, which are a combination of ethylene glycol, and terephthalic acid (made from xylene) are the most widely used. They account for about half of all synthetic fibers. The second synthetic fiber is nylon, its most important raw material being benzene. Acrylic fibers, in which the major raw material is the propylene derivative acrylonitrile, make up most of the remainder of the synthetic fibers.

9.4 Polymerization

The polymerization of ethylene under pressure (2–10 MPa) at 110–120 °C in the presence of a catalyst or initiator, such as a 1% solution of benzoyl peroxide in methanol, produces a polymer in the 2 000–3 000 molecular weight range. Polymerization at 100–300 MPa and 180–200 °C produces a wax melting at 100 °C and 15 000–20 000 molecular weight, but the reaction is not as straightforward as the equation indicates since there are branches in chain. However, considerably lower pressures can be used over catalysts composed of aluminum alkyls (R_3Al) in the presence of titanium tetrachloride ($TiCl_4$), supported chromic oxide (CrO_3), nickel (NiO), or cobalt (CoO) on charcoal, and promoted molybdenum-alumina (MoO_2-Al_2O_3), which at same time give products more linear in structure Z-N (Figure 9.3).

Polypropylene can be made in similar ways; and mixed monomers, such as ethylene-propylene and ethylene-butene mixtures, can be treated to give high-molecular-weight copolymers with good elasticity. Polyethylene has excellent electrical insulating properties; its chemical resistance, toughness, machinability, light weight, and high strength make it suitable for many other uses.

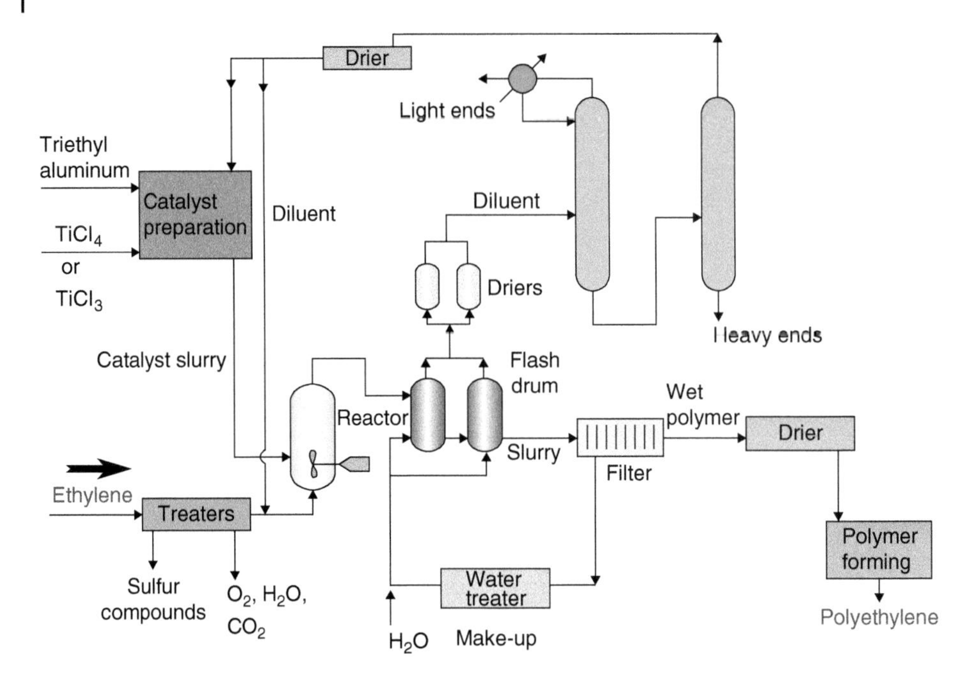

Figure 9.3 Polymerization of ethylene.

Lower-molecular-weight polymers, such as the dimers, trimers, and tetramers, are used as such in motor gasoline. The materials are normally prepared over an acid catalyst. Propylene trimer (dimethylheptenes) and tetramer (trimethylnonenes) are applied in alkylation of aromatic hydrocarbons for the production of alkylaryl sulfonate detergents and also as olefin-containing feedstocks in the manufacture of C_{10} and C_{13} oxoalcohols. Phenol is alkylated by trimer to make nonylphenol, a chemical intermediate for lubricating oil detergents and other products.

Iso-butylene also forms several series of valuable products; the di- and tri- iso-butylenes make excellent motor and aviation gasoline components, and they can be used as alkylating agents for aromatic hydrocarbons and phenols and as reactants in the oxo-alcohol synthesis. Polyisobutylenes in the viscosity range of 55 000 SUS (38 °C) have been used as viscosity index improvers in lubricating oils. 1-Butene and 2-buten participate in polymerization reactions by the way of butadiene, the dehydrogenation product, which is copolymerized with styrene (23.5%) to form GR-S rubber and with acrylonitrile (25%) to form GR-N rubber.

Derivates of acrylic acid (butyl acrylate, ethyl acrylate, 2-ethylhexyl acrylate, and methyl acrylate) can be homopolymerized using peroxide initiators or copolymerized with other monomers to generate acrylic or acrylic resins.

10

Pillar B of Petrochemistry

Production of BTX Aromatics

CHAPTER MENU

10.1 Introduction

In the petroleum refining and petrochemical technology, *BTX* refers to mixtures of benzene, toluene, and the three xylene isomers, all of which are aromatic hydrocarbons. If ethylbenzene is included, the mixture is sometimes referred to as *BTEX*.

The BTX aromatics are very important petrochemical materials. According to recently conducted research about the global benzene market, it is projected to reach 56 304 thousand tons by the end of 2023, increasing at a CAGR of around 3% per year in the period 2017–2023. In particular, the largest share of benzene consumption is for the production of ethylbenzene, which accounts for around 51% of the total in terms of volume. Meanwhile, the chemical's application for maleic anhydride has been growing at about 5.54% per year.

Regionally, the largest global benzene market is North East Asia, which accounts for about 40% of the total in terms of volume. China is the largest consumer and producer globally, with capacity dramatically increasing in recent years.

Toluene is also a valuable petrochemical for use as a solvent and intermediate in chemical manufacturing processes and as a high-octane gasoline component. Benzene, toluene, and xylenes can be made by various processes. However, most BTX production is based on the recovery of aromatics derived from the catalytic reforming of naphtha in a petroleum refinery (Section 5.5.2).

Petrochemistry: Petrochemical Processing, Hydrocarbon Technology, and Green Engineering,
First Edition. Martin Bajus.

Catalytic reforming usually utilizes a feedstock naphtha that contains non-aromatic hydrocarbons with 6–11 or 12 carbon atoms and typically produces a *reformate* product containing C_6–C_8 aromatics (benzene, toluene, xylenes) as well as paraffins and heavier aromatics containing 9–11 or 12 carbon atoms.

Another process for producing BTX aromatics involves the steam cracking of hydrocarbons, which typically produces a cracked naphtha product commonly referred to as *pyrolysis gasoline*, *pyrolysis gas*, or *pygas*. The pyrolysis gasoline typically consists of C_6–C_8 aromatics, heavier aromatics containing 9–11 or 12 carbon atoms, and non-aromatic cyclic hydrocarbons (naphthenes) containing 6 or more carbon atoms.

Table 10.1 compares the BTX content of pyrolysis gasoline produced at standard cracking severity or at medium cracking severity with the BTX content of catalytic reformate produced by either a continuous catalytic regenerative (CCR) reformer or a semi-regenerative catalytic reformer. About 70% of the global production of benzene is by extraction from either reformate or pyrolysis gasoline.

BTX aromatics can be extracted from catalytic reformate or from pyrolysis gasoline by many different methods. Most of those methods, but not all, involve the use of a solvent either for liquid–liquid extraction or for extractive distillation. Many different solvents are suitable, including sulfolane ($C_4H_8O_2S$), furfural ($C_5H_4O_2$), tetraethylene glycol ($C_8H_{18}O_5$), dimethylsulfoxide (C_2H_6OS), and N-methyl-2-pyrrolidone.

One of the strong trends observed in the North American benzene market is linked with the discovery of shale gas. The North American aromatic market has been particularly impacted by the shale gas boom. BTX production and supply have been reduced because of steam crackers, particularly in the United States, which are increasingly using ethane rather than naphtha as feedstock. One of the distinct trends for the European petrochemical industry is the benefit from a prolonged golden age for naphtha crackers amid low oil prices. Petrochemical revenues have grown versus oil-based European cracker margins.

The increasingly competitive landscape globally forces chemical companies to focus on innovation and particularly on the development of green chemistry, which is becoming crucial for the companies' overall future development. Several dynamic research areas can be identified, such as nanotechnology (Section 16.5), fuel cells (Chapter 13), biotechnology (Chapter 14), the development of biofuels as supplements to oil supplies, as well as the expansion of water-based paints and replacements for chlorofluorocarbons (CFCs). In order

Table 10.1 BTX content of pyrolysis gasoline and reformate.

	Pyrolysis gasoline		Reformate	
	Standard severity	Medium severity	CCR	SR
BTX content, wt. %	58	42	51	42
Benzene, wt. % of BTX	48	44	17	14
Toluene, wt. % of BTX	33	31	39	39
Xylenes, wt. % of BTX	19	25	44	47

CCR = Continuous catalytic regenerative reforming
SR = Semi-regenerative catalytic reforming

to increase their profits, it is important for hydrocarbon technology and petrochemistry manufacturers not only to cut their costs, but also to ensure that petrochemical processes and products conform to pressing environmental issues.

10.2 Alkylation

10.2.1 Ethylbenzene

Ethylbenzene is one of the largest-volume petrochemicals intermediate in use today. Nearly all ethylbenzene production is used for manufacturing styrene monomer. Ethylbenzene production technology has evolved considerably in recent decades. Prior to 1980, ethylbenzene was produced almost exclusively via Friedel-Crafts liquid-phase alkylation with an aluminum chloride catalyst. Despite the lower quality of the ethylbenzene produced and the need to dispose of the spent catalyst as a waste stream, many of the aluminum chloride-based ethylbenzene plants built in the 1970s are still operating today in a satisfactory way.

Starting in the 1980s and until the middle of the 1990s, a new gas-phase alkylation process based on zeolite catalyst was applied for most new installations, overcoming the main drawbacks of aluminum chloride technology. Starting at the beginning of the 1990s, the development of zeolite catalysts has led to the new liquid-phase alkylation processes, which are the only ones today applied for new installations due to best product quality and lower catalyst consumption.

Global styrene production since 2005 has been driven by polystyrene, EPS, ABS, and SAN. Total world capacity is 27 984 thousand tons. Regional capacity (in thousand tons) is as follows: North America 7 266, South America 640, West Europe 6 312, Middle Europe 405, Eastern Europe 938, Middle East 1 145, and Africa and Asia Pacific 11 278.

10.2.1.1 Process Chemistry

Alkylation of benzene with ethylene is an exothermic reaction that occurs in the liquid phase and in the presence of a zeolite catalyst. The zeolite works as a solid acid, which includes both Brönsted and Lewis acid sites. The reaction is commonly considered an electronic substitution on the aromatic ring, proceeding via a carbonium ion type mechanism. Ethylene is protonated by the acid site to form the active intermediate (see Section 1.2.8.1).

The main alkylation reaction leads to production of ethylbenzene according to the following scheme:

Alkylation No.1

$$C_6H_6(l) + C_2H_4(g) \rightleftarrows C_6H_5 - CH_2CH_3(l) \qquad \Delta H = -113.6\ \text{kJ/mol}$$

zeolitic catalyst

Moreover, due to the presence of catalyst acid sites, ethylene can further react with the ethylbenzene already produced, leading to the production of diethylbenzenes, which are the main by-product of the process; and higher-ethylated benzenes, collectively termed *polyethylated benzene* (PEB). Very small levels of other coupling reactions also occur, yielding materials such as diphenylethane (DPE) and high-boiling compounds.

Alkylation No. 2

$$C_6H_5CH_2CH_3(l) + C_2H_4(g) \underset{\text{catalyst}}{\rightleftarrows} C_6H_4 - (CH_2CH_3)_2(l) \qquad \Delta H = -107.9\ \text{kJ/mol}$$

Polyethylbenzenes produced by successive alkylations are transalkylated (transfer of ethyl groups) with benzene to produce additional ethylbenzene. These reactions are slower than alkylation and are limited in extent by equilibrium. Higher-ethylated benzenes also participate. The heat of reaction is essentially zero, and the reaction conditions are effectively isothermal. Polyethylbenzenes produced (mainly diethylbenzene) are not losses of material for the whole process because they can be recovered by a transalkylation reaction carried out in a dedicated plant section according to the following scheme:

Transalkylation

$$C_6H_6(l) + C_6H_4 - (CH_2CH_3)_2(l) \underset{\text{Catalyst}}{\rightleftarrows} 2\ C_6H_5 - CH_2CH_3(l) \qquad \Delta H = -5,4\ \text{kJ/mol}$$

Also, this reaction is catalyzed by the acid sites of a zeolite. The main reaction parameters are the same as for alkylation: temperature, residence time, and ethyl-phenyl group molar ratio, which typically ranges between 6 and 3. Diethylbenzene conversion increases if temperature or residence time increases until it reaches a value close to chemical equilibrium. In the same conditions, ethylbenzene selectivity generally decreases, even if it strongly depends on the whole set of operating conditions.

10.2.1.2 New Eco-Friendly Catalyst

The relevant mechanism seems to be different depending on the zeolite structure. For medium-pore zeolites, the operating mechanism involves the dealkylation of diethylbenzenes followed by realkylation benzene to give ethylbenzene. On the other hand, for large-pore zeolites, the reaction involves a hydride-transfer mechanism and a bulky diarylethane intermediate.

Zeolites are crystal aluminosilicates (Section 1.2.7) with a regular microporous structure that can be different types depending on the chemical composition and preparation procedure. The zeolite structure is responsible for very high catalyst selectivity. In fact, we speak about *reagent selectivity*, respectively, when a reagent or an intermediate has a kinetic diameter that is too big compared to the micropore dimensions. In both cases, reaction does not occur. On the other hand, product selectivity is related to incompatible kinetic diameter of the product that has to isomerize before going out the pore.

Other important parameters for zeolite catalytic activity are the silica-to-alumina ratio and particle size. Catalytic activity generally decreases with an increase in the silica-to-alumina ratio – in other words, decreasing the number of acid sites. The same effect on catalytic activity is obtained when the mean particle diameter is increased.

The acid strength required for transalkylation is higher than that required for alkylation; thus, for a given catalyst, the best conditions for an acceptable rate for the transalkylation reaction should be more severe than for alkylation. More conveniently, a different and specifically designed catalyst should be adopted for transalkylation.

Zeolitic materials can be deactivated either by poisoning of the active sites or by blockage of the pores. The first deactivation mode is caused by poisons that, even in small quantities, can contaminate the raw materials. The second is a consequence of coke formation, which is an undesirable side phenomenon commonly associated with the reaction of organic compounds over heterogeneous catalysts.

10.2.1.3 Environmental Protection of the Described Process

Benzene is alkylated with ethylene to yield a mixture of alkylated benzenes. This mixture is distilled to recover product ethylbenzene, excess benzene for recycle, and higher-ethylated benzenes (PEBs). The PEBs are transalkylated with benzene to form additional ethylbenzene.

Lummus and UOP were the first to introduce the concept of separate alkylation and transalkylation reactors for a zeolite-based process. This permits operation at lower benzene recycles than an alkylation-only process. Thus the process can operate at an optimal benzene recycle rate with consequent saving in energy input and equipment cost. Figure 10.1 is a simplified flow diagram of the EBOne process.

10.2.1.4 CDTECH EB Process

The CDTECH EB process-flow scheme differs only in the alkylation reactor system. At the heart of the CDTECH EB process is the patented catalytic distillation concept that combines catalytic reaction and distillation in a single operation. Ethylene is introduced as a vapor

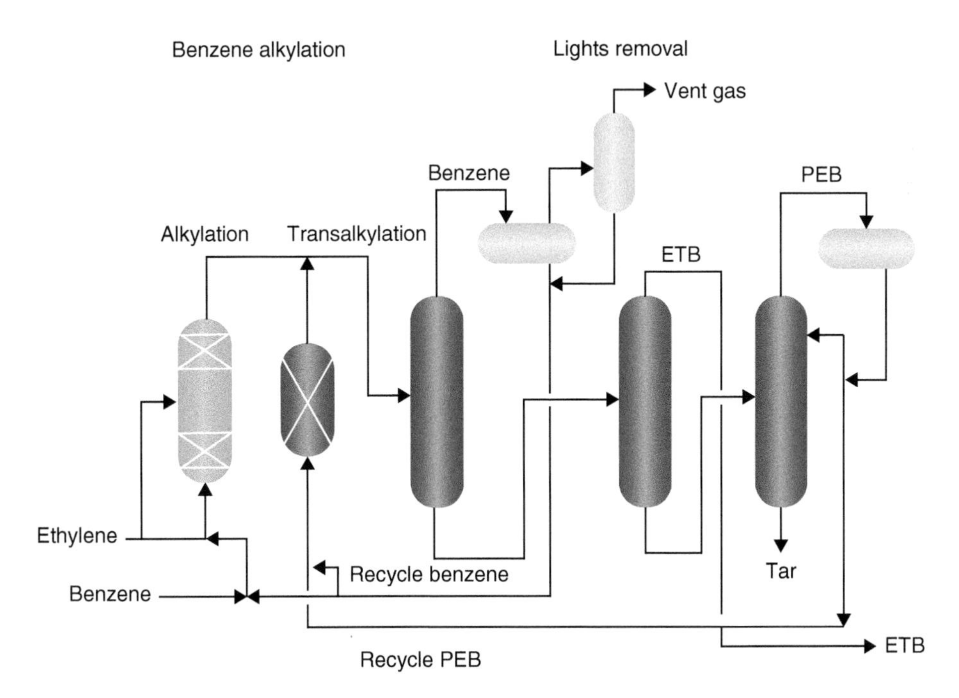

Figure 10.1 EBOne process flow scheme.

at the bottom of the reaction zone. Reaction occurs in the catalytic zone of the column, and distillation occurs throughout the column, resulting in a countercurrent flow of vapor and liquid throughout the column. Reaction products are removed continuously from the catalytic zone as a result of distillation, whereas any unconverted reactants and other lights are taken overhead.

The heat of reaction is removed immediately by vaporizing benzene, which is significant in that it provides near-isothermal operation at the optimal reaction temperature. As ethylene vapor is injected at multiple points in the reactor, it comes in contact with liquid benzene coming from above and is absorbed in the liquid phase. At equilibrium, most of the ethylene is present in the vapor phase. As the small amount of ethylene in the liquid comes in contact with the catalyst, it immediately reacts to form ethylbenzene. This moves the ethylene concentration in the liquid phase away from equilibrium. The vapor–liquid equilibrium then causes ethylene to be "injected" from the vapor phase, thus restoring equilibrium.

The high-activity zeolite catalyst is contained in glass fibers, which are rolled in a wire mesh into a spiral configuration. The unique structure of the bales gives them the required void fraction for the vapor to flow upward through the reactor. The glass-fiber packing acts as a barrier and prevents the vapors from coming in direct contact with the catalyst. The bales are about 30 cm in diameter and height and are handled easily during loading operations (Figure 2.1).

The typical range of operating parameters for the two processes is shown in Table 10.2. The Lummus/UOP EBOne and CDTECH EB processes are simple and use all-carbon-steel equipment. The two technology options can process the complete range of ethylene feedstocks. Both technologies are equally capable of processing polymer-grade ethylene, which has a typical specification of 99.9 mol% ethylene, 50 ppm maximum of other unsaturated, and 10 ppm maximum of propylene.

The CDTECH EB process is also able to handle dilute ethylene feedstock with ethylene compositions ranging from 10% to polymer grade (100%).The dilute ethylene source can be refinery FCC offgas or dilute ethylene streams from steam cracking with ethylene concentrations in the range of 30–85 mol%.

The EBOne and CDTECH EB alkylation and transalkylation reactions are highly selective. Maximum residue formation and 100% ethylene conversion result in high yields that approach theoretical. The processes have an overall yield of 99.7 length, and complete ethylene conversion is maintained throughout the catalyst run length.

Table 10.2 Typical range operating parameters.

	EBOne	**CDTECH EB**
Alkylation B/E, mol	2.0–4.0	2.0–4.0
TA P/E, mol	2.0–4.0	2.0–2.0
Alkylation temperature window, °C	185–270	200 °C top, 240 °C bottom
Alkylation pressure, MPa	3.0–4.0	2.0–2.5
Ethylene consumption, kg/kg EB	0.264	0.264
Benzene consumption, kg/kg EB	0.738	0.738

The ethylbenzene product is an ideal feed for downstream (DS) styrene monomer (SM) and propylene oxide/styrene monomer (PO/SM) plants. PO/SM technology is used for about a third of the world's production of propylene oxide and involves reaction of propylene and ethylbenzene to produce propylene oxide and styrene. Because of the low-temperature alkylation/transalkylation reactors and selective zeolite catalyst, insignificant amounts of xylenes are produced.

The proprietary zeolite catalysts used in both the alkylation and transalkylation reactors are extremely stable. In addition to extended run lengths, commercial plants also have demonstrated that operating conditions remain constant throughout the run with no changes in selectivity patterns.

10.2.1.5 EBMAX Process

ExxonMobil/Badger offers the EBMax process for the production of ethylbenzene by alkylation of benzene with ethylene. In the EBMax process, ethylbenzene is produced under full or partial liquid-phase conditions over proprietary ExxonMobil zeolite catalysts contained in fixed-bed reactors. A simplified flow diagram is presented in Figure 10.2. For an EBMax unit integrated with a cumene process.

Most of the aluminum chloride technology ethylbenzene units currently operating are more than 50 years old. It is not unusual for maintenance costs to be high in the reaction section of these aging plants due to the corrosive nature of the aluminum chloride catalyst. Disposal of the aluminum chloride waste also can be a concern. These factors, along with a desire to improve yield, provide significant incentives for reaction technology replacement.

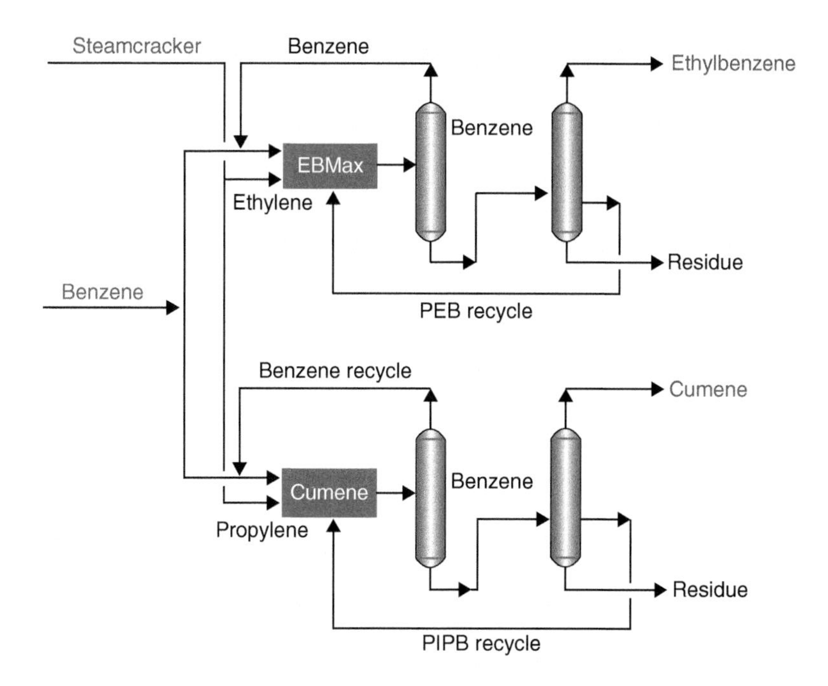

Figure 10.2 EBMax process flow diagram.

Most aluminum chloride technology units operate with only a small excess of benzene in the feed to the reaction system – benzene-ethylene molar feed ratios typically are in range of 2.0–2.5 : 1. As noted previously, the EBMax alkylation catalyst can operate with complete stability at very low B/E ratios Therefore, it is possible to replace the existing aluminum chloride reaction section equipment with EBMax reaction equipment without the addition of a second benzene column (sometimes referred to as a *prefractionator*). In most plants, no modifications to the existing benzene column are required. The existing ethylbenzene and polyethylbenzene columns also can be reused without modification.

10.2.2 Cumene

Cumene is formed by the catalytic alkylation of benzene with propylene. CDTECH's patented CD Cumene process uses a proprietary zeolite catalyst in its patented catalytic distillation structures. The catalyst is non-corrosive and environmentally sound. This modern process features higher product yields with much lower capital investment than the environmentally outdated acid-based processes. The exceptional quality of the cumene product from the CD Cumene process easily surpasses current requirements of phenol producers and may well define tomorrow's more stringent quality standards.

The unique catalytic distillation column combines reaction and fractionation in a single-unit operation. The alkylation reaction takes place isothermally and at a low temperature. Reaction products are continuously removed from the reaction zones by distillation. These factors limit the formation of by-product impurities, enhance product purity and yields, and result in expected reactor run lengths in excess of two years. Low operating temperatures result in energy efficiency and lower equipment design and operating pressures, which help to decrease capital investment, improve safety of operations, and minimize fugitive emissions. All waste heat, including the heat of reaction, is recovered for improved energy efficiency.

The CD Cumene technology can process chemical- or refinery-grade propylene. It can also use dilute propylene streams with purity as low as 10 mol%, provided the content of other olefins and related impurities are within specification.

The Q-Max process (UOP) converts benzene and propylene to high-quality cumene by using a regenerative zeolitic catalyst. This process represents a substantial improvement over older cumene technologies and is characterized by its exceptionally high yield, superior product quality, low investment and operating costs, reduction in solid waste, and corrosion-free environment.

10.2.2.1 Process Chemistry

The manufacture of cumene by the CD Cumene and Q-Max processes start with alkylation of the benzene with propylene to yield a mixture of alkylated and polyalkylated benzenes. The mixture is then processed to recover cumene (isopropylbenzenes) in a distillation train. The polyalkylated benzenes are recovered and transalkylated with benzene for maximum cumene yield.

Alkylation reactions:

$$\underset{\text{Benzene}}{C_6H_6} + \underset{\text{Propylene}}{C_3H_6} \rightleftarrows \underset{\text{Cumene}}{C_6H_5CH(CH_3)_2}$$

$$\underset{\text{Cumene}}{C_6H_5CH(CH_3)_2} + \underset{\text{Propylene}}{C_3H_6} \rightleftarrows \underset{\text{Diisopropylbenzene}}{C_6H_4[CH(CH_3)_2]_2}$$

Transalkylation reaction:

$$\underset{\text{Diisopropylbenzene}}{C_6H_4[CH(CH_3)_2]_2} + \underset{\text{Benzene}}{C_6H_6} \rightleftarrows \underset{\text{Cumene}}{C_6H_5CH(CH_3)_2}$$

Side reactions:

In addition to the principal alkylation reaction of benzene with propylene, all acid catalysts promote the following undesirable side reactions to some degree:

- *Oligomerization of olefins*, to form C_6, C_9, C_{12}, or heavier olefins
- *Alkylation of benzene with heavy olefins*, to form through oligomerization hexylbenzene and heavier alkylated benzene by-products
- *Polyalkylation*, to make diisopropylbenzene (DIPB) and heavier alkylates
- *Hydride-transfer reaction*, to form diphenyl propane

All the alkylation reactions are highly exothermic. The heat of reaction is efficiently removed by the vaporization of benzene so that the alkylation can proceeds at isothermal conditions while maintaining high catalyst productivity. As a result of the isothermal operating conditions, selectivity to cumene is high.

10.2.2.2 Environmental Protection of the Process Description

Figure 10.3 is a simple flow schematic of the CD Cumene process. The CD alkylation reaction system consists of two reaction stages: the alkylator and the finishing reactor. Propylene is introduced as a vapor at the bottom of the reaction zone in the upper section of the CD column; benzene is fed in as a liquid above the reaction zone.

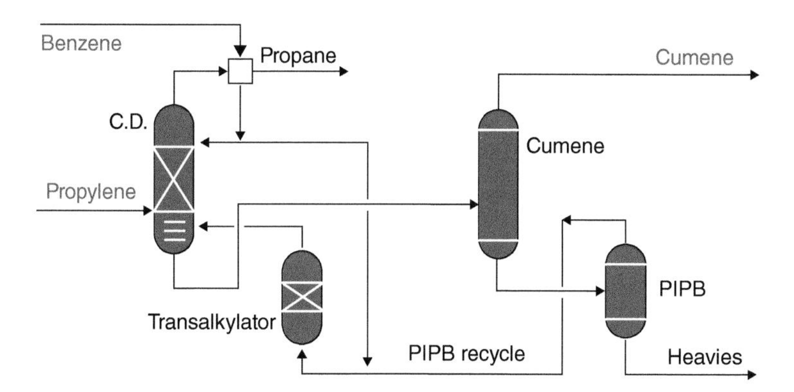

Figure 10.3 CD Cumene process flow schematic.

The proprietary CD Cumene catalyst bales containing the alkylation catalyst are stacked on top of each other inside the alkylator like structured packing, enabling the alkylation reaction and the distillation of reactants and products to take place simultaneously.

More recently, UOP has introduced QZ-2001 as a new catalyst, also based on *beta-zeolite*; it offers improved stability and operation at a very low B/P feed ratio. Because QZ-2001 is a strong acid, it can be used at a very low temperature. The low reaction temperature reduces the rate of competing olefin oligomerization reactions, resulting in higher selectivity to cumene and lower production of heavy by-products.

Further Reading

Kostrab, G., Mravec, D., Bajus, M. et al. (2006). *Appl. Catal. A Gen.* 299: 122–130.

Kostrab, G., Lovič, M., Janotka, I. et al. (2007). *Appl. Catal. A Gen.* 323: 210–218.

Kostrab, G., Lovič, M., Janotka, I. et al. (2008). *Appl. Catal. A Gen.* 335: 74–81.

Kostrab, G., Lovič, M., Janotka, I. et al. (2012). *Catal. Commun.* 18 (February): 176–181.

Global Benzene Market Analysis & Qutlook 2017–2023 (2018). www.globenewswire.com/news-release/2018/11/26/1656481/.

www.slideshare.net>Hazalztan>styrene-prod.

https://www.chemicals-technology.com/projects/lyondell.

11

Pillar B of Petrochemistry

Chemicals from BTX Aromatics

CHAPTER MENU

11.1 Chemicals from Aromatic Hydrocarbons

Briefly, *aromatic compounds* are those containing one or more benzene rings or similar ring structures. The majority are taken from refinery streams that contain them, and separated into fractions, of which the most significant are benzene, toluene, and xylenes (BTX); the two-ring condensed aromatic compound naphthalene is also a source of petrochemicals.

Phenol and acetone are key components in the cumene value chain, as illustrated in Figure 11.1.

Petrochemistry: Petrochemical Processing, Hydrocarbon Technology, and Green Engineering,
First Edition. Martin Bajus.

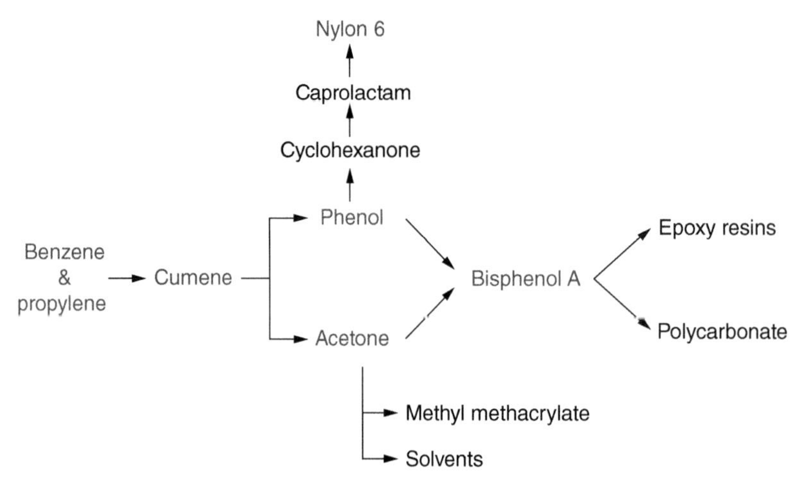

Figure 11.1 The cumene value chain.

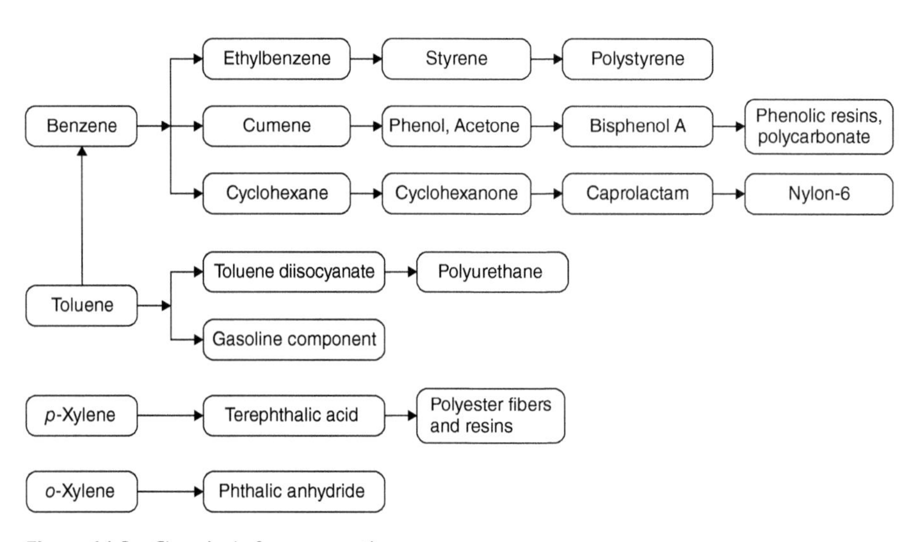

Figure 11.2 Chemicals from aromatics.

Aromatic compound such as BTX are major sources of chemicals (Figure 11.2). For example, benzene is used to make styrene, the basic ingredient of polystyrene plastics, as well as paints, epoxy resins, glues, and other adhesives. The process for the manufacture of styrene proceeds through ethylbenzene, which is produced by reaction of benzene and ethylene in the presence of an acid zeolitic catalyst.

11.2 Styrene

Styrene monomer (SM), or vinylbenzene, is an important industrial chemical used in the manufacturing of a variety of polymers. The activity of the vinyl group makes SM easy to polymerize on its own or with many other monomers. Production capacity for styrene in 2005 was nearly 28 million tons per annum (MTPA).

Over 60% of all styrene currently produced is used in the manufacture of polystyrene. The remainder is used for production of a wide range of liquid and solid copolymers, such as styrene acrylonitrile (SAN), acrylonitrile butadiene styrene (ABS), styrene butadiene (SB) latex, and styrene butadiene rubber (SBR).

11.2.1 Process Chemistry

Nearly 90% of the world styrene production comes from gas-phase ethylbenzene dehydrogenation. Ethylbenzene dehydrogenation may be carried out by a simple thermal reaction. The use of catalysts, however, allows a much higher selectivity to styrene. The reaction is highly endothermic ($\Delta H°_{610°C}$ = 124.56 kJ/mol) and occurs with an increased number of moles, as shown:

$$\underset{\text{EB}}{C_6H_5\,CH_2\,CH_3} \rightleftarrows \underset{\text{SM}}{C_6H_5CHCH_2} + H_2$$

It is an equilibrium reaction and for this reason is thermodynamically limited. The equilibrium constant K_p is calculated as follows

$$K_p = \frac{(p_{\text{sty}} \cdot p_{\text{H}_2})}{p_{\text{eb}}} \tag{11.1}$$

where p_i is the partial pressure of the component *i*, and it can be written also as

$$K_p = \left[\frac{(n_{\text{sty}} . n_{\text{H}_2})}{n_{\text{eb}}}\right] \cdot \left[\frac{P}{\sum n_{\text{i}}}\right] \tag{11.2}$$

where n_i is the number of moles of component *i* and *P* is the total pressure of the reactor system.

K_p is constant at constant temperature. According to Eq. (11.2), it is possible to increase the products (n_{sty} and n_{H_2}) by decreasing the total operating pressure, the other numerator term, in the reaction system. This way, not only the yields of styrene, but also hydrogen, are enhanced.

Before entering reaction, steam is added to the ethylbenzene. It has the essential role of supplying the heat needed by the reaction, above all in an adiabatic condition. Looking at Eq. (11.2), it is possible to increase ethylbenzene conversion also by increasing the total number of moles in the system ($\sum n_i$), reducing the ethylbenzene partial pressure by dilution. Thus it is clear that steam also has a direct thermodynamic effect on the reaction equilibrium: the ethylbenzene conversion increases if the steam-to-ethylbenzene ratio increases. Steam has one other positive effect: it cleans the catalyst by reacting with carbon to produce carbon dioxide and hydrogen.

11.2.2 Process Descriptions

The Lummus/UOP classic SM dehydrogenation technology uses commercial iron oxide-based dehydrogenation catalysts in two high-temperature (>600 °C) adiabatic reaction stages. Interstage reheating is required because of the endothermic nature of the dehydrogenation reaction. Reactor temperatures are maintained by heat exchange with superheated (up to 900 °C) steam, which is used subsequently as cofeed to the reactor. The typical range of operating parameters for the two processes is shown in Table 11.1.

Table 11.1 Typical range of operating parameters for the for the Lummus/UOP processes.

	Classic SM	Smart SM
Expected consumption	1.055–1.060	1.065–1.075
Steam-to-oil ratio	1.0–1.15	1.15–1.50
Catalyst run length	2.5 yr	2.5 yr
Liquid hourly space velocity	$0.45\,h^{-1}$	$0.45\,h^{-1}$
Conversion per pass	63–65%	65–75%
Reactor temperatures, SOR/EOR	622/627 °C	620/620 °C
Last reactor outlet pressure, kPa	35.2	36.4–48.5

The Lummus/UOP Smart SM process uses in situ oxidative reheat technology to eliminate the need for some of the heat-exchange equipment. The oxidative reheat also eliminates a portion of the hydrogen produced in the dehydrogenation reaction, releasing the constraint of thermodynamic equilibrium. In conventional dehydrogenation units, selectivity and conversion are linked so that high conversion and high selectivity cannot be achieved simultaneously. In order to achieve high selectivity, the conversion is limited to less than 70%. However, the Smart SM technology permits both high selectivity and high conversion and also results in considerably less flow through the plant for a given SM capacity, thereby increasing efficiency and reducing capital cost. The 130 000 MTPA SM plant of Kaucuk A.S. in the Czech Republic was expanded to 170 000 MTPA with the addition of a Smart reactor.

Badger Licensing LLC, now a joint venture between affiliates of Technip and ExxonMobil, jointly licenses the technology for the manufacture of SM with Atofina Petrochemicals. In the early 1960s, Versalis (at that time Montedison, then EniChem, and then Polimeri Europa) started the production of styrene in Mantova, Italy.

11.3 Hydrogenation

Small quantities of cyclohexane are from naphtha, but most is produced by hydrogenation of benzene. The process is operated in refinery complexes using hydrogen from catalytic reforming. The consecutive reaction is highly exothermic and occurs with a decreased number of moles, as shown here:

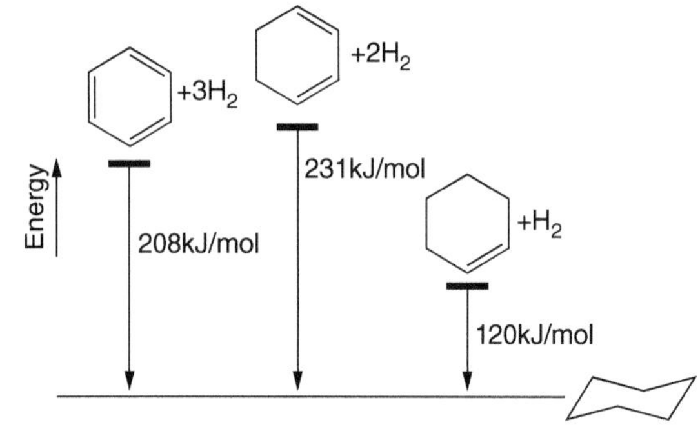

+3H2
+2H2
+H2
Energy
208kJ/mol
231kJ/mol
120kJ/mol

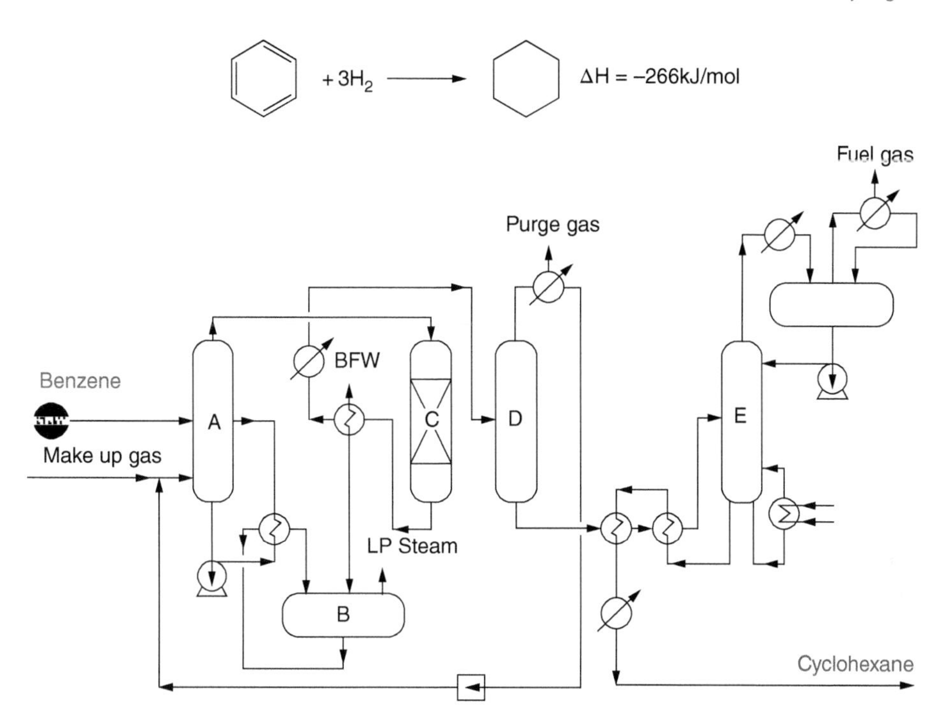

Figure 11.3 The Institut Francais du Petrole process for the hydrogenation of benzene to cyclohexane: (a) liquid-phase reactor; (b) heat exchanger; (c) catalytic pot: acts as a finishing reactor when conversion of the main reactor drops below the required level; (d) high-pressure separator; (e) stabilizer.

The hydrogenation of benzene produces cyclohexane. Many catalyst systems, such as Ni/alumina and Ni/Pd, are used for the reaction. General reaction conditions are 160–220 °C and 2.5–3.0 MPa. Higher temperatures and pressures may also be used with sulfide catalysts: Older methods use a liquid-phase process (Figure 11.3). New gas-phase processes operate at higher temperatures with noble metal catalysts. Using high temperatures accelerates the reaction (faster rate). The hydrogenation of benzene to cyclohexane is characterized by a highly exothermic reaction and a significant decrease in the product volume.

11.3.1 Partial Hydrogenation of Benzene to Cyclohexene

Asahi Kasei developed a technology for highly selective partial hydrogenation of benzene to cyclohexene and succeeded in the commercialization of a production process for producing cyclohexanol from benzene through cyclohexene. The process was considered difficult for a long time in the industry, but several innovative technologies made the partial hydrogenation reaction possible. A catalyst that consists of specific metallic ruthenium particles exhibited excellent high selectivity of cyclohexene. The catalyst was obtained by reducing a ruthenium compound, which contains a zinc compound. The use of a zinc, which forms specific sites on the metal surface as a co-catalyst, exhibited a remarkable effect to enhance selectivity. Dispersing agents of metal oxides were found to extend the life of the metallic ruthenium catalyst, and some dispersing agents had the effect of enhancing selectivity in the partial hydrogenation reaction. One of the remarkable features of the reaction was the reaction field, which comprised four phases: vapor (hydrogen), oil, aqueous, and solid (ruthenium catalyst). The catalyst was used in the aqueous phase, and the reactants (benzene and hydrogen) were dissolved in the aqueous phase, where the reaction proceeded. Therefore, the products and reactants transferred between four phases through dissolution, diffusion, and extraction: rendering quick transfer was a very important factor in enhancing reaction selectivity. The catalyst system and the reaction field described made the selectivity for cyclohexene very high, and a yield of 60% for cyclohexene was obtained.

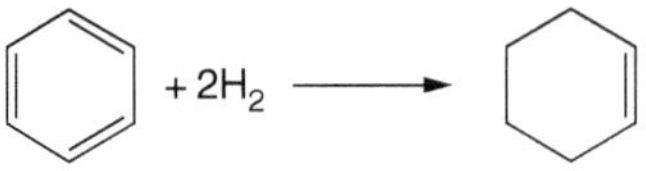

$+ 2H_2 \longrightarrow$

A Ru–Zn/*m*-ZrO_2 nanocomposite catalyst for benzene partial hydrogenation to cyclohexene was prepared by coprecipitation of ruthenium trichloride and zirconium oxychloride with ammonia, followed by hydrogen reduction in an aqueous zinc sulfate solution. By use of transmission electron microscopy (TEM), powder X-ray diffraction (XRD), and X-ray photoelectron spectroscopy (XPS), the reduction of zinc cation to metallic zinc and the transformation of zirconium hydroxide to monoclinic zirconia were proposed and explained. The hydrogenation parameters, such as the amount of zinc sulfate in the pretreatment process, the hydrogenation temperature, the stirring rate, and the hydrogen pressure, influenced the yield of cyclohexene. The pronounced positive effect of metallic zinc in enhancing the cyclohexene yield was discussed.

The cyclohexylbenzene starting material for the present process can be produced by the alkylation of benzene with cyclohexene according to the following reaction:

$+ \longrightarrow$

Cyclohexene can be supplied to the reaction zone as a separate feed from benzene, but normally it is produced in situ by the selective hydrogenation of benzene in the presence of a bifunctional catalyst. In the case of cyclohexene being produced by in situ hydrogenation of benzene, the combined reaction is generally called *hydroalkylation* and may be summarized as follows:

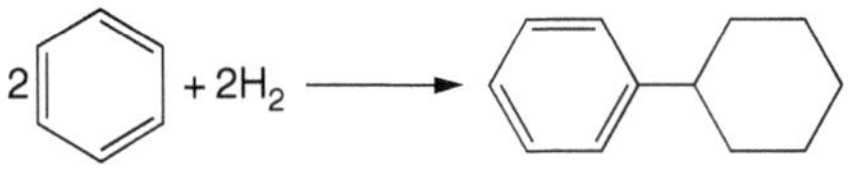

$2 \quad + 2H_2 \longrightarrow$

11.4 Hydrodealkylation of Toluene

In situ hydrodealkylation during catalytic reforming was preceded as an additional source of benzene in that reaction. Hydrodealkylation in a dedicated plant may be purely thermal or may be catalyzed by metals or supported metal oxides. Hydrogen is always present. Typical reaction conditions are 600 °C and 4–6 MPa over oxides of chromium, platinum, molybdenum, or cobalt, supported on alumina. The uncatalyzed reaction requires a higher temperature of up to 800 °C and pressure up 10 MPa.

The reaction may be thermal or catalytic:

$$C_6H_5CH_3 \underset{}{\overset{H_2}{\rightleftharpoons}} C_6H_6 + CH_4$$

11.5 Isomerization

Xylenes are produced mainly by catalytic reforming (Chapter 5). However, in Europe, the majority of benzene and about half of toluene come from pyrolysis gasoline (Chapter 7). The xylenes are isolated only with difficulty from pyrolysis gasoline because it contains 50% ethylbenzene with a similar vapor pressure.

Chemical technology needs to separate the isomers, isomerize the unwanted ones to an equilibrium mixture, and repeat the process to extinction. The separation of three xylene isomers and ethylbenzene from each other is a remarkable task. Their physical constants are shown in Table 11.2: the boiling points of all four compounds are within 9 °C. *o*-Xylene boils more than 5 °C above the others, however, and can be fractionally distilled on a huge column with 150–200 plates and a high-reflux ratio. The mixture at the top of the column contains about 40% *m*-xylene, 20% *p*-xylene, and 40% ethylbenzene. If required, the low-boiling ethylbenzene can be removed by an involved extractive distillation. Energy costs are high.

The isomers in the intermediate fraction differ markedly in melting point and, in the oldest process, are separated by low-temperature crystallization. The mixture is carefully dried to avoid icing of the equipment, and then it is cooled. Crystallization of *p*-xylene starts at −4 °C and continues until −68 °C, at which point the *p*-xylene/*m*-xylene eutectic concentration in the crystalline mass is 70%; but a series of melting, washing, and recrystallization steps eventually raises this to 99.5%.

Table 11.2 Physical constants of the C_8-stream.

	Melting point (°C)	**Boiling point (°C)**
o-Xylene	−25.2	144.4
m-Xylene	−47.9	139.1
p-Xylene	13.2	138.3
Ethylbenzene	−95.0	136.2

If *m*-xylene is required, it may be extracted from the C_8 stream by complex formation. Treatment of the stream with HF-BF_3 gives two layers. The *m*-xylene selectively dissolves in the HF-BF_3 layer as complex $C_6H_4(CH_3)_2.HBF_4$. The phases are separated, and the *m*-xylene is regenerated by heating.

The drawbacks of low-temperature crystallization are the high energy requirements for cooling and problems of handling solid *p*-xylene – which, for example, deposits on the walls of the cooling vessel and reduces the rate of heat transfer. Nonetheless, it is still being used.

More recent processes for separating *m*- and *p*-xylenes make use of molecular sieves for which the feed components show small differences in affinity. The UOP Parex process is the most widely used (Figure 5.4).

The exit streams from both the low-temperature crystallization and the adsorption processes contain unwanted products, primarily *m*-xylene, ethylbenzene (if it was not removed earlier), and the portion of *o*-xylene that the market does not require. They are catalytically isomerized in the presence of acid catalysts to provide another equilibrium mixture, which is somewhat more favorable in that it contains about 48% *m*-xylene, 22% each of *o*- and *p*-xylenes, and 8% ethylbenzene. Acid catalysts include silica-alumina and silica with HF-BF_3, the same material that complexes with *m*-xylene. The drawback of silica-alumina is that it promotes disproportionation and transalkylation. If platinized alumina is added to the silica-alumina, as a dual-function catalyst, the system will isomerize ethylbenzene as well as the xylenes.

The most important isomerization catalyst today in hydrocarbon technology is the zeolite ZSM-5. This is the same zeolite catalyst useful for toluene disproportionation to benzene and *p*-xylene. The silica-alumina/platinized alumina catalyst operates at 2.3–3.3 MPa in a hydrogen atmosphere with substantial recycle (Tables 5.5 and 5.6). The ZSM-5 process has a major economic advantage because it operates at low pressures in either vapor or liquid phase and requires less or even no hydrogen and much less recycling than does the noble metal catalyst. Its one disadvantage is that it does not isomerize ethylbenzene but rather dealkylates it to benzene. Because the ZSM-5 process requires less capital investment and has lower operating costs, it has been widely accepted. A process that was announced but not commercialized makes use of non-zeolite molecular sieves, primarily silicoaluminophosphates. This process is claimed not only to isomerize ethylbenzene to xylenes but also to provide a higher level of the desired p-xylene.

11.6 Disproportionation of Toluene

Disproportionation is important as a source of mixed xylenes from which the *para* isomer, the one most in demand, can be isolated. Two molecules of toluene react to give one of benzene and one of mixed xylene.

The reaction is carried out in the vapor phase in the presence of a non-noble metal catalyst. Even a casual estimate of the energetics of the reaction indicates that the equilibrium constant is close to unity, so large volumes of reactants must be recycled from the product-recovery section of the plant to the reactor. The advantage of the process, compared with catalytic reforming, is that the reactor effluent is free from ethylbenzene. This makes the separation of the xylene isomers easier.

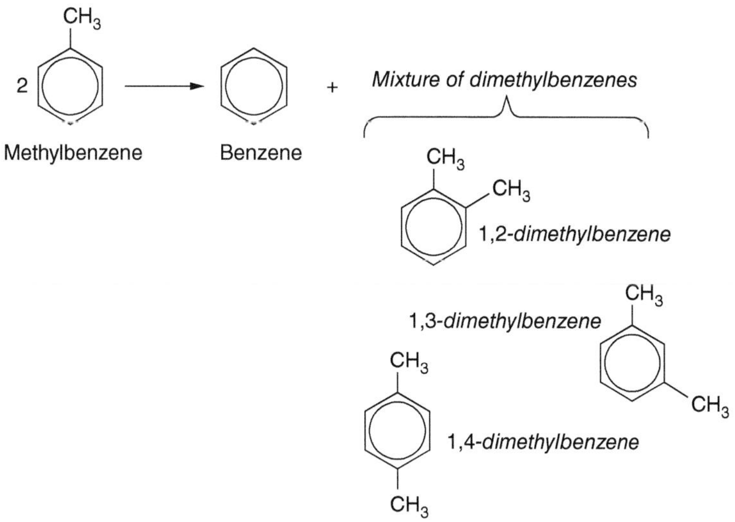

2 Methylbenzene → Benzene + Mixture of dimethylbenzenes

1,2-dimethylbenzene

1,3-dimethylbenzene

1,4-dimethylbenzene

A dramatic improvement is ExxonMobil's development of a liquid-phase process based on an antimony-doped zeolite ZSM-5 catalyst at 300 °C and 4.5 MPa. A xylene fraction results, with about 80% of the desired *p*-xylene. The disproportionation takes place inside a zeolite cage whose "mouth" is shaped such that benzene, toluene, and *p*-xylene have ready access, but *o*- and *m*-xylenes cannot escape. Accordingly, these isomers remain in the cage to undergo further isomerization and to supply the equilibrium quantities required each time 2 moles of toluene disproportionate.

Another disproportionation, called *transalkylation*, involves the interaction of toluene with trimethylbenzene (pseudocumene) or ethyltoluene, a compound produced during catalytic reforming.

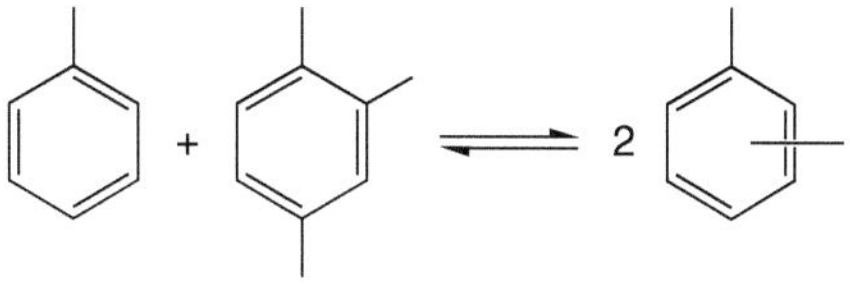

A methyl group migrates from a trimethylbenzene molecule to a toluene molecule, giving two molecules of mixed xylenes. In practice, benzene is also produced, but its volume can be kept low by using a high ratio of trimethylbenzene to toluene.

11.7 Oxidation Processes

11.7.1 Cumene → Phenol + Acetone

Cumene has become the major feedstock for the production of phenol, so the cumene hydroperoxidation process currently accounts for about 95% of phenol produced worldwide. The mechanisms for the formation and degradation of cumene hydroperoxide require a closer look and are provided following the figure.

Benzene —(H_3C–CH=CH_2 Propene, Zeolite/Δ)→ Cumene —(Air [O_2], Δ)→ Cumene hydrogen peroxide —(HCl, H_2O)→ Acetone + Phenol

11.7.2 Process Chemistry

The phenol process is extremely complex, with several recycles and by-product upgrading to increase overall efficiency. Three main sections can be defined.

11.7.2.1 Cumene Oxidation to Cumene Hydroperoxide

Cumene oxidation proceeds in the liquid phase with air under pressure at a temperatures ranging from 90 to 120 °C. It is a free-radical chain reaction initiated by the presence of small amounts of cumene hydroperoxide (CHP). The radicals so generated produce CHP but also several by-products, such as dimethylphenylcarbinol, acetophenone, dicumylperoxide, methanol, and formic and acetic acids. The acidity of the mixture must be neutralized, usually with sodium carbonate or sodium hydroxide solutions, to avoid CHP cleavage and phenol formation even to a limited extent (tenths of ppm) that would trap the radicals and reduce the oxidation efficiency.

11.7.2.2 Cumene Hydroperoxide Cleavage to Phenol and Acetone

After the oxidized product has been concentrated to recycle back the unreacted cumene, the resulting CHP solution is cleaved in the presence of small amounts of an acid catalyst, usually concentrated sulfuric acid, to give phenol and acetone at a temperature ranging from 70 to 90 °C. In addition to the main reaction, other reactions take place, such as cleavage of dicumylperoxide, dehydration of dimethyphenylcarbinol to α-methylstyrene, addition of dimethylphenylcarbinol to phenol to form cumylphenol, and coupling and dehydration of acetone mesityloxide.

11.7.2.3 Distillation Section

After neutralization of the solution, acetone and phenol are recovered in separate sections, where specific impurities are removed by distillation. α-Methylstyrene is recovered to be sold in part and to be hydrogenated in part to cumene and recycled back to oxidation.

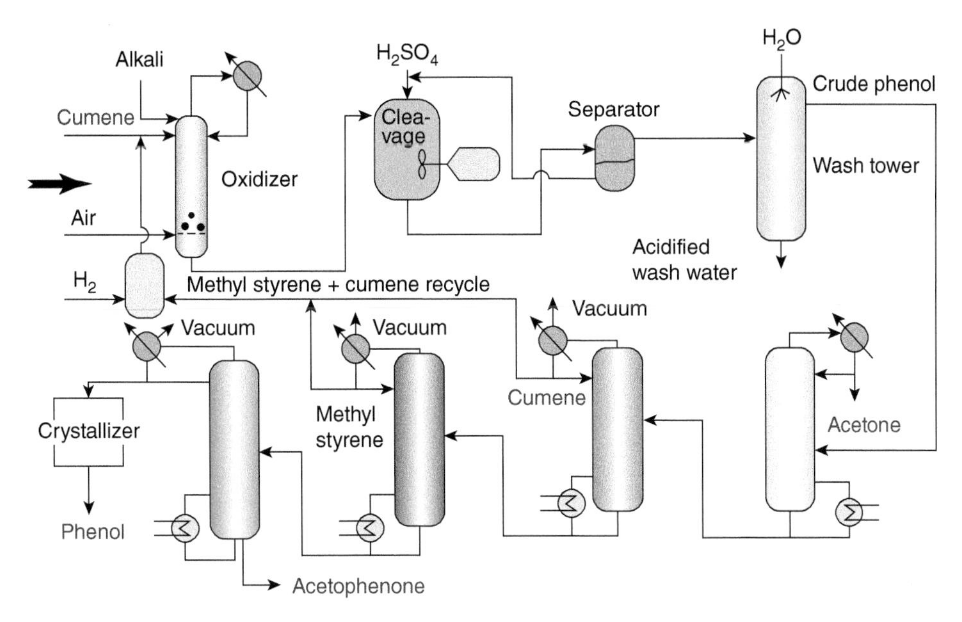

Figure 11.4 Simplified scheme of the phenol process.

11.7.3 Process Description

The air oxidation of cumene (isopropyl benzene) leads to the production of both phenol and acetone, as shown in the Figure 11.4. The fresh cumene, along with recycled cumene coming from the cumene hydroperoxide concentration and α-methylstyrene hydrogenation section, is fed, together with air, into the cumene oxidation section, where CHP is produced; the CHP reaches a concentration of about 30 wt % in the cumene mixture. Small amounts of a basic compound in aqueous solution are fed to the oxidizers to neutralize acidic by-products.

The offgases from the oxidation reactors pass through an offgas treatment section, where the hydrocarbons are recovered before this stream is sent into the atmosphere. The CHP mixture produced in the oxidation section is sent to the CHP washing section to eliminate traces of basic compounds, and then it is concentrated to about 80 wt % in the CHP concentration section while the overhead product (mainly cumene) is recycled back to the cumene oxidation section.

The concentrated CHP mixture is cleaved in the CHP cleavage section, with mineral acid as catalyst, to acetone, phenol, and by-products. The acid cleft product is neutralized in the cleft neutralization section to eliminate the salt content and then is stocked as feed to the distillation section.

The distillation section is composed, after a first raw phenol-acetone separation, of two subsections: the acetone purification section and the phenol purification section. In the acetone section, pure acetone (purity higher than 99.8 wt %) and an aqueous phase are produced. In the phenol section, pure phenol (purity 99.96 wt % on dry product) via extractive distillation and a hydrocarbon-rich cut are produced.

11.7.4 Benzene → Maleic Anhydride

Most maleic anhydride is made by the oxidation of butane or, less frequently, 1- and 2-butenes. An older method still used in Europe is based on the oxidation of benzene and is analogous to the process for phthalic anhydride from naphthalene. The vapor-phase oxidation is carried out with a supported vanadium pentoxide catalyst at 400 °C. A mixture of maleic acid and maleic anhydride results.

$$\text{C}_6\text{H}_6 + 4.5\ \text{O}_2 \longrightarrow \text{(maleic anhydride)} + 2\ \text{CO}_2 + 2\text{H}_2\text{O}$$

(ΔH = ~1875 kJ/mol)

A mixture of maleic acid and maleic anhydride results, and the acid may be dehydrated to the anhydride directly without separation. The product is purified by batch vacuum distillation.

Maleic anhydride, which is produced at a level of about 227 MTPA in the United States, finds its greatest market in unsaturated polyester resins. Its other important uses are conversion to fumaric acid for use as a food acidulant and in the formulation of agricultural chemicals and lube oil additives. Its newest use is for conversion to 1,4-butanediol.

11.7.5 Cyclohexane → Cyclohexanol + Cyclohexanone → Adipic Acid

Benzene is the source of the two most important nylons, nylon 6,6 and nylon 6. Nylon 6,6 is the polymer formed when adipic acid condenses with hexamethylene diamine. Nylon 6 is the self-condensation product of caprolactam, which is the dehydration product of 6-aminocaproic (6-aminohexanoic) acid. The number used to label nylons refers to the number of carbon atoms in the diamine and dibasic acid, in that order. A single number indicates that the amino and carboxyl functions are in one molecule. Thus $[NH(CH_2)_6NHCO(CH_2)_4CO]_n$ is nylon 6.6, $[NH(CH_2)_5CO]_n$ is nylon 6, and $[NH(CH_2)_6NHCO(CH_2)_8CO]_n$ is nylon 6.10, a specialty nylon made with sebacic rather than adipic acid.

The first step in the manufacture of the adipic acid needed for nylon 6.6 is the hydrogenation of benzene to cyclohexane (Section 11.3). Thereafter the cyclohexane may be oxidized directly to adipic acid with nitric acid or with air over cobalt acetate, but yields are low, production-valueless by-products are high, and large amounts of nitric acid consumed. Instead, a two-stage process is used. The initial oxidation at about 150 °C and 1.0–1.5 MPa over cobalt or manganese naphthenate or octanoate gives a cyclohexanol and cyclohexanone "mixed oil." In the first stage of the reaction, cyclohexyl hydroperoxide forms, and this is converted catalytically to cyclohexanol/cyclohexanone:

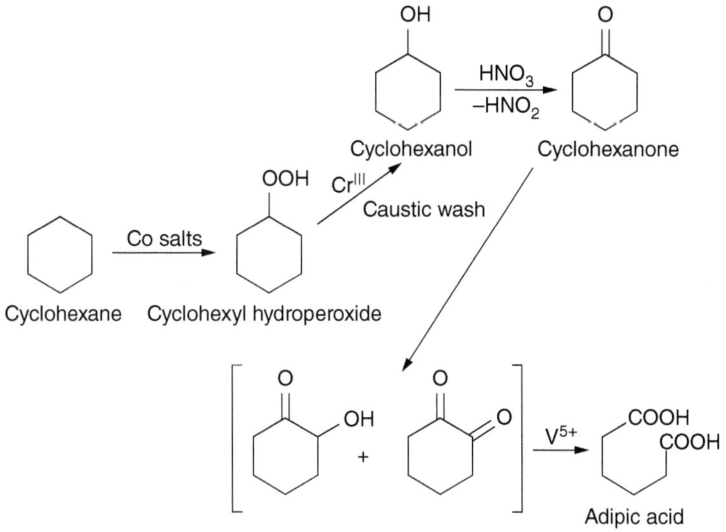

OH
HNO$_3$
–HNO$_2$
O
Cyclohexanol
Cyclohexanone
OOH
Cr^{III}
Caustic wash
Co salts
Cyclohexane
Cyclohexyl hydroperoxide
O
OH
+
O
O
V^{5+}
COOH
COOH
Adipic acid

11.7.6 P-Xylene → Terephthalic Acid / Dimethyl Terephthalate

Mixed xylenes are used as solvents, particularly in the paint industry, and are valued components of the gasoline pool because of their high octane number. The major chemical use for the individual xylenes is oxidation to terephthalic acid, isophthalic acid, and phthalic anhydride.

The major use for *p*-xylene is oxidation to terephthalic acid, which, as such or as its methyl ester, is reacted with ethylene glycol to give polyester resins for fibers, films, molding resins, and, most recently, for biaxially oriented bottle resins. Terephthalic and dimethyl terephthalate is the most widely used xylene-based chemical by a wide margin, compared with phthalic anhydride, the next largest xylenes derivate.

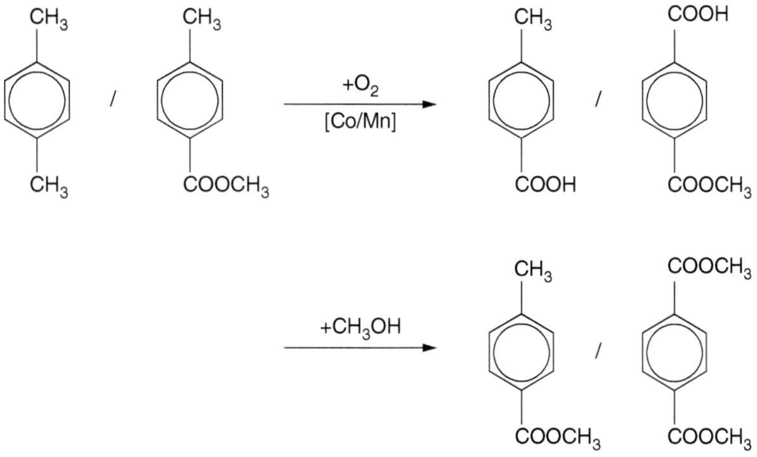

CH$_3$
CH$_3$
/
CH$_3$
COOCH$_3$
+O$_2$
[Co/Mn]
CH$_3$
COOH
/
COOH
COOCH$_3$
+CH$_3$OH
CH$_3$
COOCH$_3$
/
COOCH$_3$
COOCH$_3$

Although *o*-xylene oxidizes readily, *m*- and *p*-xylenes present a problem. The *m*- and *p*-toluic acids formed in the first stage of oxidation contain a methyl group that defies further oxidation, because the carboxyl group is electron withdrawing. There are several ways to overcome this, the most important being the Amoco Mid-Century process. The oxidation is carried out in acetic acid solution with a catalyst comprising a manganese or cobalt salt with a bromine promoter, which may actually be bromine itself but is usually manganous or cobaltous bromide. The bromide converts the recalcitrant methyl group to a free radical, which is then much more susceptible to oxidation.

11.8 Condensation Processes

11.8.1 Aniline

Next in line of benzene-based chemicals is aniline, 75% of which goes into isocyanates (Sections 11.8.2 and 11.8.3). In the traditional process, benzene is first nitrated with mixed acids (H_2SO_4/HNO_3). These form a nitronium ion (NO_2^+), which attacks the benzene ring. The reaction is exothermic, and the mixture must be cooled to maintain a temperature of about 50 °C. An adiabatic process has been described, in which 65 rather than 98% sulfuric acid is used. The water in the acid absorbs the heat, eliminating the need for external cooling. In the nitration of benzene, a small amount of *meta* isomer is obtained.

H_2SO_4, HNO_3 → NO_2

Nitrobenzene is reduced to aniline in almost quantitative yield by vapor-phase hydrogenation at 270 °C and 12.5 kPa in a fluidized bed of a copper-on-silica catalyst. Also feasible is vapor-phase hydrogenation over a fixed bed of nickel sulfide on alumina.

NO_2 + 3 H_2 —Catalyst→ NH_2 + 2 H_2O

An older process employed iron turnings and hydrochloric acid as a source of hydrogen. This liquid-phase process, in which the iron was converted to Fe_3O_4, useful as a pigment, is little used today. Another process involving the ammonolysis of chlorobenzene was wasteful of chlorine and has not been used since 1967. The newest method for aniline preparation involves the ammonolysis of phenol.

11.8.2 4,4′-Diphenylmethane Diisocyanate

4,4′-Diphenylmethane diisocyanate is produced by reaction of aniline hydrochloride with formaldehyde to form the 4,4′, 2,4′, and 2,2′ isomers of diaminodiphenylmethane. The equation shows only the 4,4′ isomer:

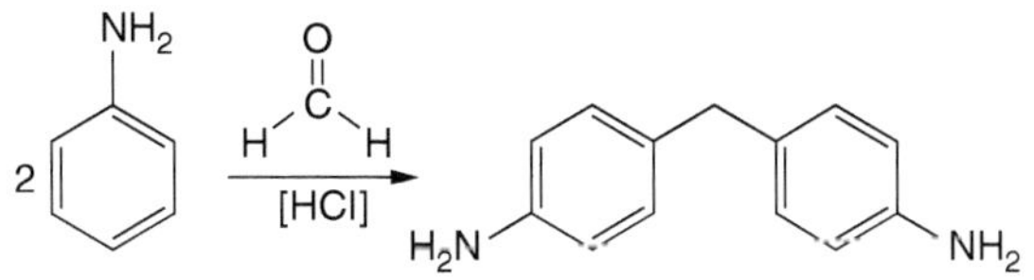

The diamine reacts with additional formaldehyde to give trimers, tetramers, and higher oligomers. The diisocyanate from the diamine is known as MDI (standing for methylene diphenylene diisocyanate), whereas the isocyanate from the oligomers is known as poly MDI or PMDI. When MDI is required, the diamine is removed from the mixture by distillation and phosgenated. Conversely, the entire mixture may be phosgenated and the MDI separated from the mixed isocyanates by distillation.

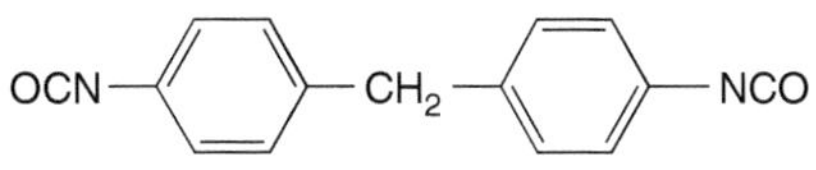

Chemical structure of MDI

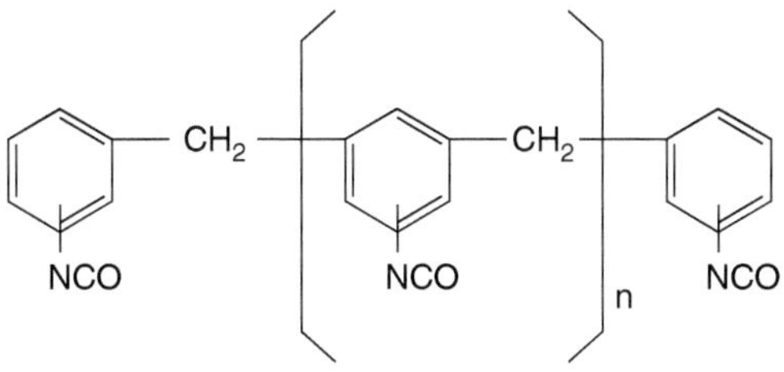

Chemical structure of PMDI

For the condensation of aniline and formaldehyde, aniline is treated with a stoichiometric amount of hydrochloric acid, and the hydrochloride is reacted with 37% formaldehyde for a few minutes at 70 °C. The condensation is completed at 100–160 °C for one hour. The mixture of di- and higher amines is recovered and phosgenated by reaction with a phosgene in chlorobenzene solution. The carbamoyl chloride forms at 50–70 °C, and this is decomposed to the isocyanate at 90–130 °C.

11.8.3 Toluene → Dinitrotoluene and Toluene Diisocyanate

Toluene is the major product from catalytic reforming (Chapter 5) and pyrolysis gasoline (Chapter 7). The chemical technology requires considerably more benzene than toluene. Toluene's second largest use is as a solvent, mainly for coatings. The finding that benzene is carcinogenic has increased the toluene's solvent usage at the expense of benzene, although the latter is in any case small.

The largest outlet for toluene in which its chemical properties are of value in their own right is as a raw material for a mixture of 2,4- and 2,6-toluene diisocynate (tolylene diisocynate, diisocyanatotoluene, TDI) used for polyurethane resins. TDI is made by chemistry similar to that used for 4,4′-MDI (Section 11.8.2). Some of the non-phosgene routes desirable there are applicable to TDI, although less development work has been done.

Commercial TDI contains about 80% of 2,4-(*o*, *p*) isomer and 20% of the 2,6-(*o*, *o*) isomer. The three-step process comprises the dinitration of toluene, the hydrogenation of the nitro compounds to diaminotoluenes, and the reaction of these with phosgene to commercial-grade TDI.

Toluene is nitrated with a mixture of nitric and sulfuric acids in two stages. The first produces the three isomers of mononitrotoluene. These are subsequently nitrated further to obtain the six possible dinitrotoluene isomers. The 2,4 isomer makes up 74–76% of the mixture and the 2,6 isomer about 19–21%. The concentrations of the other four isomers are minimal. The 3,4-compound forms at a level of 2.4–2.6%. The 2,3, 2,5, and 3,5 isomers are present at a level of no more than 1.7%. The concentration of the acids is carefully controlled so that very little of the trinitro isomers is formed.

The dinitrotoluenes are dissolved in methanol and hydrogenated continuously to diaminotoluenes by reaction with hydrogen in the presence of Raney nickel at 150–180 °C and 6.5–13.0 MPa. Numerous other catalysts such as supported platinum or palladium may be used.

The possibility of eliminating the nitration step by an intermolecular amination of toluene with ammonia is of some interest. Similar chemistry has been discussed for the possible preparation of aniline (Section 11.8.1) and *m*-aminophenol.

$$C_6H_5CH_3 + HNO_3 \xrightarrow{H_2SO_4} CH_3C_6H_3(NO_2)_2 + H_2O$$

$$CH_4 + H_2O \longrightarrow CO + H_2$$

$$CH_3C_6H_3(NO_2)_2 \xrightarrow{\text{cat.}} CH_3C_6H_3(NH_2)_2$$

$$COCl_2 + CH_3C_6H_3(NH_2)_2 \longrightarrow CH_3C_6H_3(NCO)_2 + 4\ HCl$$

The carbonylation with phosgene occurs in two stages, the first yielding a carbamoyl chloride and the second the isocyanate. In both reactions, hydrogen chloride is liberated. The reaction is carried out with a 10–20% solution of the diamine mixture in chlorobenzene. This is combined with a chlorobenzene solution of phosgene.

11.8.4 Bisphenol A

Bisphenol A is the basic monomer for the production of two polymers epoxy resins and polycarbonates. As such, it is a fast-growing end user of phenol and acetone, with a worldwide grow rate of 7% per year.

Dow's high-purity bisphenol A QBIS technology is in an advanced process producing BPA suitable for polycarbonate and epoxy production at a low manufacturing cost. This process employs a proprietary QCAT resin catalyst system that has demonstrated the capability to

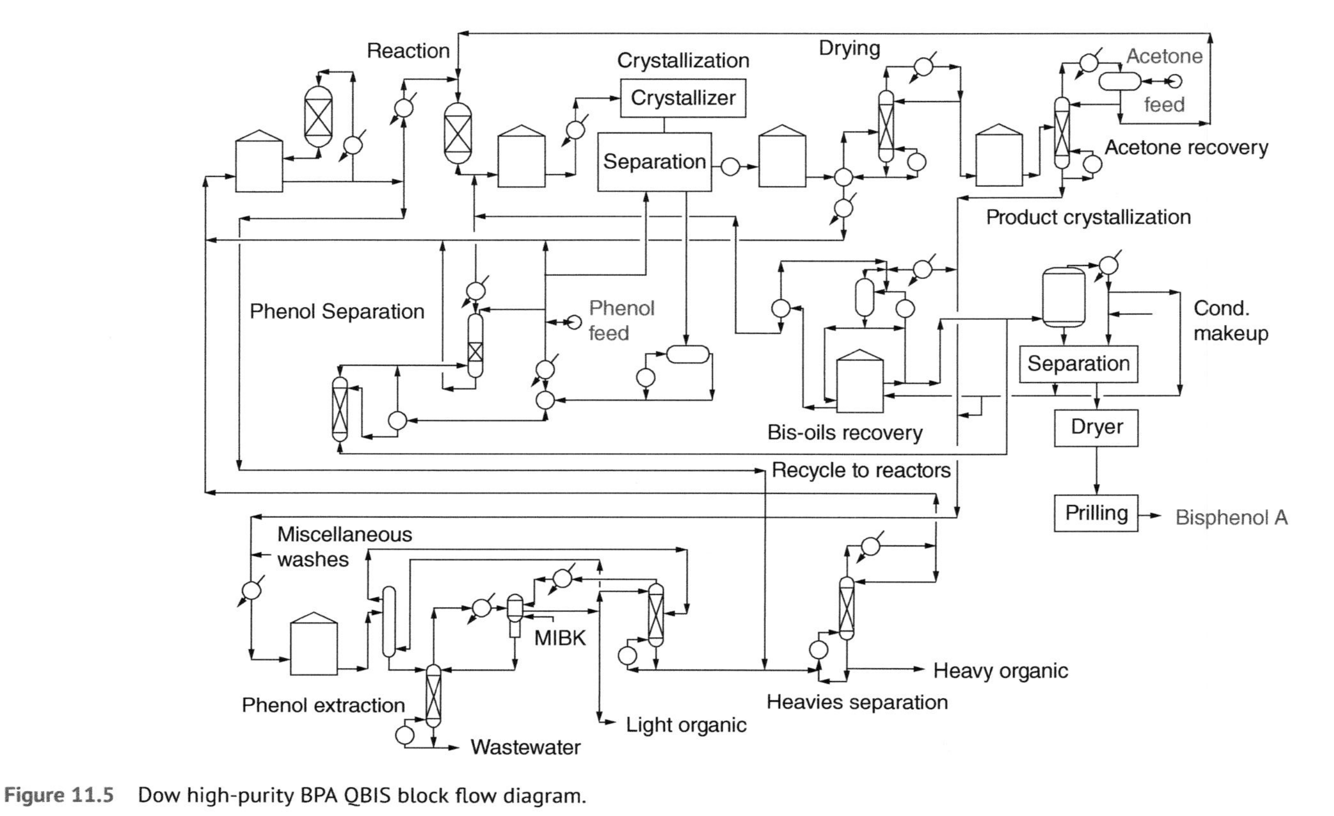

Figure 11.5 Dow high-purity BPA QBIS block flow diagram.

consistently achieve greater than 98% efficiency of raw materials utilization. The QCAT catalyst used in the main reactors is a unique catalyst providing high asset utilization and long life. The proprietary QCAT catalyst used in the rearrangement reactor has been designed for improved selectivity for isomerization and color absorption.

11.8.4.1 Bisphenol Reaction

BPA is produced by reacting 1 mol acetone with 2 mol of phenol in an acid-catalyzed condensation reaction. The reaction is exothermic, and water is the main reaction by-product.

Phenol + Acetone + Phenol $\xrightarrow{H^+}$ p,p - Bisphenol A + H_2O

Phenol Acetone Phenol QCAT p,p - Bisphenol A

A QCAT cation-exchange resin catalyzes the BPA reaction. The catalyst is back-promoted via a proprietary system to achieve the proper selectivity to the para, para isomer. The reaction is carried out in a significant excess of phenol, and the reactors run adiabatically.

11.8.4.2 Process Description

The process minimizes the formation of rearrangeable and nonrearrangeable by-products. By using a recirculation process, the rearrangeable impurities reach equilibrium such that the reaction of phenol and acetone primarily produces high-purity *p,p*-bisphenol.

The process scale is easily adjustable: facilities using the Dow BPA process are designed for 90 000, 100 000, and 120 000 MTPA. Figure 11.5 shows a process block flow diagram. The plant has multiple fixed-bed reactors. When a reactor's catalyst activity becomes too low and the reactor is taken offline, the spent catalyst is removed and replaced with the QCAT catalyst before the reactor is brought back online.

12

Pillar C of Petrochemistry

C_1 Technologies

CHAPTER MENU

Petrochemistry: Petrochemical Processing, Hydrocarbon Technology, and Green Engineering, First Edition. Martin Bajus.

12.1 Introduction

In today's petrochemical industry, most of the available technologies are considered "mature," and even considerable research effort is not expected to yield significant improvements. However, as profit margins can sometimes be increased by improving selectivity by one percentage point or less, it is still desirable to invest in improving catalyst performance. Moreover, polymer chemistry continues to demand monomers that are not only less expensive but also increasingly pure.

Petrochemistry is based in large part on the conversion of alkenes because of the ease and economy with which they can be obtained from petroleum; and as they are easily functionalized, they are versatile raw materials. However, petrochemical technology is moving toward the direct use alkanes, which can be obtained from both petroleum and natural gas and are even more economical. One of the most important applications of selective oxidation catalysis is the functionalization of hydrocarbons. Of selective oxidation processes, only the oxidation of *n*-butane to maleic anhydride has been commercialized.

Optimization of the catalytic performance in terms of reactant conversion, yield, productivity, and selectivity to the desirable product is related not only to a thorough knowledge of the nature of the catalyst and surface-active phases, reaction mechanism, thermodynamics, and kinetics but also to the development and use of a suitable reactor configuration where all features can be successfully exploited. Catalytic distillation is a unique unit operation that combines a solid heterogeneous catalyst and a single mass-transfer column, providing a simultaneous reaction and distillation (Section 3.4.3, Figure 3.4). Catalytic distillation is being applied to various applications and promises to play a major role in the refining and petrochemical industries. The first catalytic distillation was started up the production of methyl *tertiary* butyl ether (MTBE), and later catalytic distillation (CD) was applied for the production of ETBE, TAME, and TAEE (Sections 12.2.8 and 12.2.8.1).

Significant developments have been achieved in recent years in membrane reactor technology, where the capability of these in the selective permeation of particular components in a mixture is combined with reactive application. The selective transport of oxygen or hydrogen in membranes can be used in oxidation and dehydrogenation processes.

In any case, hydrogen will become more and more of a focal point and a basic necessity in the petrochemical technology of tomorrow. Hydrogen's production potential resides in hydrocarbons whose H/C ratio is greater than motor fuel requirements, i.e. the alkanic C_1–C_4 fractions have specific uses. Two compounds will be critical for this new trend. Hydrogen is the first, as even regenerative reformers will no longer suffice. Partial oxidation, in particular of gases or heavy residues by stream, is needed to supplement production. The second is selective, efficient catalysts so as to limit severity of operating conditions.

The evolution of tomorrow's refinery will not be confined to strictly petroleum and petrochemical processes. The utilization of waste energy production, whether electricity or steam, is an important item in plant cost and reliability. The major consequence will be a much more environmentally friendly product quality. These problems are solved in the refinery of the future, with the arrival of deep conversion processing such as residue hydrocracking, carbon rejection, and gasification processes that can lead to the elimination of heavy fuel production if need be; supplementary processes for deep treatment of distillates coming from conversion or deep conversion; and synthesis of compounds from the light ends of the same conversion processes that led to the advanced flow schemes themselves. This type of refinery approaches that of a petrochemical complex capable of supplying traditional refining products, but also meeting much more severe specifications, and petrochemical intermediates such as olefins, aromatics, hydrogen, and methanol.

12.2 Synthesis Gas

Synthesis gas (syngas) is the name given to a variety of mixtures of carbon monoxide (CO) and hydrogen (H_2), or nitrogen and hydrogen. It is made from methane from natural gas if the latter is available, as it is in the United States and most of Europe. Countries lacking natural gas (e.g. Japan and Israel) make synthesis gas from naphtha. It may, however, be made from virtually any hydrocarbon (e.g. so-called *reside*, the residue from petroleum distillation). Indeed, it may be made from almost any carbonaceous material including coal, peat, wood, biomass, agricultural residues, and municipal solid waste. Synthesis gas is the beginning of a wide range of chemicals (Figure 12.1).

Coal was an important feedstock prior to 1960. During World War II, it was used in Germany to provide syngas for the manufacture of fuel and chemicals by the Fischer–Tropsch (FT) process (see Section 12.2.10). The same technology is applied today in South Africa and makes coal the basis for at least half that country's energy needs. Coal is also gasified in China.

Different applications for syngas require different mixtures. The FT reaction and methanol manufacture require CO/H_2 = 1:2; hydroformylation (the oxo process [Section 12.2.12]) requires CO/H_2 = 1:1, and the Haber process for ammonia requires N_2/H_2 = 1:3 without any CO at all. Other organic chemical syntheses require pure hydrogen. Properly adjusted, nonetheless, the basic syngas processes can give all these mixtures.

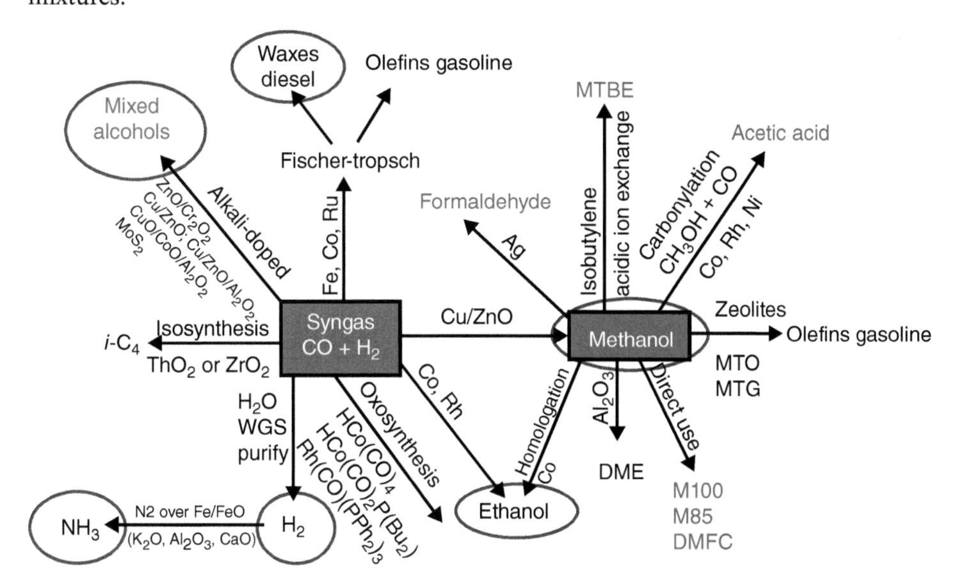

Figure 12.1 Production of chemicals from syngas.

12.2.1 Steam Reforming of Methane – Stringent Greenhouse Gas

The most widely used syngas process is the steam reforming of hydrocarbons, with partial oxidation of hydrocarbons as another possibility. Coal-based routes are also of high significance. The dominant syngas process is the steam reforming of methane.

Before steam reforming can take place, the methane feedstock is desulfurized by passage over a zinc oxide catalyst at 360–400 °C. Sulfur levels less than 2 ppm are required if the nickel steam-reforming catalyst is to have an adequate life. It is then mixed with steam in a molar ratio of between 2.5 and 3.5 of steam to methane and passed over an alkali-promoted nickel catalyst, supported on potassium oxide or alumina. The reaction is carried out at about 800 °C and 3.5 MPa; it is endothermic and the heat is supplied externally. The principal reaction is

$$CH_4 + H_2O \rightleftarrows CO + 3H_2 \qquad \Delta H_{800°C} = 227\ kJ/mol$$

The exit gases contain about 7% unchanged methane plus some CO_2 and traces of nitrogen present in the natural gas. The excess steam minimizes coking of the catalyst. The higher the steam/carbon ratio, the higher the CO_2/CO ratio in the products:

$$C + H_2O \rightleftarrows CO + H_2$$

$$C + 2H_2O \rightleftarrows CO_2 + 2H_2$$

If the syngas is to be used to make ammonia, nitrogen must be added and CO removed. This is done by addition of an amount of air calculated to provide a N_2/H_2 ratio of 1 : 3. The oxygen in the air reacts with some of the CO to give CO_2:

$$2CO + O_2 \rightleftarrows 2CO_2$$

The gases then pass to a second reforming unit at 370 °C, possibly with additional steam. In the presence of an iron oxide catalyst, the water-gas conversion or shift reaction takes place:

$$CO + H_2O \rightleftarrows CO_2 + H_2 \qquad \Delta H_{370°C} = -38\ \text{kJ/mol}$$

The names of the processes may be somewhat confusing. The term *steam reforming* is used to describe the reaction of hydrocarbons with steam in the presence of a catalyst. This process should not be confused with the *catalytic reforming* process for the improvement of the gasoline octane number (Section 5.5.5.2). In the gas industry, the term *reforming* is commonly used for the conversion of a hydrocarbon by reacting it with O-containing molecules, usually H_2O, CO_2, and/or O_2. *Partial oxidation* (also called *steam/oxygen reforming*) is the non-catalytic reaction of hydrocarbons with oxygen and usually also steam (catalysis is possible, and the process in that case is referred to as *catalytic partial oxidation*). This process may be carried out in an autothermic or allothermic way. *Coal gasification* is the more common term to describe partial oxidation of coal. A combination of steam reforming and partial oxidation, in which endothermic and exothermic reactions are coupled, is often referred to as *autothermic reforming*. These reactions are summarized in Table 12.1.

The product gases contain hydrogen, nitrogen, CO_2, and traces of methane, CO, and argon. They are compressed and scrubbed with aqueous monoethanolamine and diethanolamine (a variety of other processes is available) to remove CO_2. Some of the

Table 12.1 Reactions during methane conversion with steam and/or oxygen.

Reactions	ΔH°_{298} (kJ/mol)	
$CH_4 + H_2O \rightleftarrows CO + 3H_2$	206	(1)
$CO + H_2O \rightleftarrows CO_2 + H_2$	−41	(2)
$CH_4 + CO_2 \rightleftarrows 2CO + 2H_2$	247	(3)
$CH_4 \rightleftarrows C + 2H_2$	206	(4)
$2CO \rightleftarrows C + CO_2$	−41	(5)
$CH_4 + 1/2O_2 \rightarrow CO + 2H_2$	−36	(6)
$CH_4 + 2O_2 \rightarrow CO_2 + 2H_2O$	803	(7)
$CO + 1/2O_2 \rightarrow CO_2$	284	(8)
$H_2 + 1/2O_2 \rightarrow H_2O$	242	(9)

CO_2 dissolves in the water – at high pressure, the solubility is high – and some reacts with the amine to give an unstable salt. The ethanolamine may then be recovered by steam stripping, which decomposes the salt. A further shift reaction may improve yields, and other processes are employed to reduce CO to a very low level. The product is then delivered to an ammonia plant.

12.2.1.1 Reactions and Thermodynamics

Although natural gas does not solely consist of methane (Section 4.2) for simplicity we will assume that this is the case. Table 12.1 shows the main reactions during methane conversion.

When converting methane in the presence of steam, the most important reactions are the steam-reforming reaction (1) and the water-gas shift reaction (2). Some processes, such as the reduction of iron ore and the hydroformylation reaction (Section 12.2.12) require syngas with a high CO content, which might be produced from methane and CO_2 in a reaction known as *CO_2 reforming* (3). The latter reaction is also referred to as *dry reforming*, obviously because of the absence of steam.

The main reactions may be accompanied by coke formation, which leads to deactivation of the catalyst. Coke may be formed by decomposition of methane (4) or by disproportionation of CO, the Boudouard reaction (5).

In the presence of oxygen, methane undergoes partial oxidation to produce CO and H_2 (6). Side reactions such as the complete oxidation of methane to CO_2 and H_2O (7), and oxidation of the formed CO and H_2 (8) and (9), may also occur.

The reaction of methane with steam is highly endothermic, while the reactions with oxygen are moderately to extremely exothermic. Operation can hence be allothermic (steam and no or little oxygen added, required heat generated outside the reactor), or autothermic (steam and oxygen added, heat generated by reaction with oxygen within the reactor), depending on the steam/oxygen ratio.

Stoichiometric amounts of steam and oxygen were added for reactions (1) and (6), respectively. This is most obvious at high temperatures where only H_2 and CO are present.

The H_2/CO ratios resulting from steam reforming and partial oxidation are 3 and 2, respectively, at these high temperatures.

The H_2 and CO contents of the equilibrium gas increase with temperature, which is explained by the fact that the reforming reactions (1) and (3) are endothermic.

The CO_2 content goes through a maximum. This can be explained as follows: CO_2 is formed in exothermic reactions only, while it is a reactant in the endothermic reactions. Hence, at low temperature CO_2 is formed, while with increasing temperature the endothermic reactions in which CO_2 is converted (mainly reaction (3)) become more important.

Both steam reforming and partial oxidation of methane are hindered by elevated pressure because the number of molecules increases due to these reactions. At a pressure of 3.0 MPa, the equilibrium conversion to H_2 and CO is only complete at a temperature of over 1400 K. However, in industrial practice, temperatures much in excess of 1200 K cannot be applied as a result of metallurgical constraints. It will be shown later that this point is of great practical significance.

12.2.2 Steam Reforming Process

It is interesting that even though steam reforming is carried out at high temperature (>1000 K), a catalyst (supported nickel) is still required to accelerate the reaction due to the very high stability of methane. The catalyst is contained in tubes, which are placed inside a furnace that is heated by combustion of fuel. The steam reformer consists of two sections. In the convection section, heat recovered from the hot flue gases is used for preheating of the natural gas feed and process steam, and for the generation of superheated steam. In the radiant section, the reforming reactions take place.

After sulfur removal, the natural gas feed is mixed with steam (and optionally CO_2) and preheated to approximately 780 K before entering the reformer tubes. The heat for the endothermic reforming reaction is supplied by combustion of fuel in the reformer furnace (allothermic operation). The hot reformer exit gas is used to raise steam. In a *knock-out drum* (a vessel in which liquid is separated from vapor by gravitation), water is removed and the raw syngas can be treated further, depending on its use.

A second route to syngas is partial oxidation of carbonaceous materials reacted with steam. If the feedstock can be vaporized, the reactions are carried out simultaneously; but if it is a solid, they must be carried out sequentially.

A vaporizable hydrocarbon feedstock (e.g. methane, propane, or naphtha) is burned in a flame in the presence of about 35% of the stoichiometric amount of oxygen:

$$CH_4 + O_2 \rightleftarrows CO_2 + 2H_2 \qquad \text{Rapid} \qquad \Delta H = 318\ \text{kJ/mol}$$

A little water is also formed by conventional combustion reactions. Excess hydrocarbons can now react:

$$CH_4 + 1/2O_2 \rightleftarrows CO + 2H_2 \qquad \text{Rapid} \qquad \Delta H = -36\ \text{kJ/mol}$$

$$CH_4 + CO_2 \rightleftarrows 2CO + 2H_2 \qquad \text{Slow} \qquad \Delta H = 247\ \text{kJ/mol}$$

$$CH_4 + H_2O \rightleftarrows CO + 3H_2 \qquad \text{Slow} \qquad \Delta H = 227\ \text{kJ/mol}$$

The flame temperature is 1300–1400 °C at a pressure of 6.0–8.0 MPa with a residence time of 2–5 seconds. The rapid initial reactions provide the heat required to drive the subsequent endothermic reactions. These reactions are slower than the initial reactions, however, and as incomplete combustion is always accompanied by carbon formation, a finely divided carbon is always present in the products and must be removed by washing.

Desulfurization is unnecessary. Sulfur in the feedstock is converted primarily to hydrogen sulfide but also to carbonyl sulfide (COS). Nitrogen compounds end up as elemental nitrogen or ammonia. A plant must be built, however, to produce the pure oxygen for the process. This is generally an economic drawback but confers a slight advantage in ammonia manufacture in that the nitrogen coproduced with the oxygen can be added to the syngas to provide the correct composition for ammonia production. Some of the shift reaction and CO_2 removal stages can be avoided.

12.2.3 Hydrogen

Hydrogen is manufactured by steam reforming and partial oxidation, but large amounts are also obtained in the refinery as byproducts of cracking and catalytic reforming reactions (Chapters 5 and 7). Minor sources include coke oven gases, as well as the electrolysis of water, brine (which gives principally sodium hydroxide and chlorine), and hydrogen chloride and hydrogen fluoride (to give chlorine and fluorine).

When hydrogen is obtained from syngas or refinery processes, it is purified by washing successively at 180 °C and about 2.0 MPa with liquid methane, to remove nitrogen and CO, and with liquid propane to remove methane.

About 60% of all hydrogen is used for the production of ammonia. The second largest use is in refinery processes in hydrotreating, hydrocracking, hydrodesulfurization, and toluene hydrodealkylation (Sections 11.3 and 11.4). Most of this hydrogen is produced internally in other refinery processes. Among organic chemicals made outside the refinery, methanol is the largest consumer of hydrogen. Other applications include the conversion of benzene to cyclohexane (Section 11.3), nitrobenzene to aniline (Section 11.8.1), and unsaturated fats to saturated or hard fats.

12.2.4 MegaMethanol Technology

Vast natural gas, shale gas, and oil-associated gas reserves are available in remote areas at low and stable cost. Combining low-cost feedstock with large single-train synthesis technology will be the strategy of the next decades in order to achieve a remarkable production cost reduction.

Lurgi developed its MegaMethanol technology on the basis of the syngas technologies available in the 1990s, i.e. conventional steam reforming and combined reforming together with a new synthesis concept. Lurgi's MegaMethanol process is an advanced technology for converting natural gas to methanol at low cost in large quantities. It permits the construction of highly efficient single-train plants of at least double the capacity of those built to date. This paves the way for new downstream (DS) industries that can use methanol as a competitive feedstock.

The MegaMethanol technology has been developed for world-scale methanol plants with capacities larger than 1 MMTA. To achieve such a capacity in a single train, a special process design is needed, incorporating advanced but proven and reliable technology, cost-optimized energy efficiency, low environmental impact, and low investment cost.

The main process features to achieve these targets are:

- Oxygen-blown natural gas reforming, either in combination with steam methane reforming or as pure autothermal reforming.
- Two-step methanol synthesis in water- and gas-cooled reactors (GCRs) operating along the optimal reaction route.
- Adjustment of syngas composition by hydrogen recycles.

The configuration of the reforming process – autothermal or combined reforming – mainly depends on the feedstock composition, which may vary from light natural gas (nearly 100% methane content) to oil-associated gases. The aim is to generate an optimal syngas, characterized by the stoichiometric number (SN) given as follows:

$$\mathsf{SN} = \frac{H_2 - CO_2}{CO + CO_2} = 2.0 - 2.1$$

The Lurgi MegaMethanol process, based on reforming of gaseous hydrocarbons, especially natural gas, consists of the following essential process steps:

- Desulfurization
- Saturation
- Prereforming (optional)
- Pure autothermal or combined reforming
- Methanol synthesis
- Methanol distillation

The syngas generation section of a MegaMethanol plant using combined reforming is shown in Figure 12.2.

12.2.4.1 Process Description

The syngas production section of a conventional methanol plant accounts for more than 50% of the capital cost of the entire plant. Thus, optimizing this section yields a significant cost benefit. Conventional steam methane reforming is economically applied in small and medium-sized methanol plants only, with the maximum single-train capacity being limited to about 3000 metric tons per day (MTD). Oxygen-blown natural gas reforming, either in combination with steam reforming or as pure autothermal reforming, is today considered to be the best-suited technology for large syngas plants. The reason for this appraisal is that the syngas generated through oxygen-blown technology becomes available in stoichiometric composition and under very high pressure. Hence very high quantities can be produced in a single train using reasonably small equipment.

Desulfurization Catalyst activity is seriously affected even by traces of catalyst poisons in the gas feedstock. Among others, sulfur compounds in particular lower the catalyst activity considerably.

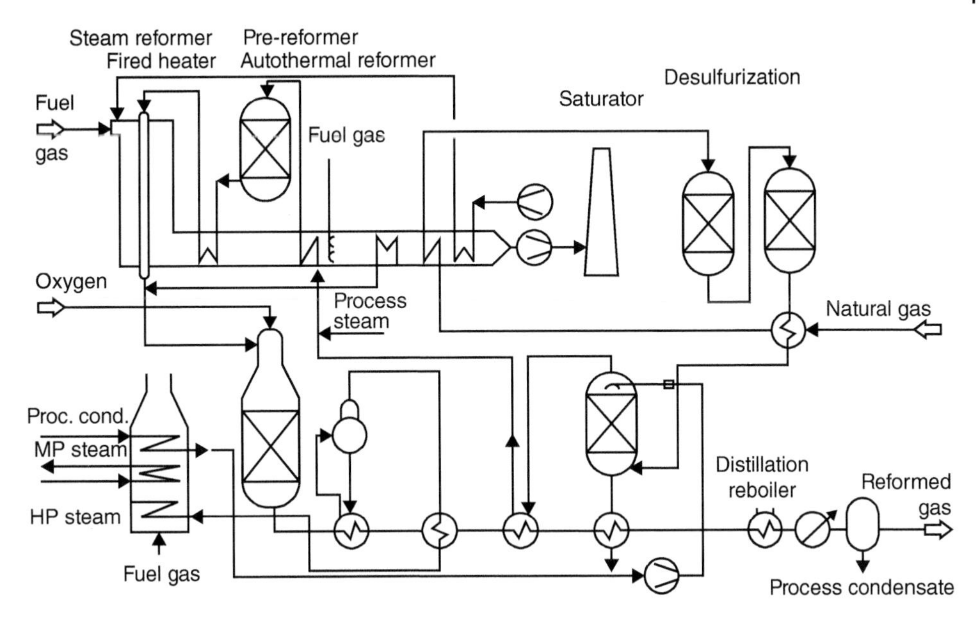

Figure 12.2 Syngas generation using combined reforming for the MegaMethanol process.

In order to protect the reformer and synthesis catalysts from sulfur poisoning, the feedstock must be desulfurized. Desulfurization operates at approximately 350–380 °C in the desulfurization reactor. The feedstock is routed through zinc oxide beds, where hydrogen sulfide is adsorbed according to the following equation:

$$H_2S + ZnO \rightleftarrows ZnS + H_2O$$

If the feedstock contains organic sulfur compounds such as mercaptans or thiophenes, hydrogenation is required prior to desulfurization. This is often accomplished in a separate reactor, where the feedstock, after adding a small amount of hydrogen-rich methanol synthesis purge gas, is hydrogenated over cobalt-molybdenum catalysts. A residual sulfur content of less 0.1 ppm is obtainable and can be tolerated for the DS processes.

Saturation After desulfurization, the natural gas feedgas is cooled and then enters the top of a saturator for saturation with water in the wet process. After makeup with process condensate and process water from distillation, hot circulation water is fed to the top section of the saturator. Circulation water is withdrawn from the bottom of the saturator by a recirculation pump and then is heated by a circulation water heater in the reformed gas cooling train before it is refed to the saturator.

Prereforming (Optional) If the feedstock contains fractions of higher hydrocarbons, the steam-reformer catalyst can be affected by carbon deposits owing to cracking reactions when operated at low steam-carbon ratios. This should be avoided by prereforming the feedstock in a prereformer. The conversion of higher hydrocarbons in an adiabatic reactor produces a gas rich in methane and hydrogen that is perfectly suitable for further steam reforming. The fixed-bed-type prereformer is arranged between the process feed superheater and the steam reformer. The desulfurized feedstock, with process steam added, is routed through the catalyst bed.

Catalyst:

Nickel 3-35% wt./ support

Support: α-modification of Al_2O_3; MgO; ceramic support with cementing, where almost all higher hydrocarbons and a small percentage of the methane are reformed with steam according to the following equations:

Steam reforming of higher hydrocarbons:

$$C_nH_m + nH_2O \rightleftarrows nCO + (m/2 + n)H_2$$

Methanation:

$$CO + 3H_2 \rightleftarrows CH_4 + H_2O$$

Water-gas shift reaction:

$$CO + H_2O \rightleftarrows CO_2 + H_2 \qquad \Delta_T H_{293} = -42.3\ \text{kJ/mol}$$

Prereformed gas is produced at about 380–480 °C, the overall heat balance being slightly endothermic or exothermic depending on the content of higher HCs in the feedstock. The prereformed gas contains only a few parts per million of hydrocarbons higher than methane and allows reduction of the steam-to-carbon ratio for the steam reformer to 1.8. A low steam-to-carbon ratio and superheating of the prereformed gas upstream (US) of the steam-reformer inlet reduces the size of the steam reformer significantly. Additionally, the amount of waste heat is reduced by saving steam-reformer underfiring when combined reforming is used.

12.2.5 Autothermal Reforming

Pure autothermal reforming can be applied for syngas production in MegaMethanol plants whenever light natural gas is available as feedstock to the process. The desulfurized and optionally prereformed feedstock is reformed with steam to syngas at about 4.0 MPa using oxygen as the reforming agent. The process offers great operating flexibility over a wide range to meet specific requirements. Reformer outlet temperatures are typically in the range of 950–1050 °C. The syngas is compressed to the pressure required for methanol synthesis in a single-casing syngas compressor with integrated recycle stage.

With the help of a proprietary and proven three-dimensional computational fluid dynamics (CFD) model, gas flows and temperature profiles are simulated with the objective of designing burner and reactor as an integrated unit. Autothermal processes produce the heat required for gasification through partial combustion of the feedgas to be converted in the reactor. Oxygen is usually added for this purpose. Suitable feedstocks for autothermal catalytic reforming are light natural gases or steam-reformed gases with high residual methane content.

The principal chemical reactions involved in the process are those of complete combustion of methane

$$CH_4 + 2O_2 \rightleftarrows CO_2 + 2H_2O \qquad \Delta_T H_{293} = -801.5\ \text{kJ/mol}$$

partial oxidation of methane

$$CH_4 + O_2 \rightleftarrows CO + H_2 + H_2O$$

as well as methane and higher hydrocarbon reforming and a CO shift reaction, as described in Section 12.2.2.

Catalyst:

Nickel 3–6% wt./ support

The combustion (partial oxidation) is highly exothermic, the reforming reaction being endothermic. In order to attain the desired product gas quality, the outlet temperature of the reactor is selected and controlled by metering in the required amount of oxygen to maintain the balance between exothermic and endothermic reactions. The autothermal reformer (ATR) is a refractory-lined pressure vessel.

12.2.6 Combined Reforming

The combination of oxygen-blown autothermal reforming and conventional steam methane reforming, the so-called *combined reforming process*, has an advantage in that it yields syngas of optimal composition and at a high pressure.

$$C_nH_m + nH_2O \rightleftarrows nCO + (m/2 + n)H_2$$

$$CH_4 + H_2O \rightleftarrows CO + 3H_2 \qquad \Delta_T H_{293} = 206.5\ \text{kJ/mol}$$

$$CO + H_2O \rightleftarrows CO_2 + H_2 \qquad \Delta_T H_{293} = -42.3\ \text{kJ/mol}$$

The overall reaction is highly endothermic, so reaction heat has to be provided externally. The syngas is characterized by a relatively low pressure (LP) and a surplus of hydrogen. By adding CO_2, the composition of the syngas can be adjusted to be more favorable for methanol production.

Methanol syngas generation by means of combined reforming is a well-proven technology. For natural gases or oil-associated gases with methane content above 80% and for methanol syntheses with capacities above 1500 MTD of methanol, this process route offers potential capital investment savings compared with the conventional steam reforming process.

The composition of the generated syngas is characterized by the SN

$$SN = \frac{H_2 - CO_2}{CO + CO_2}$$

In the steam methane reforming process, the given C : H ratio of the natural gas and the hydrogen added by steam decomposition leads to a SN that is higher than optimal for methanol production.

Therefore, Lurgi combined the two processes in such a way that only the amount of natural gas is routed through the steam reformer that is required to generate a final syngas with the desired SN of about 2.05. Thus the syngas demand per ton of methanol is reduced by approximately 25% compared with steam methane reforming. Depending on the composition of the natural gas, only about 30% of the hydrocarbons are converted in the steam methane reformer, and hence the steam methane reformer in the combined reforming is only about one-quarter of the size of a reformer in the conventional steam methane reforming process. This means considerable savings in cost and energy. Owing to the higher pressure in the reforming section, the compression energy is reduced, and compression to synthesis pressure is possible in a single-stage compressor.

12.2.7 Methanol Synthesis

In the Lurgi MegaMethanol process, methanol is synthesized from H_2, CO, and CO_2 in the presence of a highly selective copper-based catalyst. The principal synthesis reactions are as follows:

$$CO + 2H_2 \rightleftarrows CH_3OH \qquad \Delta_T H_{298} = -90.8\,\text{kJ/mol}$$

$$CO_2 + 3H_2 \rightleftarrows CH_3OH + H_2O \qquad \Delta_T H_{298} = -49.6\,\text{kJ/mol}$$

These reactions are highly exothermic, and the heat of reaction must be removed promptly from its source. This is accomplished most effectively in the Lurgi methanol reactors described shortly. The principle scheme of the methanol synthesis from natural gas is shown in Figure 12.3.

Efficient conversion in the methanol synthesis unit is essential to low-cost methanol production. In addition, optimal use of the reaction heat offers cost advantages and energy savings for the overall plant.

Nowadays, two types of catalytic fixed-bed reactors are used in industry: steam-raising reactors and GCRs. From the very beginning of the low-pressure technology era, Lurgi has been equipping its methanol plants with a tubular reactor in which the heat of reaction is transferred to boiling water (Figure 12.4).

The Lurgi water-cooled methanol reactor (WCR) is basically a vertical shell-and-tube heat exchanger with fixed-tube sheets. The catalyst is accommodated in tubes and rests on a bed of inert material. The water-steam mixture generated by the heat of reaction is drawn off

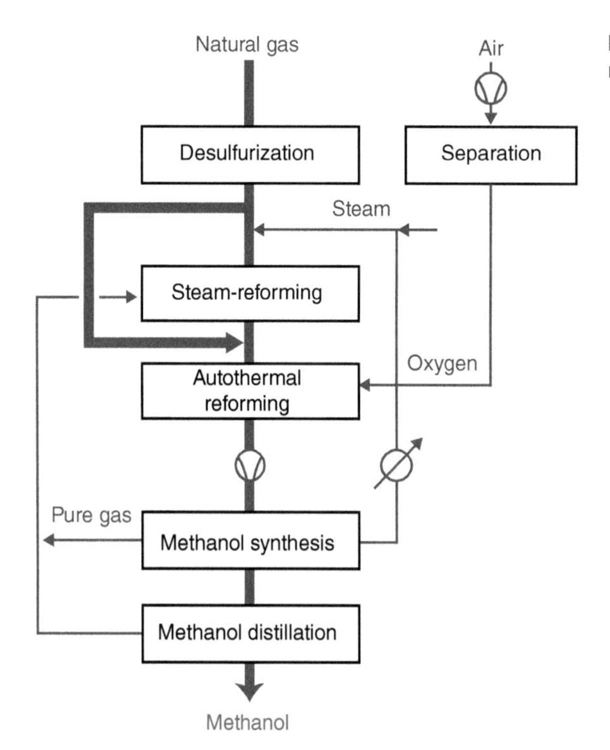

Figure 12.3 Principle scheme of methanol synthesis from natural gas.

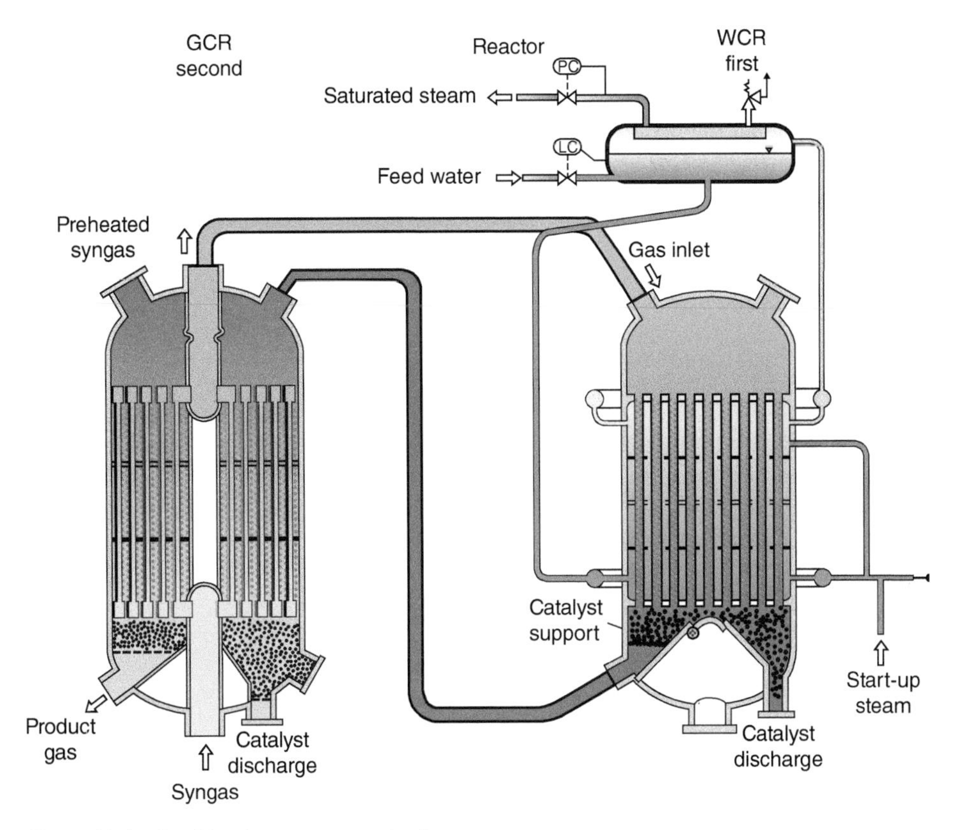

Figure 12.4 Combined converter synthesis.

below the upper tube sheet. Steam pressure control permits an exact control of the reaction temperature. The quasi-isothermal reactor achieves very high yields at low recycle ratios and minimizes the production of by-products.

A significant improvement in synthesis technology has been achieved by combining the WCR with a DS GCR. The so-called *combined converter synthesis* is shown in Figure 12.4. The excellent heat transfer in the WCR allows this reactor to operate with a high concentration of reaction components in the inlet gas. This highly concentrated gas results from a drastically reduced recycle rate. Under these conditions, a very high methanol yield is achieved in the WCR. The methanol-containing exit gas of the WCR is fed to the DS GCR. In the GCR, the reaction is accomplished at continuously reduced temperatures along the optimal reaction route. The optimal temperature profile is achieved by countercurrent preheating in the inlet gas to the WCR.

Methanol production on an industrial scale was introduced in 1923, when BASF Ludwigshafen commissioned a methanol synthesis on the basis of a chromium/zinc catalyst. Since this catalyst was not highly active, it was necessary to apply an operating pressure of between 30 and 40 MPa and operating temperatures of between 350 and 400 °C (high-pressure [HP] methanol catalyst):

$$ZnO - Cr_2O_3$$

The first LP methanol tests were run in 1969. Since Lurgi itself is not a catalyst manufacturer, it started cooperation with Süd-Chemie AG in 1970 for fabrication of the catalyst. In the presence of syngas, a highly selective copper-based catalyst is used. This catalyst is very sensitive to catalytic poisons, such as sulfur compounds, chlorine, and phosphorus.

The latest generation of methanol catalysts makes it possible to select an outlet temperature of the GCR of about 220 °C and pressure of about 5–10 MPa:

$$Cu - ZnO - Cr_2O_3$$

12.2.7.1 Methanol Synthesis Loop

Since economic conversion of the syngas to methanol cannot be achieved in a single reactor pass, unreacted gases are circulated in a loop, thus increasing the conversion rate. Figure 12.5 shows a typical diagram of the synthesis loop. Recycle gas and syngas are mixed and preheated in the trim heater by cooling the reactor outlet gas. Preheated recycle gas and syngas are routed to the GCR. On the tube side of the GCR, the reactor inlet gas is further heated to the inlet temperature of the WCR (approximately 240 °C).

12.2.7.2 Methanol Distillation

The raw methanol in the methanol synthesis unit contains water, dissolved gases, and a quantity of undesired but unavoidable by-products that have either lower or higher boiling points than methanol. The purpose of the distillation unit is to remove those impurities in order to achieve the methanol purity specification. A three-column methanol distillation is accomplished in the following process steps:

- Degassing
- Removal of low-boiling by-products
- Removal of high-boiling by-products

Methanol (methyl alcohol; wood alcohol) is a colorless, volatile, inflammable, poisonous alcohol.

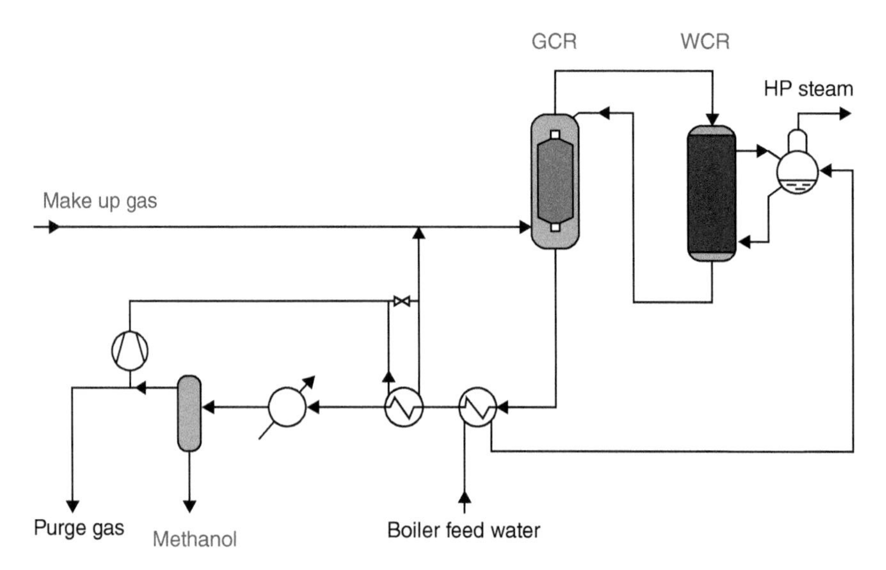

Figure 12.5 Scheme representing only the methanol synthesis unit.

12.2.8 MTBE

This brief description of past and present petrochemical developments leads to a certain number of important remarks. First of all, we are observing a gradual, continuous evolution. It could hardly be otherwise, considering the large time factors – it takes several years to build a petrochemistry – the capital investments, and the tightness of the product specification. Moreover, petrochemistry evolves around successive modifications to a basic flow scheme containing a limited number of processes. These processes have been greatly improved over the past 20 years from the technological point of view and, for catalytic processes, the level of performance of the catalysts in service. On other hand, very few new processes have appeared: as early as 1970, one could almost have built the petrochemical refinery of today, but with much lower performance with regard to energy, economics, and product quality. Among the truly new processes, one can name selective oligomerization, light olefin etherification, and very LP reforming with continuous catalyst regeneration.

Increasing amounts of methanol (currently over 35%) are used in combination with isobutene in the production of MTBE. MTBE has a high octane number (116) and is used as an octane booster in gasoline. It has been the fastest-growing chemical in the bulk chemical technology of the last decade. Recently, however, its environmental friendliness has been questioned. It appears that in case of leakage and spills, MTBE enters the groundwater because it is miscible with water.

The etherification reaction is an equilibrium-limited exothermic reaction:

$$(CH_3)_2C{=}CH_2 + CH_3{-}OH \longrightarrow CH_3{-}C(CH_3)_2{-}O{-}CH_3$$

Isobutene Methanol MTBE

$$\Delta H^{\circ}{}_{298.\text{liq}} = -37.5\ \text{kJ/mol}$$

Ethanol can be substituted for methanol to make ethyl *tertiary* butyl ether (ETBE) with the same process configuration (Slovnaft). Furthermore, isopentene may be used in addition to or instead of field butenes to make *tertiary* amyl methyl ether (TAME) or *tertiary* amyl ethyl ether (TAEE).

In the etherification reactions for the production of MTBE, ETBE, TAME, and TAEE, the acid catalyst type AMBERLYST (15Wet, 35Wet) has been used. AMBERLYST is a polymeric (resin) catalyst that is resistant to breakdown by osmotic and mechanical shock. The excellent etherification catalyst with higher acid strength improves its performance and prolongs its lifetime as compared to conventional catalysts (Dow Water & Process Solutions).

The isobutene feed usually is a mixed C_4 hydrocarbon stream containing about 30% isobutene. The other constituents are mainly *n*-butane, isobutane, and *n*-butene, which are inert under typical reaction conditions:

Temperature: 320–360 K
Pressure: 2.0 MPa
Conversion of isobutene: 90–97%, CD to 99%.
Catalyst: AMBERLYST
The reaction takes place in the liquid phase.

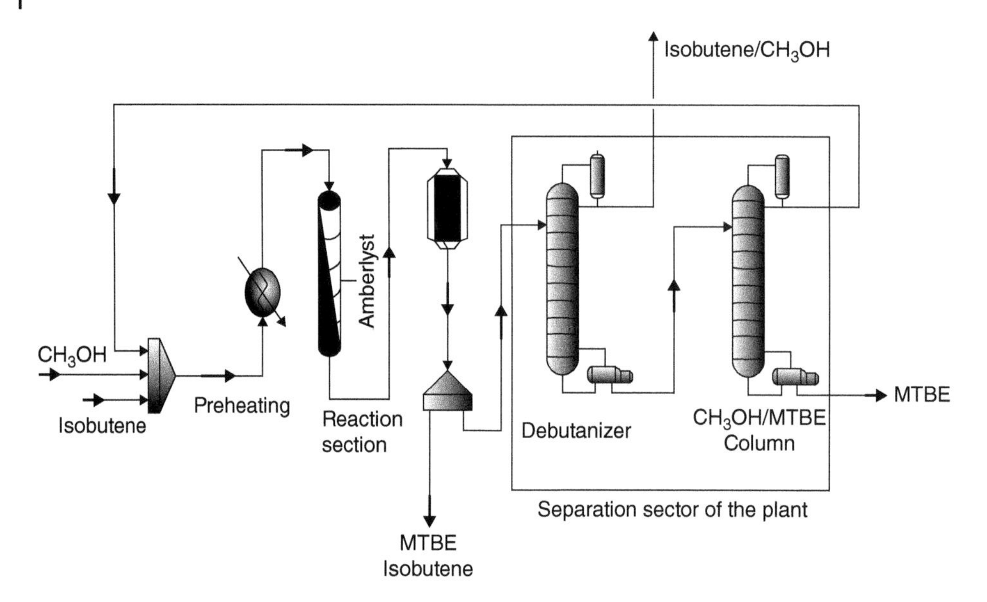

Figure 12.6 Production of MTBE by the conventional process.

With a conventional process employing fixed-bed reactors and slight excess methanol, it is possible to obtain isobutene conversions of 90–97%, which is slightly less than thermodynamic attainable conversion.

Separation of MTBE from the unconverted C_4-fraction and methanol is achieved by distillation. Because separation of isobutene from the other hydrocarbons is difficult, it usually leaves the process with the inerts. Therefore, an isobutene loss of 3–10% must be accepted (Figure 12.6). This is not the case with catalytic distillation.

12.2.8.1 Environmentally Friendly Process of Catalytic Distillation

The isobutene can be enhanced by continuous removal of the product MTBE. Catalytic distillation offers this possibility, thus increasing the isobutene conversion to up to 99%. (Figure 12.7) shows a schematic of this process. The catalyst is placed inside a distillation column. Feed methanol and isobutene come in contact with the catalyst. MTBE is formed and is continuously removed from the reaction section. Methanol, isobutene, and the inert hydrocarbons flow upward through the catalyst bed and then to a rectification section, which returns liquid MTBE.

An additional advantage of catalytic distillation is that the exothermic heat of reaction is directly used for the distillate separation of reactants and products. Hence, perfect heat integration is achieved: the heat of reaction will not increase the temperature, and operation in the reaction zone is nearly isothermal.

In practice, the catalytic distillation column is preceded by a conventional fixed-bed reactor (which is smaller than the one in the conventional process). In this reactor, the largest part of the conversion takes place at relatively high temperature (high rate, lower equilibrium conversion), while the remainder takes place in the catalytic distillation column at lower temperature (lower rate, high equilibrium conversion).

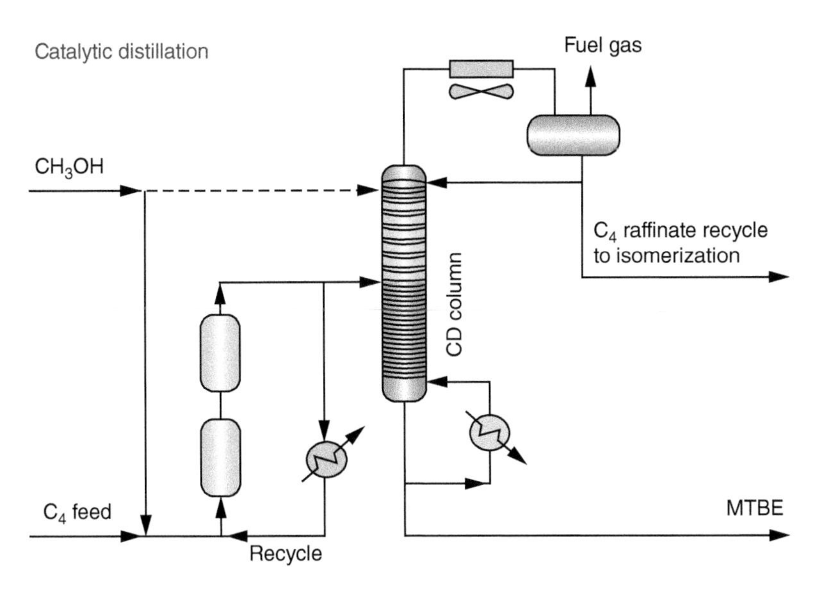

Figure 12.7 Production of MTBE by combine reaction and distillation.

A catalytic distillation column is a nice example of a so-called *multifunctional reactor*. This technology is very promising.

In many cases, considerable effort has been directed toward improving performance in all stages of system, from pretreatment of reagents to storage of the product. Such research is generally toward maximizing operating flexibility, increasing efficiency in the use of raw materials utilities, and minimizing waste. Catalysts have been improved, computer simulations have been developed to optimize parameters, and catalyst beds have been redesigned using a single bed with a catalyst of varying activity (structured reactors). Almost all oxidation plants have been renovated along these lines, and it is not unusual for actual productivities to be two to three times or higher than the plant's original productivity. Thus, although innovation requires improvements (in the sense of a radical change in production), much industrial research effort is still being devoted to incremental improvements, which requires thorough knowledge of the basic of the catalytic reaction and catalyst/reactor system. Reducing the number of steps in a chemical transformation is advantageous not only in terms reduced capital cost, especially to: (i) avoid storage of possibly dangerous products, (ii) reduce the risks and the cost of environmental control in complex plants, and (iii) increase process flexibility.

Sometimes, simplification of a process just involves eliminating a purification step, so that the product of one reactor goes straight into another without isolation of an intermediate. We should therefore conclude that petrochemistry will witness a very important evolution, without revolution, but which will affect both the processes and procedures utilized, the objective being to produce "clean" petrochemicals.

12.2.9 Etherification of Glycerol by Isobutylene

Technological processes and reaction conditions used for ethers, which are currently commercially produced as gasoline additives (MTBE, ETBE, TAME, TAEE) were presented in the previous section. The next two ethers used as additives are applied in diesel fuel. The application of these ethers in diesel fuel could have a highly positive effect on the decrease of diesel engine emissions. From the ethers of polyhydroxy alcohols, the main interest is focused on the ethers of glycerol. Glycerol is etherified with an appropriate O-alkylation agent (*tert*-butyl alcohol, isobutylene, and mainly C_4 fraction) using acid catalysts and temperatures in the range of 50–120 °C. The glycerol formed as a by-product with biodiesel production can be utilized for ether preparation.

The literature search is focused on measurement and evaluation of the different parameters and their influence on glycerol etherification (temperature, molar ratio, pressure, time), selection of the alkylation agent, investigation of the effect of water present in the catalyst, estimation of kinetics parameters for all the reactions occurring in the reaction system, as well as preparation of ethers from preferred reagents in the case of commercial production of tert-butyl ethers of glycerol (e.g. purified glycerol from biodiesel production and C_4-fraction without 1,3-butadiene) (Figure 12.8).

For mixtures of higher ethers and diesel (and/or biodiesel), motor emission tests were performed. From the results of the tests it can be concluded that there is no negative influence of glycerol ethers on fuel properties. Furthermore, the tests indicated a positive

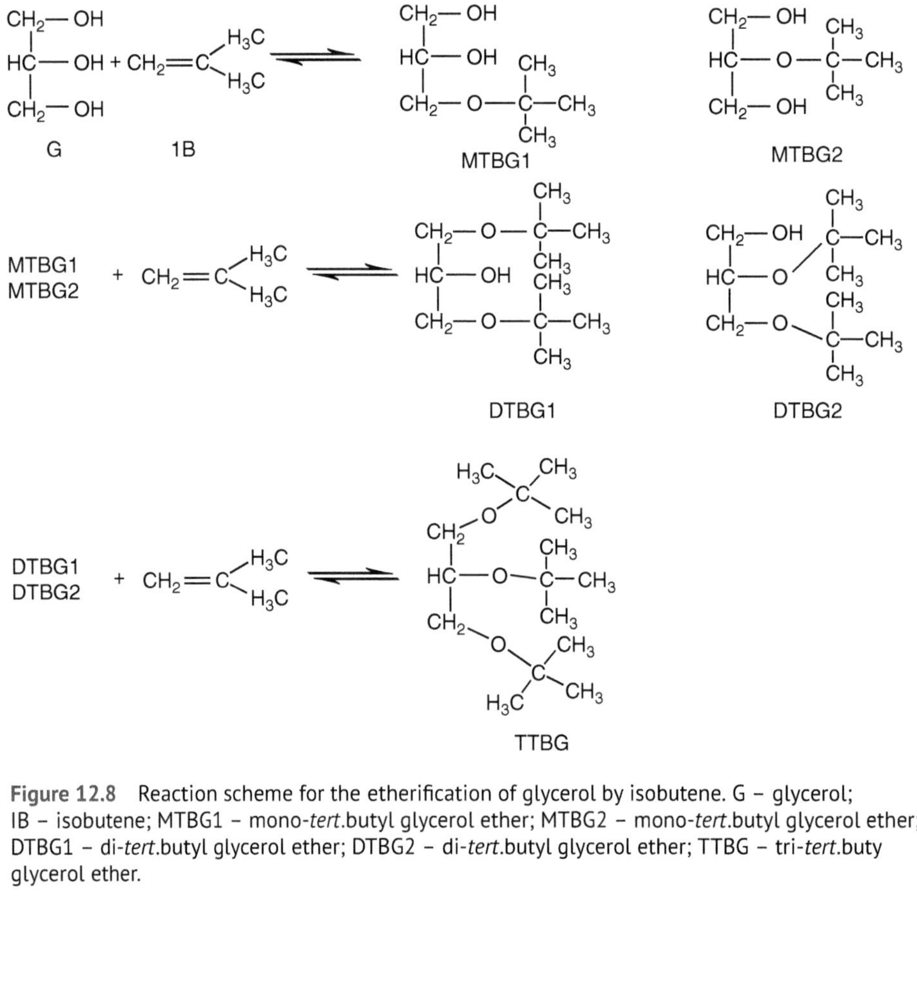

Figure 12.8 Reaction scheme for the etherification of glycerol by isobutene. G – glycerol; IB – isobutene; MTBG1 – mono-*tert*.butyl glycerol ether; MTBG2 – mono-*tert*.butyl glycerol ether; DTBG1 – di-*tert*.butyl glycerol ether; DTBG2 – di-*tert*.butyl glycerol ether; TTBG – tri-*tert*.buty glycerol ether.

influence on decrease of exhaust emissions, especially CO and particulate matter (PM) (Section 3.3.3.1).

Various types of heterogeneous catalysts (AMBERLYST 35, 36, 39, 31, and 119, zeolites H-Y, CBV 720, H-Beta, CP814E, Zeolyst Int.) and homogeneous catalysts (sulfuric, methanesulfonic and para-toluenesulfonic acids) with different properties were used. The research is mainly concentrated on ion-exchange resins, which are currently the most applied catalysts for etherification reactions. Their benefits, in addition to high catalytic activity, are simple separation from the reaction mixture, low toxicity, and relatively low price.

The most suitable etherification agent is isobutene, with which no reaction water is formed, as is the case for the etherification agent *tert*-butyl alcohol. Experiments are performed to eliminate the external mass transfer by varying stirring intensity. Stirring intensity is one of the most important factors having a major influence on the etherification reaction, due to the fact that the reaction mixture consists of two liquid phases with different polarity at the start of the etherification reaction and due to the use of a heterogeneous catalyst.

The most appropriate reaction temperature is in the range between 60 and 90 °C, depending on the type of etherification agent and glycerol purity. At higher temperatures, side reactions are preferred (oligomerization of isobutene and dealkylation of ethers).

With increasing molar ratio isobutene/glycerol, the amount of higher ethers (primarily triether) was rising, but at the same time the amount of formed diisobutenes increased. The molar ratio of IB/G = 4 is appointed as optimal. At this molar ratio, the amount of mono-ethers in the reaction mixture was low.

In the O-alkylation with *tert*-butyl alcohol, water is formed as a by-product. Water has an inhibiting influence on the catalyst and entire reaction system, which caused a drop in reaction rate. Due to these phenomena, *tert*-butyl alcohol is not considered as a direct alkylation agent for industrial utilization.

The reaction rate increases with time, which is related to the cross-solubility of products formed in the two liquid phases, until the formation of one phase only. The formation of one phase occurs at a conversion of glycerol of approximately 70%. After reaching the total conversion of glycerol, only the etherification reaction to higher ethers occurs (the selectivity to higher ethers increases). At higher temperatures the reaction time is shorter, but selectivities of different products at equilibrium are lower.

The water content in ion-exchange resin catalysts is shown by inhibition of the reaction rate of the etherification reaction (water occupies the active centers of the catalyst). It was demonstrated that on highly crosslinked ion-exchange resins (AMBERLYST 15 and 35), the catalytic behavior decreased only slightly compared to less cross-linked ion-exchange resins (e.g. A 36 and A 39). This difference in decrease in catalytic behavior can be attributed to the cross-linking of the catalyst: the macropores of highly crosslinked ions. Exchange resins do not shrink in the presence of water, whereas the macropores of less-crosslinked ion exchange resins do. The best results were reached when AMBERLYST 35 in dry form (water <3 wt.%) was used.

The kinetics of the main and side reactions of etherification of glycerol and ethylene glycol with isobutylene were measured. The results represent the most complete kinetic parameters summary for these reactions compared to date in the literature. Reactions were performed in solvents (dioxane, sulfolane, or dimethylsulfoxide) to increase the solubility of all reagents in the reaction mixture. From the performed experiments, the pseudohomogenous kinetic model with neglected influence of diffusion was determined as the consistent model for describing the behavior of glycerol and ethylene glycol etherification reactions.

Other results from literature describe results of the preparation of tert-butyl ethers from purified glycerol from existing biodiesel production plants. Several types of refined glycerol are used. As an alkylation agent, isobutylene and C_4-fractions were tested. Refined glycerol from biodiesel production etherified with C_4-fraction seems to be the most feasible approach to commercial production of tert-butyl ethers of glycerol.

The research points out that *tert*-butyl ethers of glycerol satisfy the demands (e.g. effects of addition of *tert*-butyl ethers of glycerol on fuel properties and the environment) to become a high potential diesel and biodiesel additive. When produced from the glycerol waste stream from biodiesel production, the *tert*-butyl ethers of glycerol can be regarded as a product from renewable sources, which introduces the "green" aspect of the fuel additive for their consideration as a viable commercial fuel additive.

12.2.10 Fisher–Tropsch Synthesis

Fischer–Tropsch Synthesis (FTS) is another technology that converts syngas into basic petrochemicals and mainly fuel-range hydrocarbons. Like methanol to olefins (MTO), it offers the possibility of market diversification to remote natural gas resource holders. Syngas is catalytically converted into a broad range of products that mainly consists of linear alkanes, 1-alkenes, and a small amount of oxygenates (alcohols, ketones, and aldehydes).

Iron (Fe)- and cobalt (Co)-based catalysts are mainly used as the industrially relevant catalysts. In the mid-1920s, Franz Fischer and Hans Tropsch reported the catalytic formation of higher hydrocarbons and other organic compounds from syngas mixtures on a nickel (Ni) and Co catalyst. The process itself has been shown to be a satisfying alternative for the production of fuels in regions where crude oil is scarce while other carbonaceous sources are abundant. Today, FTS is industrially operated in only a few countries. Sasol operates four plants in South Africa, and Shell has one operational plant in Malaysia. Two plants are operational in Qatar: one, Oryx GTL (gas to liquids), is a joint venture between Qatar Petroleum and Sasol; the other, Pearl GTL, is a joint venture between Qatar Petroleum and Shell. The latter is the largest FTS plant in the world, with a capacity of 18 000 tons of kerosene per day and a total investment cost of $18 billion USD – which was a much higher cost than initially estimated. This emphasizes the rather high investment cost of the technology and project cost risks. In addition, the viability of the FTS process is highly dependent on both oil and gas price fluctuations. FTS is clearly a mature technology and is typically operated at two distinct operational modes, as described by Dry: the low-temperature and the high-temperature mode. The product spectrum depends mostly on the temperature mode. At a low temperature, the product distribution is shifted toward high-molecular linear waxes; at a higher temperature, the average molecular weight is less and larger amounts of 1-alkenes are formed, which is of interest for the production of light olefins. For the lower-temperature mode, a supported Co-catalyst is more suitable, as it has higher activity. The main design specification for a FT reactor is heat removal, as the occurring reaction is highly exothermic. Four reactor technologies have been designed and commercially used: the circulating fluidized-bed (CFB) reactor, the fixed-fluidized-bed (FFB) reactor, the multi-tubular fixed-bed reactor, and the slurry phase reactor. Process intensification has led to the Sasol advanced Synthol-FFB reactors, which replaced the older CFB reactor configuration. The reaction section is bigger and has a higher heat-removal capacity; in addition, lower pressure drops and higher throughputs are possible without affecting the isothermal operation of the reactor. This reactor technology is used for high-temperature FTS. The most advanced reactor design is the slurry bed. This reactor is designed to operate at low temperatures, which increases the selectivity to longer hydrocarbons and suppresses high methane yields. The selectivity is rather low and the product spectra cover a broad carbon range, as described by the Anderson-Schulz-Flory distribution. As a result, further US processing and refining are essential. For process improvement, a better fundamental understanding is required in order to improve the catalyst design and, in particular, to increase the poor resistance to sulfur poisoning.

12.2.11 Acetic Acid

Apart from food applications, acetic acid is used for the production of vinyl acetate, for cellulose acetate (thermoplastic and fiber constituent), and as a solvent. It can be produced from a number of raw materials. Commercially, nearly all non-food acetic acid is produced by one of the three processes shown in (Table 12.2).

All three reactions are based on homogeneous catalysis. The first two processes proceed through radical reactions. The chemistry of methanol carbonylation is fundamentally different.

12.2.11.1 Background Information

Until late in the nineteenth century, acetic acid was manufactured by the age-old process of sugar fermentation to ethanol and subsequent oxidation to acetic acid. Table vinegar occasionally is still produced by fermentation. In 1916, the first process for production of acetic

Table 12.2 Processes for the production of acetic acid.

Reaction	ΔH°_{298} (kJ/mol)
Oxidation of acetaldehyde	−292
$H_3C{-}C(=O)H + 1/2O_2 \rightarrow CH_3C(=O){-}OH$	
Direct oxidation of hydrocarbons (mainly naphtha and n-butane)	«O
$CH_3CH_2CH_2CH_3 + O_2 \rightarrow CH_3C(=O){-}OH$ + by-prodocts	
Carbonylation of methanol	−138
$CH_3OH + CO \rightarrow CH_3C(=O){-}OH$	

acid on an industrial scale was commercialized. This process was based on the liquid-phase oxidation of acetaldehyde. The acetaldehyde process is still widely applied, but since the development of the Monsanto methanol carbonylation process it is progressively losing ground.

12.2.11.2 Principal Reaction

The reaction unit of the Chiyoda CT-ACETICA process operates at a moderate temperature range of 170–190 °C and pressure range of 3.0–4.5 MPa in the presence of a heterogeneous rhodium (Rh) catalyst and methyl iodide as a promoter in a bubble-column reactor. The usual acetic acid yields based on methanol and CO consumption are greater than 99 and 92%, respectively.

The basic chemistry of carbonylation is similar to that of a homogeneous catalyst in the conventional processes. The net reaction in the carbonylation of methanol is Carbonylation:

$$CH_3OH(l) + CO(g) \rightarrow CH_3COOH(l);\ \Delta H^\circ_{298} = -138\ \text{kJ}/_{\text{gmol methanol}}$$

CT-ACETICA requires no additional water to stabilize the active rhodium complex. Therefore, the reaction solution contains less than 8% water according to the following equilibrium reactions:

$$\text{Esterification}: CH_3OH + CH_3COOH \rightleftarrows CH_3COOCH_3 + H_2O$$

$$\text{Etherification}: 2CH_3OH \rightleftarrows CH_3OCH_3 + H_2O$$

Hydrogen iodide is formed in the reaction solution by hydrolysis of the metal iodide:

$$CH_3I + H_2O \rightleftarrows CH_3OH + HI$$

Certain amounts of methanol and acetic acid may react with the methyl iodide as shown here:

$$CH_3I + CH_3COOH \rightleftarrows CH_3COOCH_3 + HI$$

$$CH_3I + CH_3OH \rightleftarrows CH_3OCH_3 + HI$$

12.2.11.3 Catalyst Preparation Reactions

Promoter (CH_3I). The CT-ACETICA process uses methyl iodide as the promoter for the carbonylation reaction, as described earlier. The method of producing CH_3I in a methyl iodide generator is

$$I_2 + CO + H_2O \rightarrow 2HI + CO_2$$

$$HI + CH_3OH \rightarrow CH_3I + H_2O$$

Heterogeneous catalyst (rhodium immobilized on resin). The CT-ACETICA process is based on a heterogeneous rhodium catalyst. The nitrogen atoms of the resin pyridine groups become positively charged after quaternization with methyl iodide.

The active rhodium complex, $[Rh(CO)_2I_2]$, is immobilized by the ion exchange on the quaternized polyvinylpyridine resin. Because the ion-exchange equilibrium favors the solid phase, almost all Rh in the reaction mixture is immobilized on the resin support.

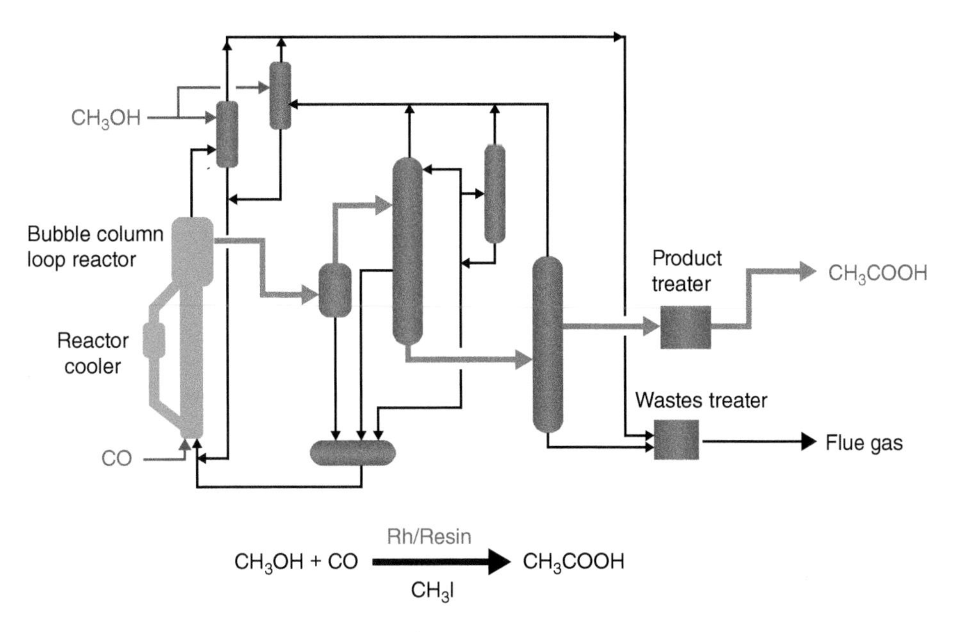

Figure 12.9 Flowsheet of the CT-ACETICA process.

12.2.11.4 Process Description

A simplified diagram of the process flow of the CT-ACETICA process is shown in Figure 12.9. The CO compressor compresses CO received by pipeline at the battery limit to the reaction level. After the moisture is removed, the CO is spared into the carbonylation reactor.

To enhance absorbing efficiency, fresh methanol is split into two streams, each of which is fed to a separate countercurrent HP absorber and LP absorber. Part of the methanol feed contacts the reactor offgas in the HP absorber, which mostly contains unconverted CO, methyl iodide, methyl acetate, and so on. The other part of the methanol feed stream contacts the light gases in the LP absorber, which were released at LP from the distillation unit. The main purpose of this absorption system is to maximize recovery of valuable methyl acetate and methyl iodide, which otherwise would exit the system with the vented gas, resulting in unnecessarily high chemical consumption and yield loss. The methanol feed streams exiting the absorbers are recombined and mixed with the recycled liquid from the recycle vessel and makeup methyl iodide from the MI generator unit. The combined and recycled stream is then charged to the bottom of the carbonylation reactor riser section. The carbonylation reactor, using a three-phase bubble-column system, consists of a riser, separator, downcomer, and reactor cooler. The reaction conditions are as follows:

Temperature: 170–190 °C
Pressure: 3.0–4.5 MPa

12.2.12 Hydroformylation

Hydroformylation, also referred to as *oxo synthesis*, was discovered in Germany before World War II in one of the many programs aimed at application of syngas from coal (CO/H_2)

mixtures. It was discovered that alkenes react with syngas, providing an appropriate catalyst is present. Formally, formaldehyde (CH_2O) is added to the double bond. Therefore, the reaction was called *hydroformylation* (analogous to *hydrogenation* being the addition of hydrogen). The most important application is the hydroformylation of propene, yielding two isomers:

$$\underset{\text{propene}}{CH_3CH=CH_2} + CO + H_2 \rightarrow \underset{\text{iso-butyraldehyde}}{CH_3\underset{\displaystyle CH_3}{\underset{|}{C}}H\overset{\displaystyle O}{\overset{\|}{C}}H} + \underset{n\text{-butyraldehyde}}{CH_3CH_2CH_2\overset{\displaystyle O}{\overset{\|}{C}}H}$$

$$\Delta H^{\circ}_{298} = -125\ \text{kJ/mol}$$

In most cases, the linear aldehyde is the preferred product because it enjoys a much larger market than the branched aldehyde. The major reason is that the aldehydes are for a large part used for the production of alcohols, and linear alcohols show a relatively high biodegradability.

Hydroformylation is not limited to propene. Linear and branched C_3–C_{17} alkenes are common feedstocks.

The products of hydroformylation usually are intermediates for the production of several types of alcohols. For instance, an important practical process is the production of 2-ethyl-1-hexanol from *n*-butyraldehyde (formed by hydroformylation of propene), as shown in Figure 12.10.

The alcohols produced by the oxo synthesis are commonly known as oxo alcohols. *n*-Butanol and 2-ethyl-1-hexanol are the most important products of hydroformylation. *n*-Butanol has three main uses: as a solvent in surface coatings, for the manufacture of butyl esters, and for the additivation of fuels.

2-Ethyl-1-hexanol is mainly used in plasticizers. Other oxo alcohols in the C_6–C_{11} range are also applied in the synthesis of plasticizers (in particular for PVC). A trend in this field of application is the use of longer chain alcohols in order to reduce the volatility of the plasticizers. The oxo alcohols in the C_{12}–C_{18} range are applied in detergents.

12.2.12.1 Thermodynamics

Hydroformylation requires low temperature and elevated pressure. As mentioned earlier, normal aldehydes are the preferred products.

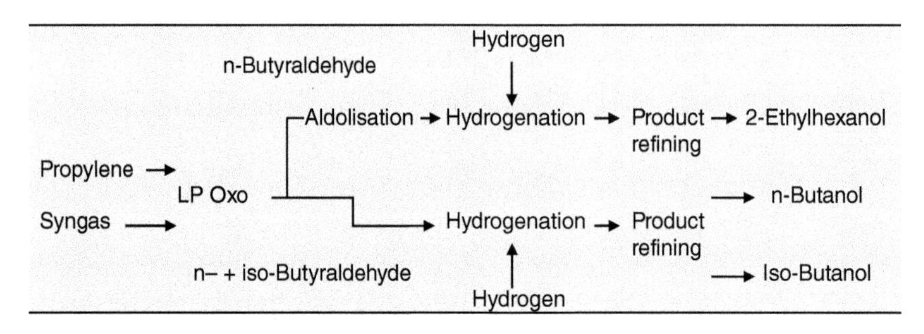

Figure 12.10 Reaction sequence for the production of 2-ethyl-1-hexanol.

The iso-aldehyde is the thermodynamically favored product. During hydroformylation, other reactions also can occur, in particular hydrogenation to propane. Thermodynamically, the latter reaction is the preferential reaction. Therefore, a catalyst is required that has a high selectivity.

12.2.12.2 Catalyst Development

The most important catalysts for hydroformylation are based on Rh and Co, which are introduced as carbonyls. Rh and Co are not the only metals showing catalytic activity, but they are the most active ones. The metal activity for hydroformylation (relative units) is shown in (Table 12.3). From this list, it is understandable that in practice only Rh and Co are used.

12.2.12.3 Catalytic Cycle

Polyorganophosphite-modified rhodium catalysts are very reactive and show good regioselectivity (selectivity to the straight chain aldehyde) in comparison with phoshine-modified catalyst such as triphenylphosphine (TPP). This process is often called the *low-pressure oxo process* (LPO). The probable sequence of the reaction with propylene to form normal butyraldehyde is shown in Figure 12.11; the figure shows the catalytic cycle for hydroformylation of propene by $HRh(CO)L_2$.

Table 12.3 Relative activities of metals for hydroformylation.

Rh	Co	Ru	Mn	Fe	Cr, Mo, W, Ni
10^3–10^4	1	10^{-2}	10^{-4}	10^{-6}	0

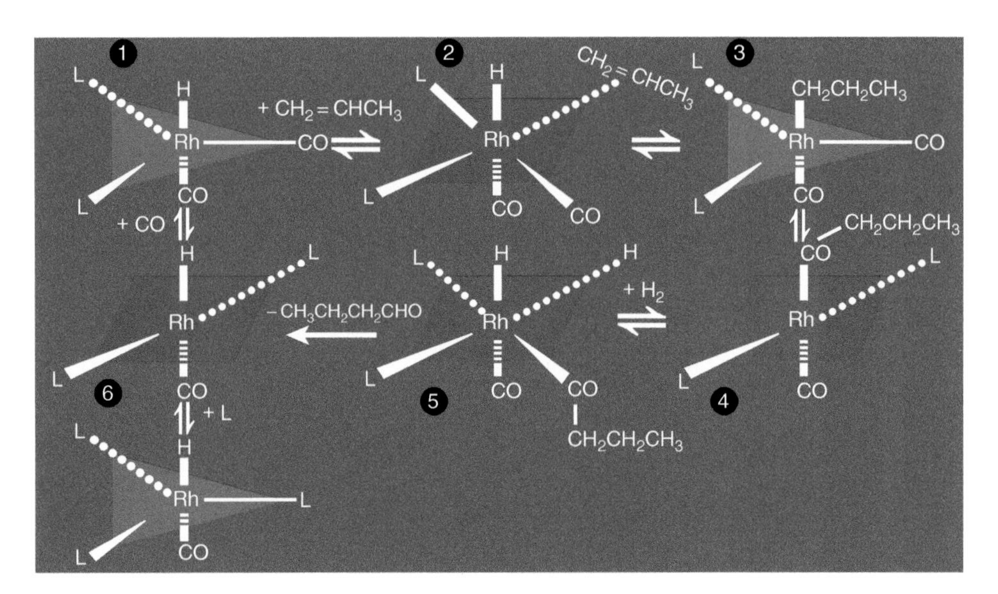

Figure 12.11 Catalytic cycle in the hydroformylation of propene. Probable reaction cycle for formation of normal butyraldehyde from propylene (L = triphenylphosphine (TPP); (1): product of reaction of ROPAC with CO and TPP; (2): addition product of (1) and propylene; (3): alkyl complex resulting from rearrangement of (2); (4): acyl complex resulting from CO insertion to (3); (5): dihydroacyl complex resulting from oxidative addition of hydrogen to (4); (6): product of elimination of butyraldehyde from (5)).

Rhodium is introduced to the oxo reactor in the form of a solution of ROPAC (a stable crystalline compound) in butyraldehyde. Complex (1) in Figure 12.11 is formed from fresh Rh in the presence of CO and TPP. In this coordination complex, the Rh atom carries five labile-bonded ligands: two TPP, two CO, and one hydrogen. In the first reaction step, a propylene ligand is added to form complex (2), which rearranges to the alkyl complex (3). This undergoes CO insertion to form the acyl complex (4). Oxidative addition of hydrogen gives the dihydroacyl complex (5). Finally, hydrogen transfers to the acyl group, and normal butyraldehyde is formed together with complex (6). Coordination of (6) with CO regenerates complex (1).

Some *iso*-butyraldehyde is produced along with the normal butyraldehyde, but a high selectivity to the latter is ensured by exploiting a steric hindrance effect as follows. The reaction is carried out in the presence of a large excess of TPP. Under the low-pressure conditions of the reaction, the high TPP concentration suppresses the dissociation of complex (1) into one containing only a single phosphine ligand. If largely undissociated complex (1) is present, with its two bulky TPP ligands in contact with propylene, then a high proportion of primary alkyl is favored – if fewer such ligands were present, then more propylene would form secondary alkyl groups, leading to more *iso*-butyraldehyde.

12.2.12.4 Kinetics

Kinetic studies have been performed for the hydroformylation reaction. For a simple system (the only ligands were CO), the following rate expression has been found:

$$r_{\text{hydroformylation}} = k[\text{Rh}][\text{H}_2][\text{CO}]^{-1}$$

For a Rh system containing triphenylphosphine ligands, the following rate equation was reported:

$$r_{\text{hydroformylation}} = k\,[\text{Rh}][\text{PPh}_3]^{-0.7}[\text{alkene}]^{-0.6}[\text{H}_2]^{0.05}[\text{CO}]^{-0.1}$$

12.2.12.5 Process Flowsheet

Figure 12.12 shows a simplified flow diagram of the basic process of rhodium-catalysis hydroformylation of propene (LPO process). (The catalyst ligands are triphenylphosphines, approximately 75 mol/mol Rh).

A feed of alkene (fresh feed or recycle light ends) is reacted with a 50/50 mixture of CO and hydrogen at:

Pressure of 3.0 MPa (g)
Temperature in the range 95–180 °C
Conversion per pass ca. 30%

This produces an aldehyde containing one carbon atom more than the feed alkene.

The gaseous reactants are fed to the reactor through a sparer. The reactor contains the catalyst dissolved in product butyraldehyde and by-products (e.g. trimers and tetramers). Rhodium stays in the reactor, apart from a small purification cycle. The reaction is exothermic and is carried out in a train of equal-sized water-cooled reactors. The number of reactors is determined by the required plant capacity.

The gaseous reactor effluent passes through the demister in which fine droplets of catalyst contained in the product gases are removed and fed back to the reactor. The reaction products and unconverted propene are condensed and fed to the gas–liquid separator.

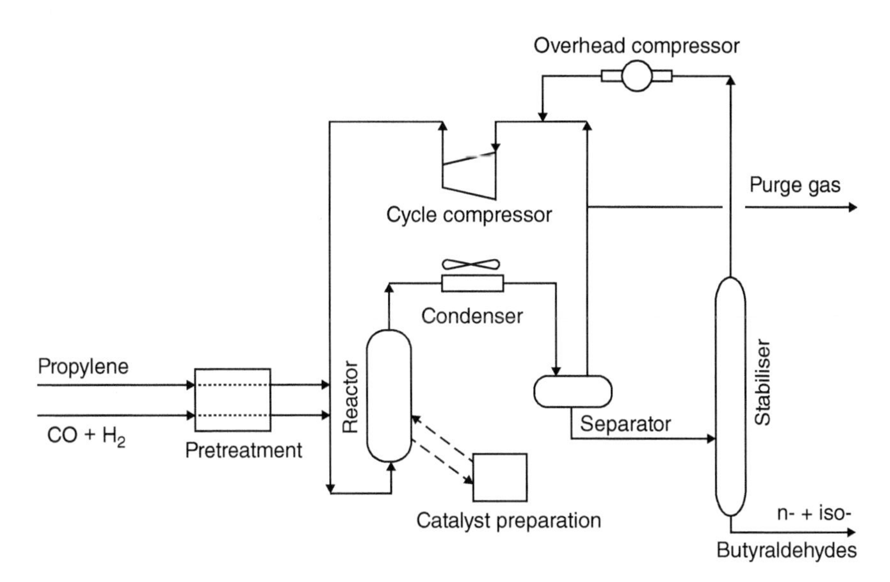

Figure 12.12 Hydroformylation of propene (LPO process) flowsheet.

The first industrial oxo process was based on an unmodified cobalt catalyst: $HCo(CO)_4$. This catalyst system requires high pressure and temperatures. Hydroformylation of higher alkenes to produce plasticizers and detergent range alcohols still uses this type of catalyst (Kuhlmann hydroformylation process).

12.2.12.6 Comparison of the Hydroformylation Process

Table 12.4 summarizes the characteristics of the major hydroformylation processes. The rhodium catalyst is attractive for the conversion of propene to the linear aldehyde.

Table 12.4 Comparison of hydroformylation processes.

Catalyst	Rh/phosph.	Rh/phosph., water-sol.	$HCo(CO)_4$	Co/phosph.
Pressure, MPa	2	5	20	7
Temperature (K)	370	390	410	440
Alkene	Terminal C_3	Terminal C_3	Intern. C_3–C_{10}^+	Intern.C_3–C_{10}^+
Product	Aldehyde	Aldehyde	Aldehyde	Alcohol
Linearity (%)	70–95	95	60–80	70–90
Alkane by-product (%)	0	0	2	10.15
Metal deposition	No	No	Yes	Yes
Heavy ends	Little	Little	Yes	Yes
Poison sensitivity	High	High	Low	Low
Metal costs	1000	1000	1	1
Ligand costs	High	?	Low	High
Company	Union Carbide, Davy Powergas, Johnson Matthey	Ruhrchemie, Rhône-Poulenc	Kuhlmann	Shell

Selectivity is approximately 95%. The branched product also has a market, although a much smaller one than the linear product, and its production is, therefore, desired in these quantities. Hydroformylation of higher alkenes with internal double bonds requires isomerization activity and, as a consequence, in that case cobalt catalysts are preferred. In all processes except one, aldehydes are the primary products. These can be hydrogenated to alcohols in a separate reactor. The modified cobalt catalyst is used for the direct production of alcohols rather than aldehydes, because this catalyst promotes hydrogenation.

Further Reading

de Almeida, E.L.F., Dunham, F.B., Bomtempo, J.V., and Bicalho, R.G. *The Renewal of Gas to Liquids Technology: Perspectives and Impacts*. Brazil: Instituto de EconoUFRJ edmar@ http://ie.ufrj.br.

Amin, N.A.S. and Angoro, D.D. (2004). *Fuel* 83: 487–494.

Arata, K., Matsuhashi, H., Hino, M., and Nakamura, H. (2003). *Catal. Today* 81: 17–30.

Bajus, M. (2007). *Petroleum and Coal* 49 (1): 1–9.

Chou, L., Cai, Y., Zhang, B. et al. (2003). *Appl. Catal. A General* 238: 185–191.

Energetika (2004). Štatistický úrad Slovenskej republiky, Bratislava, 2005. www.uspto.gov

Iglesia, E., Spivey, J.J., and Fleisch, T.H. (eds.) (2001). Studies in surface science and catalysis. In: *Natural Gas Conversion VI*, vol. 136, 561. Amsterdam: Elsevier.

Kawabe, T., Tabata, K., Suzuki, E. et al. (2001). *Catal. Today* 71: 21–22.

Klepáčová, K., Mravec, D., Hájeková, E., and Bajus, M. (2003). *Petroleum and Coal* 45 (1–2): 54–57.

Klepáčová, K., Mravec, D., and Bajus, M. (2005). *Appl. Catal. A Gen.* 294: 141–147.

Klepáčová, K., Mravec, D., and Bajus, M. (2006). Etherification of glycerol with tert.-butyl alcohol catalysed by ion-exchange resins. *Chemical Papers*: 224–230.

Know, S.H., Shin, J.W., Oh, J.K. et al. (1999). MTBE – a chemical under fire. *Hydrocarbon Processing* (January): 73–75.

Liu, X., Linghu, W., Li, X. et al. (2006). *Appl. Catal. A* 303: 251–257.

Rahmin, I.I. (2003). Gas to liquid technology. Paper presented at the 26th IAEE Annual International Conference, Prague, June.

Rahmin, I.I. (2005). Paper presented at AchemAmerica 2005: International Exhibition and Conference on Process Technologies, Mexico City.

Ratan, S. and Vales, C.F. (2002). *Hydrocarbon Processing*: 57–64.

Takemoto, T., Tabata, K., Teng, Y. et al. (2001). *Catal. Today* 11: 47–53.

Teng, Y., Tabata, K., Yamaguchi, Y. et al. (2001). *Catal. Today* 71: 37–45.

Wolcox, E.M., Roberts, G.W., and Spivey, J.J. (2003). *Catal. Today* 88: 83–90.

Wood, D., Towler, B., and Nwaoha, C. (2012). Gas-to-liquids (GTL): a review of an industry offering several routes for monetizing natural gas. *Journal of Natural Gas Science and Engineering* 9: 196–208.

Zaman, J. (1999). *Fuel Process. Technol.* 58: 61–68.

13

Hydrogen Technologies

CHAPTER MENU

13.1 Introduction

Hydrogen and catalysts became powerful tools in the natural gas and petroleum sectors in the last century. Catalysis with focus on hydrocarbon conversion and formation today covers a broad range of processes related to the upgrading of crude oil and natural gas. Although it is often said that the field of gas to liquid (GTL) technologies (Fischer–Tropsch synthesis [FTS]) is mature and there are not many directions for research, the increasing demand for natural gas has made gas-conversion technologies a challenging task for refiners as well as for researchers. Therefore, the importance of catalysts and hydrogen is not only focused on

Petrochemistry: Petrochemical Processing, Hydrocarbon Technology, and Green Engineering,
First Edition. Martin Bajus.

development; in addition, tremendous work has to be done for catalyst formulation and hydrogen production and storage. This chapter reviews basic aspects dealing with actual and future trends in hydrocarbon technologies for:

- Hydrogen/biohydrogen production
- On-board fuel reforming
- Hydrogen storage
- Fuel cells
- Natural gas conversion/gas conversion (GTL)

Thus, specific processes or a sequence of them can be selected or approved. The major objective is to cover the following elements of enormous interest in today's hydrocarbon technologies:

- Recent advances in hydrogen technologies and GTL technologies based on the literature
- Limitations when using hydrogen technologies

By *hydrogen production*, we mean extracting and isolating hydrogen in the form of independent molecules, at the level of purity required for a given application. The process is naturally depend on the starting point, and the currently dominant scheme of production from methane only makes sense if the energy is initially contained in methane or can easily be transformed to methane. Thus, in the case of fossil fuels, the transformation of natural gas into hydrogen is relatively easy and that of oils a bit more elaborate, while transformation of coal requires an initial step of high-temperature gasification.

To reduce dependence on imported oil, a number of strategies are under consideration including the increased use of gasoline hybrid vehicles in the near term. For the long term, however, petroleum substitution is required, and that necessitates the development of alternative energy carriers. Hydrogen has the potential to be an attractive alternative energy carrier, particularly for the transportation sector. It can be clean, efficient, and derived from diverse domestic resources, such as renewables (biomass, hydro, wind, solar, and geothermal) as well as fossil fuels and nuclear energy. In the case of fossil fuels, natural gas is likely to be used for the distributed production of hydrogen in the near term, before the infrastructure required for centralized hydrogen production and hydrogen delivery is developed. In the long term, centralized production, using coal with carbon sequestration or nuclear energy (through high-temperature water splitting or thermochemical cycles) could be employed to produce hydrogen using a number of delivery options. Hydrogen can then be employed in high-efficiency power-generation systems, including internal combustion engines or fuel cells for both vehicular transportation and distributed electricity generation.

There are three primary barriers that must be overcome to enable industry commercialization of hydrogen fuel cell vehicles: (i) on-board hydrogen storage systems are needed that allow a vehicle driving range of greater than 500 km while meeting vehicle packaging, cost, and performance requirements; (ii) fuel cell system cost must be lowered to $30 per kilowatt while meeting performance and durability requirements; (iii) the cost of safe and efficient hydrogen production and delivery must be lowered to be competitive with gasoline independent of production pathway and without adverse environmental impact.

This part provides an overview of today's and future trends in using hydrocarbon technologies and related hydrogen technologies. This chapter is devoted to hydrogen/bio-hydrogen production, focusing on on-board fuel reforming and vehicular hydrogen storage. Hydrogen storage is widely recognized as a major technological barrier to the successful implementation of fuel cells for transportation and portable applications. The chapter shows that GTL technology is undergoing a renewal of its innovative process. An overview is first provided, to give a general understanding of on-board fuel reforming and hydrogen storage.

13.2 Hydrogen as an Alternative Fuel

Pollution from automobiles is, particularly in urban areas, becoming increasingly unacceptable to people living in, visiting, or working in the cities of the world. Demands for zero-emissions have been voiced, and the automobile industry is facing louder and louder criticism for not addressing the problem. The simplest solution to reducing emissions is a hydrogen fuel-to-wheel efficiency of about 36%, to make the vehicle more efficient. For fuel cell cars, a hydrogen fuel-to-wheel efficiency of 36% implies a fuel-to-wheel efficiency of around 25% for the chain starting from hydrogen production from natural gas, over proton exchange membrane (PEM) fuel cells and electric motors to wheels, all for a standard mixed driving cycle.

On a weight basis, hydrogen has nearly three times the energy content of gasoline (120 MJ/kg for hydrogen versus 44 MJ/kg for gasoline). However, on a volume basis, the situation is reversed, and hydrogen has only about a quarter of the energy content of gasoline (8 MJ/l for liquid hydrogen versus 32 MJ/l for gasoline). On-board storage in the range of 5–13 kg (1 kg hydrogen = 3785 l of gasoline energy equivalent (GGE) of hydrogen is required to encompass the full platform of light-duty automotive fuel cell vehicles. Engine power plants with efficiencies less than PEM fuel cells would require a larger payload of hydrogen to achieve a comparable driving range.

The Freedom Concil for Automotive Research (CAR) partnership was expanded in 2003 to include major energy companies (BP America, Chevron Corporation, ConocoPhillips, ExxonMobil Corporation, and Shell Hydrogen) and is now known as the Freedom CAR and Fuel Partnership. The performance targets developed are system and application driven, based on achieving similar performance and cost levels as current gasoline fuel storage systems for light-duty vehicles. The storage system includes the tank, valves, regulators, piping, mounting brackets, insulation, added cooling capacity, thermal management, and any other balance-of-plant components in addition to the first charge of hydrogen and any storage media such as solid adsorbent or liquid used to store the hydrogen.

Currently, research priorities are on achieving the volumetric and gravimetric capacity targets in Table 13.1, while also paying attention to energy and temperature requirements for hydrogen release as well as kinetics of hydrogen charging and discharging. It is important to note that to achieve a system level of 2 kWh/kg (6 wt.% hydrogen) and 1.5 kWh/l (0.045 kg hydrogen/l), the gravimetric and volumetric capacities of the material/media alone must clearly be higher than the system-level targets. To restate, development of hydrogen storage material/media (e.g. metal hydride, carbon nanostructured material) that meets 6 wt.% or 45 g/l is not sufficient to meet the system targets. Depending on the material and on the system design, material capacities may need to be a factor

Table 13.1 US DOE hydrogen storage system performance targets.

Storage parameter	Units	2007	2010	2015
U.S. DOE technical targets for on-board hydrogen storage systems				
System gravimetric capacity:	kWh/kg	1.5	2	2
usable, specific-energy from H_2 (net useful energy/max system mass)	kg H_2/kg	(0.045)	(0.06)	(0.06)
System volumetric capacity:	kWh/l	1.2	1.5	2.7
usable energy density from H_2 (net useful energy/max system volume)	kgH_2/l	(0.036)	(0.045)	(0.081)
Storage system cost	$/kWh net	6	4	2
(and fuel cost)	($/kg H_2)	(200)	(133)	(67)
	$ gge at pump	–	2–3	2–3

of 1.2–2 times higher than capacity targets. Given the wide number of options for specific materials and system designs, only system-level targets are specified.

13.2.1 Production of Hydrogen

Hydrogen is the most abundant element in the universe. It is produced on a large scale mainly by steam reforming, partial oxidation, coal gasification, and electrolysis. Its current worldwide production is around 5×10^{11} Nm^3 per year. It is primarily used as feedstock in the chemical industry: for instance, in the manufacture of ammonia and methanol, and in refinery reprocessing and conversion processes. However, with environmental regulations becoming more stringent, there is now growing interest in the use of hydrogen as an alternative fuel. Its combustion does not result in any emissions other than water vapor (although under certain air hydrogen ratios, NO_x can also be produced), and hence it is the least polluting fuel that could be used in an internal combustion engine. It can also be used in a fuel cell for the production of electricity for stationary application and mobile electric vehicle operations.

Different technologies are suited for application at scale determined both by the type ⇄ of application and by the characteristic of the technology itself. If the cost of given technology exhibits an economy of scale, it is preferred to use that technology in a centralized fashion. If the technology is cheapest in smaller units (for fixed overall production), the situation is of course the opposite, but this is rare, and a more common situation is that the cost is insensitive to scale of production. This allows the kind of applications aimed for to determine the scale employed. However, there may also be specific scale requirements set by the type of usage. For example, technology for passenger cars must have a size and weight suitable for typical motor vehicles. Here one finds little flexibility, due to existing infrastructure such as roads, sizes of garages and parking spaces, etc. Generally, if a new technology requires changes in infrastructure, both the cost and the inconvenience associated with changing to it must be taken into consideration.

The identification of hydrogen production by reverse operation of low-temperature fuel cells as a technology not suffering by small scale of operation explains the interest in developing this technology for dispersed (decentralized) employment.

13.2.1.1 Dry Reforming, Methane, and CO_2 Chemical Transformation

As an alternative to conventional steam reforming, methane could be reformed in carbon dioxide steam rather than water steam:

$$CH_4 + CO_2 \rightleftarrows 2\,CO + 2\,H_2 \qquad \Delta_r H_{293} = 247.3\ \text{kJ/mol} \tag{13.1}$$

Advantages of this reaction could be the disposal of CO_2 and the possibility of operating at fairly low temperatures, for example in combination with convention steam reforming (Section 12.2.1).

Syngas production technology plays a central role in petrochemistry. The methane reforming reaction with CO_2 (dry reforming) was first studied by Fischer and Tropsch (1928) over a number of base metal catalysts. Calculations indicate that the reaction is thermodynamically favored above 913 K. It is more endothermic than steam reforming. Carbon deposition is predicted, and carbon formation over metal catalysts during the CO_2 reforming reaction has been observed. An industrial process, the Calcor process, has been developed, which uses methane and a large excess of CO_2 to make CO-rich synthesis; a nickel-based catalyst is indicated. The SPARG process is essentially the same as a conventional steam methane reformer for the addition of sulfur to the catalyst.

Recently, renewed attention in both academic and industrial research has been focused on the CO_2 reforming reaction. The main problem of the CO_2 reforming reaction is that CO_2 is at the bottom of a potential energy well, and its use in the dry reforming reaction requires a very large energy input. Two themes in particular are under investigation:

i) Development of new catalysts that are inactive toward the carbon formation reactions. The research employed the La_2NiO_4-zeolite coupled membrane catalyst to combat the problem of catalyst deactivation due to coking in the CO_2 reforming of methane over nickel catalysts.
ii) The possibility of producing syngas with CO_2 reforming reactions performed as follows, rapidly separating the products (CO and H_2) from the reactants (CH_4 and CO_2).

In recent years, the CH_4/CO_2 reforming reaction leading to the formation of syngas with low H_2/CO ratios has received renewed attention in the context of natural gas upgrading. Although noble metals have been found to be active and selective catalysts, their high costs limit their application. Other metals, such as cobalt and nickel, have been reported as alternatives, but they are easily deactivated due to metal volatilization and carbon deposition. The use of alkaline earth compounds, such as calcium and magnesium oxides, to modify nickel has been found to be satisfactory. A series of complex oxide catalysts, such as $LaNiO_3$, $La_{0.8}Ca$ or $(Sr)_{0.2}NiO_3$, and $LaNi_{1-x}Co_xO_3$ (where $x = 0.2 - 1$) with a perovskite structure have been found to be resistant to coking. Bolt et al. have reported that a catalyst with a discontinuous interfacial $NiAl_2O_4$ layer showed less sintering of nickel. Recently, Gao et al. have reported that a La_2NiO_4 mixed oxide catalyst was more stable than $LaNiO_3$ at high temperatures (500–850 °C) and exhibited better activity in the methane dissociation reaction. Due to the interaction between the rare-earth and nickel entities, both nickel stability and catalyst life time are improved by the addition of rare earth to the nickel catalyst.

The coupling of catalysis with separation through membranes is becoming feasible for practical application. The technology is rather new and novel. According to Le Chatelier-Braun's law, if an external force leads to the separation of a specified component from a system at equilibrium, the system adjusts so as to minimize the effect of the applied force and thereby produces more of the removed component. Hence, by combining reaction and membrane separation in a membrane reactor, one can possibly enhance product selectivity and achieve higher conversion at lower temperatures. There are other advantages in the employment of membrane separation in a reactor, such as simplified downstream separation and recovery savings on energy consumption, and life-prolonging of reactor materials and catalysts.

It has been pointed out that with steam reforming of methane, up to 100% conversion can be accomplished in palladium-membrane reactors at temperatures as low as 500 °C, white carbon deposition can be avoided entirely. We have not come across any application of membrane reactors in CO_2 reforming of methane in the literature. To separate molecules other than H_2, palladium and its alloys are not suitable. Currently, composite membranes, consisting of thin layers of zeolite film on a porous supporting structure, are under development to cater to such needs. Molecules are separated on the basis of molecular size through these porous membranes. Both permselectivity and mechanical durability are essential factors, and the development of the sol–gel procedures for the deposition of thin microporous (<2 nm) and mesoporous (2–50 nm) layers on macroporous (>50 nm) supports has enabled the engineering of porous membranes suitable for industrial applications. Thin microporous layers have been obtained by depositing polymeric silica sols on γ-alumina supports. Yan et al. have prepared ZSM-5 membranes on porous alumina with good permselectivities (permeability at 185 °C : 10.1, 5.9, and 0.19×10^{-8} mol/m^2/s/Pa for hydrogen, n-butane, and isobutane, respectively) by in situ hydrothermal synthesis. Similarly, a continuous silicalite-1 membrane has been prepared on the inner wall of a porous alumina tube by Noble and Falconer. It goes without saying that a continuous zeolite layer without pinholes or cracks is crucial for reaching high permselectivities. Matsukata et al. have prepared a crackless zeolite layer on porous alumina. The thickness of the zeolite layers in membranes thus prepared is in the range of 40–500 μm. Further study of the mechanism of thin layer formation is likely to lead to thinner and more permeable membranes.

The use of membranes as intrinsically active catalysts or a catalyst support by depositing a catalytically active component on it has not been very successful in reactors due to low catalyst loading per unit volume of the reactor. Recently, collaboration has been established between Gao and Au in the application of the membrane techniques in the reforming of methane by CO_2. Work has been done to adopt the La_2NiO_4-Zeolite coupled membrane (thickness ~ 0.1 μm) being developed by Gao and coworkers in a microreactor for the reforming reaction. The relative permeance of the membrane is $CH_4 : CO_2 : CO : H_2 = 1 : 6 : 30 : 70$. The novel part of their approach is to combine catalysis with membrane separation: both La_2NiO_4 and zeolite membranes are deposited on the inner and outer walls of a porous alumina tube, and the catalytic reaction and product separation occur simultaneously in the reactor.

As far as the mechanism of the CH_4/CO_2 reforming reaction is concerned, there are quite a number of proposed schemes: Solymosi, Rostrup-Nielsen, Goula, and Bradford. According to the proposed schemes, the dissociative adsorption of CH_4 is considered to be the initial and rate determining step and CO_2 activation can be promoted by CH_x and H species on the surface. However, the CH_4/CD_4 results of Au et al. over Ni/SiO_2 and Rh/SiO_2 catalysts indicate that the reaction pathways for the formation of CO in the CO_2 reforming of methane can be represented as two different pathways for the formation of CO, i.e. the oxidation of CH_x species and the dissociation of CO_2; and the latter takes place prior to the former in the process of CO_2 reforming of methane over the SiO_2-supported nickel and rhodium catalysts. Similar conclusions have been drawn in studies of CH_4/CO_2 reforming over Ni/SiO_2 by Kroll et al.

Lee et al. have studied tri-reforming of methane to synthesize syngas with desirable H_2/CO ratios by simultaneous oxy-CO_2-steam reforming of methane. Results of tri-reforming of CH_4 by three catalysts (Ni/Ce-ZrO_2, Ni/ZrO_2, and Haldor Topsoe R-67-7H) showed that coke on the reactor wall and the surface of catalyst were reduced dramatically. It was found that weak acidic sites, basic sites, and redox ability of Ce-ZrO_2 play an important role in tri-reforming of methane conversion.

The mechanism and the rate-determining steps of CO_2 reforming of methane were investigated over the typical Ni/α-Al_2O_3 catalyst in a wide temperature range of 550–750 °C using steady-state and transient kinetic methods. The methane dissociation reached equilibrium with Ni—H species above 650 °C. The surface oxygen species originating from CO_2 became removable and reacted with CH_x species above 575 °C. The reaction of CH_x with CO_2 was slower than that of CH_4 dissociation above 650 °C, leading to durative carbon deposition on the catalyst. The formation of hydrogen is a rapid or equilibrium step in the reforming reaction.

13.3 Vehicle On-Board Fuel Reforming

The question of suitability for decentralization hydrogen production becomes more critical in the case of vehicle-integrated production systems, which must be economic at small scale. The basis for on-board production of hydrogen in principle could be fuels such as fossil fuels or biofuels, notably gasoline, as well as methanol, ethanol, and similar intermediate stages between fuels produced from natural organic or from more artificial industrial primary materials. Of these, only methanol can be obtained by a reforming process similar to that of natural gas, at moderate temperatures of 200–300 °C. Reforming of other hydrocarbons usually requires temperatures above 800 °C. Methanol is also interesting because of its similarity to gasoline in terms of fueling infrastructure. The methanol energy content of 21 MJ/kg or 17 GJ/m^3 is lower than that of gasoline, but because fuel cell cars are more efficient than gasoline cars, the fuel tank size will be similar.

Fuel cells (FCs) are electrochemical devices that convert the chemical energy of a fuel and an oxidant directly into electricity and heat on a continuous basis. A FC consists of an electrolyte and two electrodes. A fuel such as hydrogen is continuously oxidized at the negative anode while an oxidant such as oxygen is continuously reduced as the positive cathode. The electrochemical reactions take place as the electrodes to produce a direct electric current. FCs use hydrogen as a fuel, which results in the formation of water vapor only, and thus they provide clean energy. FCs offer high conversion efficiency and hence are promising. The current status of fuel technology for mobile and stationary applications has recently been discussed.

Among the various types of FCs, proton exchange membrane FCs (PEMFCs), solid oxide FCs (SOFCs), and molten carbonate FCs (MCFCs) have attracted considerable interest. SOFCs and MCFCs operate at high temperatures (around 973 K) and are used for stationary power generation. PEMFCs are primarily used for automotive applications. PEMFCs can be characterized into two categories: reformed and direct systems. Reformed systems required the use of an external reformer to reform fuel (methane, methanol, ethanol, gasoline, etc.) into hydrogen for use in the FC. In direct systems, the fuel is oxidized at the surface of the electrode without treatment. They have a low operating temperature (353 K), high current density, and low CO tolerance (10 ppm). They use hydrogen as fuel, and this can be supplied as pure hydrogen. Thus, FC vehicles can be equipped with pressurized hydrogen tanks, thereby ensuring a continuous supply of fuel. Alternately, hydrogen can be stored as a liquid in cryogenic tanks at 20 K. However, these ways of storing hydrogen are inconvenient. Moreover, the use of compressed hydrogen involves safety issues. Also, there is no proper infrastructure for hydrogen transport and distribution. Therefore, in practice, other hydrogen-containing fuels are used.

A number of hydrogen-generation routes have been explored. Methanol, ethanol, ammonia, naphtha, diesel fuel, and natural gas are some possible sources of hydrogen for FCs. In addition, petroleum distillates, liquefied petroleum gas (LPG), oil, gasified coal, and even gas from landfills and wastewater treatment plants can be processed to supply hydrogen. For stationary applications, natural gas is the fuel of choice due to its availability and ease of distribution. For automotive applications, naphtha and diesel fuel are the most convenient fuels since they can be easily transported. However, PEMFCs are very sensitive to impurities in fuel and have a sulfur specification less than 1 ppm, Gasoline has a 30 ppm sulfur standard in the United States, while hydrogen from coal gasification may contain 100–200 ppm sulfur. Catalytic cracking of ammonia generates a CO_2-free mixture containing 75% hydrogen. However, ammonia is toxic and poses a problem of generating nitrogen oxides during catalytic combustion of the cell effluent. Methanol, which is mainly prepared by syngas conversion, has a favorable H : C ratio of 4, is largely distributed, and is available in abundance. Moreover, it can be transported and reformed more easily than natural gas. However, its main drawback is its high toxicity. Ethanol is more promising since it is less toxic. It can also be more easily stored and safely handled. Most importantly, it can be produced in large amounts from biomass such as agricultural wastes and forestry residues and hence is a renewable resource, unlike methanol and gasoline. This could prove advantageous in tropical countries with warm climates where there are large plantations of corn and sugarcane. The bio-ethanol thus produced is free from sulfur, which otherwise may poison the FC catalyst. The use of biomass as a new feedstock for hydrogen production or energy source has attracted considerable attention in recent years, because it is an environmentally friendly and renewable rich source. Biomass of all types was used in Europe to produce energy accounting for 144.087 kilotons of oil equivalent in 2017. As a way of comparison, this means biomass used for energy is on its way to surpassing the European production of coal in the same period. The most promising option to generate hydrogen from bio-oil is via steam reforming followed by a water-gas shift reaction.

13.3.1 Steam Reforming of Naphtha (Gasoline)

Generating hydrogen by the steam reforming of hydrocarbons is a well-known technology. Tailoring the process for FC-powered vehicles, however, introduces additional constraints. Here, the purity of hydrogen becomes essential to avoid poisoning the electrodes of the FC system. This could necessitate a selective membrane separation technology. This can be accomplished using a palladium-silver membrane, which can be integrated directly into the reaction system (membrane reactor) or separately employed in later stages. The first choice (membrane reactor), however, offers attractive features regarding the energy density of the whole system, which is an important criterion to judge the performance of FC-powered vehicles. Compactness is also enhanced because conversion levels in such membrane reactors are increased. A significant amount of research work on membrane reactors is underway, and there are also signs of successful technological developments.

The flowsheet for the direct generation of hydrogen using steam reforming of naphtha on-board a vehicle, which is suggested here, is shown in Figure 13.1. The process can be described as follows: naphtha feedstock is vaporized and passed through a bed of zinc oxide to remove sulfur contamination, which is a potential poison for reforming catalysts and FC electrodes. Naphtha is then passed through the reformer, where it is catalytically reacted with steam to produce a mixture of steam, H_2, CO, CO_2, and CH_4. Heat is supplied to the reformer by the combustion of either a portion of the incoming naphtha or a portion of the generated hydrogen. The gas mixture is then cooled down by the incoming process water and directed to the low temperature (LT) shift reactor, where CO is converted catalytically to CO_2. The remaining traces of CO are further converted to CH_4 in the methanator. The gas mixture at this point contains about 53% H_2, 17% CO_2, 28% H_2O, and traces of CH_4. Both the shift and methanation reactions are exothermic. Before directing the product gas mixture to the selective membrane, it is compressed to 0.3 MPa. The pure hydrogen stream is then sent to the anode of the FC, and the unused H_2 from the FC is combusted to provide

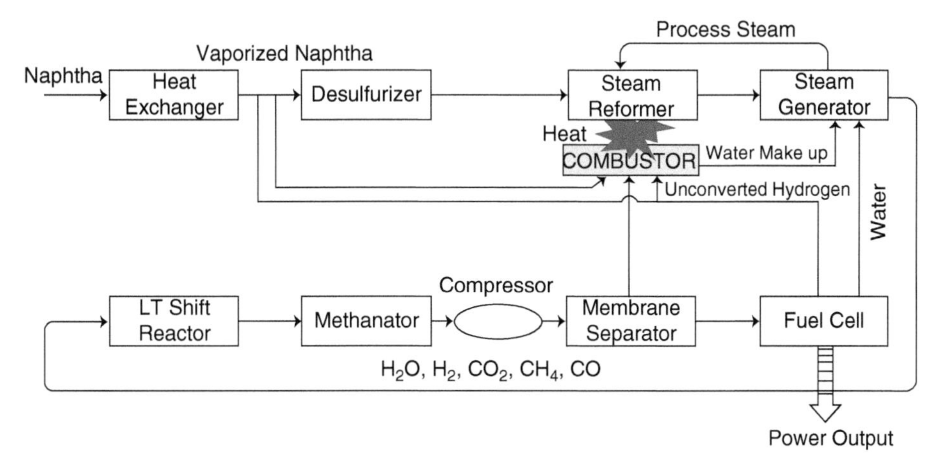

Figure 13.1 The flowsheet for the direct generation of hydrogen on-board a vehicle.

the required heat in the reformer. This unconverted H_2 is assumed to be 30% of the amount entering the FC (corresponding to 70% conversion in the FC). The other stream leaving the membrane separator, which has less H_2, undergoes a process to recover water. This water will be combined with the water recovered from the FC and circulated back to the reformer after being vaporized by the reformer products and then heated to the reformer temperature.

For the peak load of 50 kW, it is found that 14 l/h naphtha is needed, which means a 70 l fuel tank in the vehicle is sufficient for 5 hours of driving. The amount of water needed is not a critical factor, since it is generated in the FC and quantities of water can be kept at the minimum level. In the desulfurization step, it is found that about 1.6 l of a bed of ZnO is capable of handling a stream of naphtha with 1500 ppm of sulfur for 45 hours of continuous operation before regeneration or replacement based on operation at 1 MPa. Operation at a lower pressure level will increase the desulfurization catalyst requirements, maybe to a prohibitive level. Over the reformer liquid hourly space velocity (LHSV) range of 1–4 $hours^{-1}$, the amount of supported nickel catalyst varies from 14 to 4 l, respectively. For the LT shift reactor, the amount of catalyst required ranges from 4 to 60 l on going from 300 to 4000 $hours^{-1}$ typical gas hourly space velocity (GHSV). The catalyst here is CuO-ZnO supported on Al_2O_3. The last methanation step to remove traces of poisonous CO requires about 3.5 l nickel supported by various oxides. To selectively separate hydrogen, it is suggested to use a palladium–silver membrane, which is reported to give ultra-pure hydrogen.

13.3.2 On-Board Diesel Fuel Processing

The fuel gas for the SOFC stack is generated by catalytic partial oxidation (CPO) of diesel fuel, using a reformer without an additional water supply. A time near market entry is desired to use synergetic effects in FC improvement and system integration as well, and commercialization of FCs is expected to proceed first through SOFC systems for stationary and auxiliary power units (APU) application, initially using conventional fuels.

Webasto's APU consists of three major subunits. First, the fuel-processing unit has to generate a homogeneous fuel/air mixture for the reforming catalyst. Then, this catalyst converts the diesel fuel with oxygen into a hydrogen- and CO-rich fuel gas, which is fed to the SOFC stack. The reformer as well as the subsequent SOFC operate at the same temperature level of approximately 800 °C. Due to the high temperature level of the FC stack, the CO-containing gas can be fed directly to the SOFC without any shifting devices between the reformer and the FC. Downstream (DS) of the SOFC, the unconverted part of the fuel gas leaving the anode compartment is combusted in an afterburner with an integrated heat exchanger to use the thermal energy for preheating the cathode air. Auxiliary components like diesel pumps and air blowers complete the setup to a stand-alone system. All core components are located in one thermally insulated hotbox and operate near ambient pressure. This way, conventional blowers and pumps can be used, so no expensive mass-flow controllers or air compressors are necessary for operation. Figure 13.2 shows a block diagram of the system setup.

In general, different technologies can be used for the generation of a fuel gas for an SOFC from liquid fuels. The pyrolysis of hydrocarbons (Eq. 13.2) occurs at high temperatures, and besides hydrogen, solid carbon is formed, which accumulates in the reactor. The CPO with

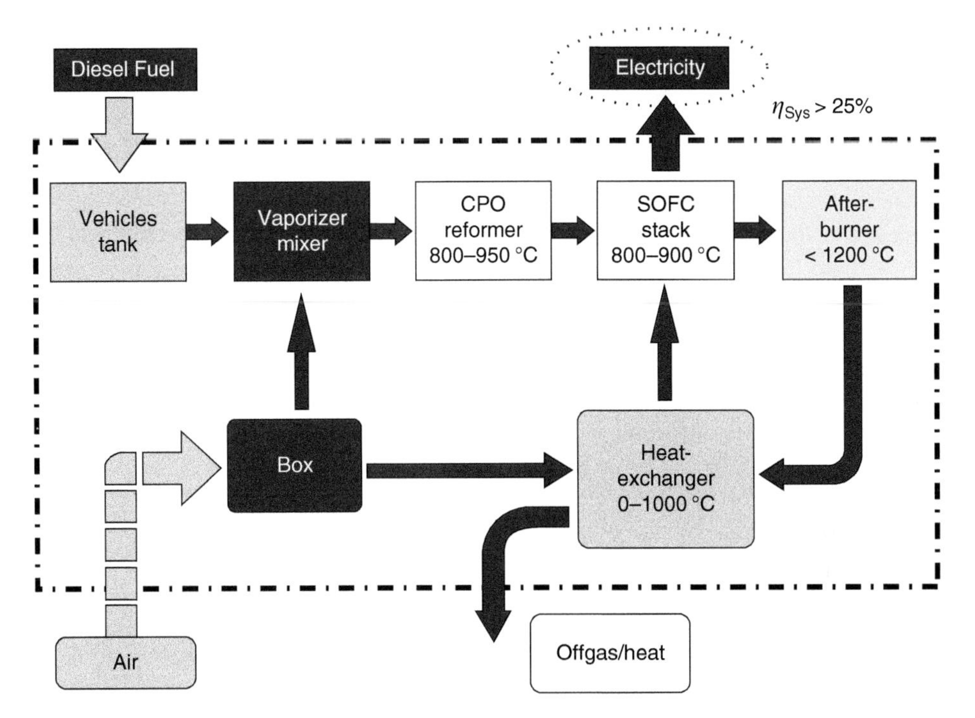

Figure 13.2 Setup of Webasto's SOFC-APU system.

a sub-stoichiometric ratio of oxygen to hydrocarbons is weakly exothermal, and H_2 and CO are the major products (Eq. 13.3). For endothermal steam reforming (SR), water is used for the conversion of the hydrocarbons (Eq. 13.4). Autothermal reforming (ATR) combines CPO and SR in such a way that a thermoneutral process results.

$$C_xH_y \rightarrow x\,C(s) + y/2\,H_2; -\Delta_RH < 0 \tag{13.2}$$

$$C_xH_y + x/2\,O_2 \rightarrow x\,CO + y/2\,H_2; -\Delta_RH > 0 \tag{13.3}$$

$$C_xH_y + xH_2O \rightleftarrows x\,CO + (x + y/2)\,H_2; -\Delta_RH < 0 \tag{13.4}$$

The CPO of hydrocarbons is chosen as reformer technology. Though the addition of water has a beneficial influence on the hydrogen content and decreases the catalyst temperatures due to the endothermal steam reforming reactions, the higher system complexity and the need for an additional reactant stream for SR and ATR are counterproductive.

The measurements at different air- umbers λ and molar system steam to carbon ratios S/C were made, to detect the most suitable ones for the CPO of diesel fuel. For this purpose, a reference fuel (C_{13}–C_{15} alkane blend, sulfur content <1 ppm per weight) was used with a boiling range between 235 and 265 °C. Experiments were carried out in a laboratory-scale reactor using a heating coil for water vaporization and a diesel fuel preheater. The fuel was heated to approximately 220 °C and mixed with the air/steam flow, which had a temperature higher than 350 °C. Thus, the fuel was completely vaporized and fed to the catalyst sample.

The fuel-processor units consist mainly of a metal-felt vaporizer with a ceramic glow plug, radial distributed air inlet holes, and downstream located pre-catalyst made from a catalytic monolith (Cordierite). For the system start, the reformer is operated in burner mode at an air number λ of approximately 1.3. After the temperature of the stack reaches 500 °C, the air number λ is reduced to values between 0.6 and 0.8 to prevent oxidation of the SOFC anode. The system is heated further until the desired operation temperature of approximately 850 °C is reached. At this time, a stack power of 340 W_e can be delivered at a stack voltage of 22 VDC.

At present, the seventh generation of a stand-alone APU system is under test, and the first results are very promising. At the same time, a further improved system is under construction that aspires to a net power of 1 kW_e and start times less than 1 hour. The system's overall efficiency is expected to be more than 20%.

13.3.3 Direct and Gradual Internal Reforming of Methane

SOFCs are promising candidates for generating power while preserving the environment. Nowadays, most SOFC developers use fuel containing a significant proportion of hydrogen. From this point of view, the most interesting fuel for SOFC systems remains methane (natural gas, biogas). Nevertheless, today natural gas is directly available whereas hydrogen must be produced, and thus direct internal reforming (DIR) is a very hopeful method. Indeed, gradual internal reforming (GIR) or DIR within a SOFC anode allows the conversion of methane into hydrogen without using a separate reformer. Such a concept is convenient for high-temperature FCs in which the steam reforming reaction can be sustained with catalysts. The reforming reaction at the anode and the water-gas shift reaction are carried out over a supported catalyst such as nickel. These reactions provide the system with the dihydrogen required by the electrochemical reaction. SOFC internal reforming is thus designed by closely coupling the catalytic reforming reaction and the electrochemical oxidation reaction within the atomic anode (Ni-YSZ) of the cell.

The anode material is a nickel and yttrium stabilized zirconia cermet (Ni-YSZ). DIR and also GIR involve adding the reforming function to the SOFC anode (Figure 13.3);

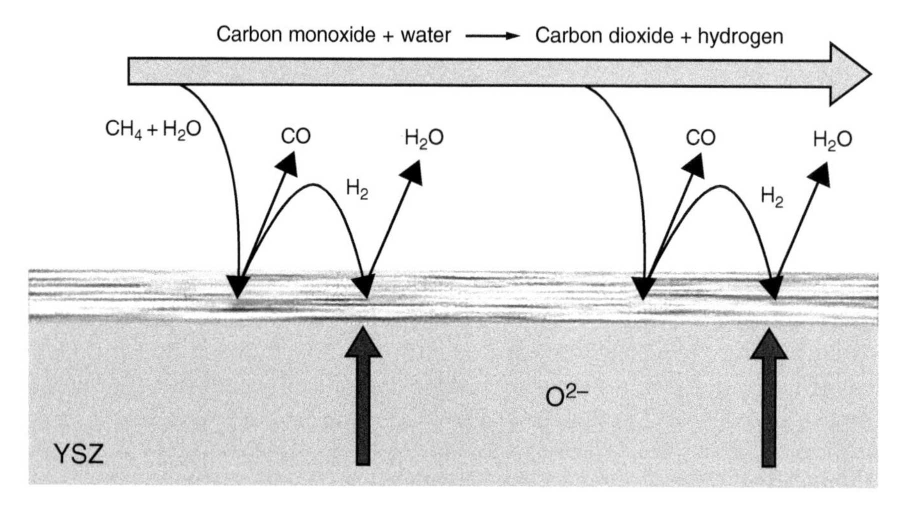

Figure 13.3 Direct internal reforming (DIR) or gradual internal reforming (GIR) of methane by steam.

direct production of hydrogen from methane is possible due to high temperature in the FC. Methane can thus be converted into H_2 and CO through the steam reforming reaction (Eq. 13.4) and the water-gas shift (Eq. 13.5), while hydrogen is electrochemically oxidized within thc anodc.

$$CO + H_2O \rightleftarrows CO_2 + H_2 \tag{13.5}$$

However, unlike DIR, where steam is fed in large amounts, the GIR process requires a very small quantity of steam since it uses the steam generated by the electrochemical reaction in the steam reforming reaction. Moreover, in DIR, because of the great difference between the reactions rates of the endothermic methane reforming reaction and the exothermic electrochemical hydrogen oxidation, cooling effects arise, resulting in a temperature drop at the cell inlet. With operation in GIR, a delocalization of the steam reforming reaction along the cell may occur; consequently, it would imply homogenization of the temperature gradient. However, at these working temperatures, Boudouard and cracking reactions can also be favored. Carbon formation is consequently possible, with the risk of carbon deposits polluting the anode. A carbon deposit on the anode surface can obviously block the fuel supply and the transfer of oxide ions, leading to a decrease in the power efficiency of the cell.

13.3.4 Methanol-to-Hydrogen Production

Compressed and liquefied hydrogen are limited for use in automobiles by low-energy new infrastructure for fueling, so there is a strong interest in converting conventional fuels to hydrogen on-board. It is primarily to avoid having to make large changes to current gasoline and diesel fuel stations that schemes based on methanol as the fuel distributed to the vehicle fuel tank have been explored. The energy density of methanol given earlier corresponds to 4.4 kWh/l, which is roughly half that of gasoline. Hydrogen is then formed on-board by a methanol reformer before being fed to the FC to produce electric power for an electric motor. The setup is illustrated in Figure 13.4. Prototype vehicles with this setup have been tested in recent years.

Methanol may eventually be used directly in a FC without the extra step of reforming to H_2. Such FCs are similar to PEM and are called direct methanol FCs (DMFCs). Over the last 40 years, there has been extensive research on DMFCs and direct ethanol FCs (DEFCs). A theoretical investigation into the comparison of direct methanol and direct ethanol FCs shows that DEFCs have higher theoretical energy densities compared to DMFCs. The energy density of DEFCs is 8.01 kW/kg, compared to 6.09 kW/kg for DMFCs. The major problem associated with using ethanol as a fuel is the low reaction kinetics of ethanol oxidation versus methanol oxidation.

As methanol may serve as a substitute for hydrogen in FCs and an intermediate fuel that can be used to produce hydrogen, its own production is of relevance and will be discussed in brief. Methanol may be produced from fossil sources such as natural gas or from biological material. Conventional steam reforming of natural gas may produce methanol.

Methanol steam reforming refers to the chemical reaction between methanol and water vapor for the production of hydrogen gas. This process is typically carried out in the

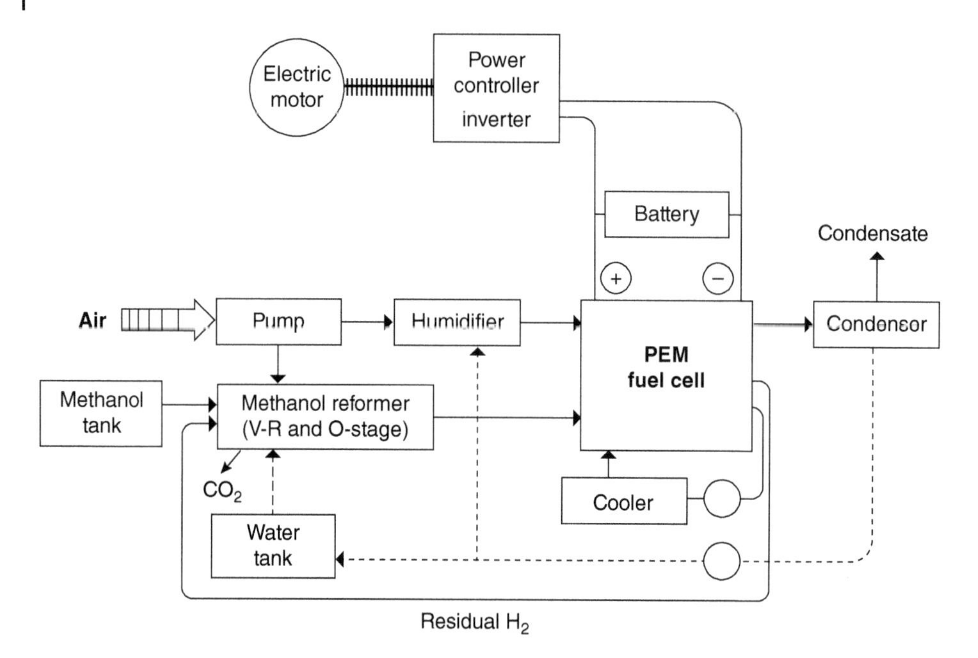

Figure 13.4 Layout of a methanol-to-hydrogen vehicle power system with a fuel cell and an electric motor.

presence of metal oxide catalysts at temperature ranging from 200 to 300 °C. The chemical reactions taking place during the reforming process are outlined as follows:

$$CH_3OH + H_2O \rightleftarrows CO_2 + 3\,H_2 \qquad \Delta H^\circ{}_f = 49.4\,kJ/mol \tag{13.6}$$

$$CH_3OH \rightleftarrows CO + 2\,H_2 \qquad \Delta H^\circ{}_f = 90.5\,kJ/mol \tag{13.7}$$

Reaction (13.6) is the main reforming reaction, which gives the stoichiometric conversion of methanol to hydrogen. It can be regarded as the overall effect of the methanol decomposition reaction to CO and H_2O and the water-gas shift reaction. The amount of CO as an intermediate product formed in the process is determined from the relative kinetics of these two reactions. An appropriate feed ratio of methanol to water, and proper control of temperature and pressure in the reformer, are required to minimize the amount of CO formed in the process. Steam reforming of methanol is an endothermic reaction. Therefore the temperature is difficult to control inside the reactor, which affects performance.

The water-gas shift reaction or its reverse may be operating as well,

$$CO_2 + H_2 \rightleftarrows CO + H_2O \tag{13.8}$$

with $\Delta H^\circ = -41$ kJ/mol. This could lead to CO contamination of the hydrogen stream, which is unacceptable for FC types such as the PEM or alkaline FCs (above ppm level) and only minimally acceptable (up to 2%) for phosphorous acid FCs. Fortunately, CO production is low at the modest temperatures needed for steam reforming, and adjusting the amount of surplus steam (H_2O) may force the reaction (Eq. 13.8) to go toward the left at a rate achieving the desired reduction of CO. At the high end of the temperature regime, this control of CO becomes more difficult. However, use of suitable membrane reactors with separate catalysts for steam reforming and water-gas shifts allows this problem to be overcome. A typical thermal efficiency of hydrogen formation by this method is 74%, with near 100% conversion of the methanol feed.

It is possible to make the process autothermal, i.e. to avoid having to heat the reactants, by adding the possibility of the exothermic partial oxidation process:

$$CH_3OH + ½\,O_2 \rightarrow CO_2 + 2\,H_2 \tag{13.9}$$

Here $\Delta H° = -155\,kJ/mol$. By suitable combination of (Eqs. 13.6 and 13.9), the overall enthalpy difference may become approximately zero. There are still problems with controlling the temperature across the reactor, because the oxidation reaction (Eq. (13.9)) is considerably faster than the steam reforming reaction (Eq. (13.6)). Proposed solutions include the use of a catalyst filament wire design leading to near-laminar flow through the reactor.

Catalysts traditionally used for the steam reforming process include a Cu—Zn catalyst containing mole fractions of 0.38 CuO, 0.41 ZnO, and 0.21 Al_2O_3, and a metallic Cu—Zn catalyst in the case of the wire concept.

A microchannel reactor with a catalyst coating in steam reforming of methanol is a promising candidate for portable electronics in order to achieve compactness in the structure of the fuel processor, along with advantages with respect to transient behavior, hydrodynamics, and heat and mass transfer characteristics. The main feature of microstructured reactors is the high surface-area-to-volume ratio in comparison to conventional chemical reactors (Chapter 16). The heat-transfer coefficient in this microchannel reactor is also significantly higher than that for traditional heat exchangers. The high heat-exchanging efficiency allows for carrying out reactions under isothermal conditions. In addition to heat transport, mass transport is also improved considerably in microstructured reactors. Heat flows in the microchannel are mostly laminar, directed, and highly symmetric. Process parameters such as pressure, temperature, residence time, and flow rate are more easily controlled in this reactor. Microstructured reactors also give opportunities for new production concepts. Depending on demand, more microstructured reactors or such subunits could be connected in parallel so that the required intermediate products and end products can be produced.

The first use of a wall-coated reactor for steam reforming of methanol was reported by de Wild and Verhaak. They used a wash-coated plate-fin type heat exchanger for the reaction. The wash-coated heat exchanger showed better performance as compared to packed beds. The problem with microchannel reactors lies in low durability of the catalyst. There is limited literature concerning the durability of the catalyst where catalyst deactivation at a faster rate has been reported. More catalyst loading and increase in porosity inside the coating are required in order to increase the durability of the catalyst. There are four types of catalyst for steam reforming of methanol in the literature under this category of reactor: $CuO/ZnO/Al_2O_3$, $CuO/Cr_2O_3/Al_2O_3$, $CuO/CeO_2/Al_2O_3$, and Pd/ZnO. In the microchannel reactor, a $CuO/ZnO/Al_2O_3$ catalyst coating was used by Bravo et al., Park et al., Lim et al., and Kawamura et al. Another research group studied $CuO/Cr_2O_3/Al_2O_3$, $CuO/CeO_2/Al_2O_3$, and Pd/ZnO catalysts, respectively. For a $CuO/ZnO/Al_2O_3$ catalyst, all the groups but one utilized a commercial catalyst, and the coatings were made with a slurry of catalyst powder and alumina/zirconia sol. Before coating of the catalyst, the microchannel sheets were undercoated with the alumina/zirconia sol in order to enhance adhesion between the catalyst powders and substrate structure. For $CuO/Cr_2O_3/Al_2O_3$ and $CuO/CeO_2/Al_2O_3$ catalysts, the alumina binder was prepared by mixing γ-alumina powder, water, polyvinyl alcohol (PVA) (binder), and acetic acid.

A novel catalyst fabricated from Al-Cu-Fe quasicrystals can also be used for steam reforming of methanol. A dense Pd/Ag membrane reactor was developed for methanol steam reforming. Hydrogen production from methanol by oxidative steam reforming was carried out in a membrane reactor.

For higher hydrocarbons, such as gasoline steam, reforming has to be performed at high temperature. Using conventional Ni catalysts, the temperature must exceed 900 °C, but addition of Co, Mo, and Re or the use of zeolites allows a reduction in the temperature by some 10%.

Small-scale applications of FCs, aimed at increasing operation time before recharge over that of current battery technology, have simulated the development of miniature reformers for both methanol and other hydrocarbons. Although the thermal efficiency of these devices aiming at use in conjunction with FCs is rated in the range of 10 mW to 100 W, which is lower than for larger units, system efficiencies compare favorably with those of existing devices in the same power range.

13.3.5 Steam Reforming of Ethanol

Ethanol is more promising because it is less toxic than methanol. It can also be more easily stored and safety handled. Most importantly, it can be produced in large amounts from biomass such as agricultural wastes and forestry residues and hence is a renewable resource, unlike methanol and gasoline. The bio-ethanol thus produced is free from sulfur, which otherwise may poison the FC catalyst. Similar, it is free from metals. Bio-ethanol, which is a dilute aqueous solution containing around 12 wt.% ethanol, could be directly subjected to steam reforming, thereby eliminating one unit operation of distillation required to produce pure ethanol. The entire process could therefore be economically attractive. Above all, ethanol is CO_2 neutral since the CO_2 that is produced in this process is consumed by biomass growth, and a closed carbon cycle is operated; the use of methanol and gasoline adds to CO_2 emissions. Thus, the use of ethanol will not contribute to global warming.

The first step in the conversion of ethanol to hydrogen is reforming. This reaction is carried out in the range of temperatures 673–1273 K. Reforming can be either by steam (steam reforming), or by humidified air (partial oxidation reforming), or by a mixture of air and steam (autothermal reforming). Here, attention is focused on the steam reforming reaction. This yields a H_2-rich gas containing CO, which is a poison for PEMFCs. Except for use in high-temperature cells, the CO concentration must be reduced to a very low level (around 10 ppm). A water-gas shift reactor is therefore used to reduce the CO content of this gas stream. After high-temperature and low-temperature water-gas shifts (HTS and LTS), the residual CO is then reduced further to the ppm level in a CO preferential oxidation (PROX) reactor. This product gas is then suitable for feeding PEMFCs.

Thermodynamics aspects of ethanol steam reforming have received a fair amount of attention in the published literature. The reaction is strongly endothermic and produces only H_2 and CO_2 if ethanol reacts in the most desirable way. The basic reaction scheme is as follows:

$$C_2H_5OH + 3\,H_2O \rightarrow 2\,CO_2 + 6\,H_2\ (\Delta H^\circ_{298} = 174\ kJ/mol) \tag{13.10}$$

However, other undesirable products such as CO and CH_4 are also usually formed during reaction. Aupretre et al. have discussed the main reactions in ethanol steam reforming that account for the formation of these by-products:

$$C_2H_5OH + H_2O \rightarrow 2\,CO + 4\,H_2\ (\Delta H^\circ_{298} = 256\ kJ/mol) \tag{13.11}$$

$$C_2H_5OH + 2\,H_2 \rightarrow 2\,CH_4 + H_2O\ (\Delta H^\circ_{298} = -157\ kJ/mol) \tag{13.12}$$

Other reactions that can also occur are: ethanol dehydrogenation to acetaldehyde, ethanol dehydration to ethylene, and ethanol decomposition to CO_2 and CH_4 or to CO, CH_4, and H_2. Coke formation may occur as per the following Boudouard reaction:

$$2\,CO \rightleftarrows CO_2 + C\ (\Delta H^\circ_{298} = -171.5\ kJ/mol) \tag{13.13}$$

Another possible route for the formation of carbon is through ethylene:

$$C_2H_4 \rightarrow \text{polymers} \rightarrow \text{coke} \tag{13.14}$$

From the thermodynamic standpoint, since reaction Eq. (13.10) is endothermic and results in an increase in the number of moles, increasing the temperature and lowering the pressure is in favor of ethanol reforming. At 500 K, steam reforming of ethanol does not occur ($\Delta G^\circ > 0$). However, ethanol decomposition can easily occur at this temperature since the value of ΔG° is sufficiently negative.

Vasudeva et al. found that H_2 yields as high as 5.5 mol/mol of ethanol in the feed can be obtained at equilibrium at temperatures around 773–873 K with water-to-ethanol molar ratios above 20. They suggested that carbon formation occurs only at low water-to-ethanol ratios (<2) and low temperatures (883 K).

The steam reforming of ethanol over Ni, Co, Ni/Cu, and noble metals (Pd, Pt, Rh) has been extensively studied. The greatest concern lies in developing an active catalyst that inhibits coke formation and CO production.

The catalytic steam reforming of ethanol for H_2 production is discussed in depth by Haga et al. The effects of process variables such as temperature, pressure, and the water-to-ethanol molar ratio in the feed on the H_2 yield at equilibrium are discussed. An overview of previous studies using Ni, Co, Ni/Cu, and noble metals (Pd, Pt, and Rh) is given. The catalyst performance characteristics suggest strong metal-support interaction. The reaction pathway is complex, and a number of undesirable side reactions occur, thereby affecting the selectivity to H_2. Catalyst coking is mainly due to the formation of ethylene by ethanol dehydration. The use of a two-layer fixed-bed reactor is therefore promising: at low temperatures, ethanol should first be converted by dehydrogenation over Cu into acetaldehyde, which has lower coking activity. The resulting mixture can then be passed at low temperatures (around 723 K) over a bed containing a mixture of Ni catalyst and a chemisorbing acetaldehyde. While Ni will be active in steam reforming of acetaldehyde thus formed, the selective removal of CO_2 from the product mixture by chemisorption will enable production of H_2-rich streams that can be fed to a PEMFC.

Haryanto et al. reviewed steam reforming of ethanol, examined the various catalysts reported to date, and presented a comparative analysis. They concluded that ethanol conversion and H_2 production vary greatly with the reaction conditions, type of catalyst, and method of catalyst preparation. The importance of process engineering–related aspects is evident, and these need to be discussed at length. The Haryanto study is aimed at fulfilling this need. It reviews the available literature on catalytic steam reforming of ethanol. All published information on this topic is analyzed and presented in a coherent manner. The role of the catalyst composition and the process conditions in determent product distribution is elucidated. The possible reaction pathways and kinetic and thermodynamic considerations are also discussed. The coupling of ethanol steam reforming with selective removal of CO_2 by chemisorption to produce high-purity H_2 at low temperatures has been discussed. It is expected that this will provide an insight into steam reforming of ethanol.

Haga et al. and Sahoo et al. studied the catalytic properties of Co among other metals and found that selectivity to H_2 was in the order Co > Ni > Rh > Pt, Ru, Cu. Metals alone may not assist hydrogen production significantly. So, performance of metal catalysts could be improved using supports. The nature of the support also plays a key role in determining selectivity to H_2; choice of support is hence crucial.

Al_2O_3 is commonly used as a support in the steam reforming reaction. However, it is acidic and promotes dehydration of ethanol to ethylene, which in turn polymerizes to form coke on the catalyst surface (Eq. (13.14)). In contrast, MgO is basic. ZnO also has basic characteristics. CeO_2 is also basic and has redox properties.

13.4 Vehicular Hydrogen Storage Approaches

Current on-board hydrogen storage approaches include compressed hydrogen gas, cryogenic and liquid hydrogen, metal hydrides, high-surface-area sorbents (such as carbon-based nanostructured materials), and chemical hydrogen storage. Compressed and cryogenic hydrogen, metal hydrides, high-surface-area sorbents, and carbon-based materials are categorized as *reversible* on-board, because these approaches may be recharged with hydrogen on-board the vehicle, similar to refueling with gasoline today. Systems that bind hydrogen with low binding energy (less than 20–25 kJ/mol H_2) can undergo relatively easy charging and discharging of hydrogen under conditions that may be applicable at refueling stations. For chemical hydrogen storage approaches as well as selected metal hydrides, the hydrogen is incorporated in much stronger bonds (e.g. with bond energies typically in excess of 60–100 kJ/mol H_2). Once the hydrogen is released for use during vehicle operation, recharging with hydrogen under operating conditions convenient at a refueling station is problematic. Such systems are referred to as *regenerable off-board*, which requires the spent media to be recovered from the vehicle and then regenerated with hydrogen either at a fueling station or at a centralized processing facility. Materials with binding energies between 25 and 60 kJ/mol H_2 may require substantial thermal management during recharging on-board the vehicle.

Both reversible on-board storage and regenerable off-board storage approaches have advantages and disadvantages. The US Department of Energy (DOE) is currently supporting research in both areas with a schedule for down-select decisions planned as materials

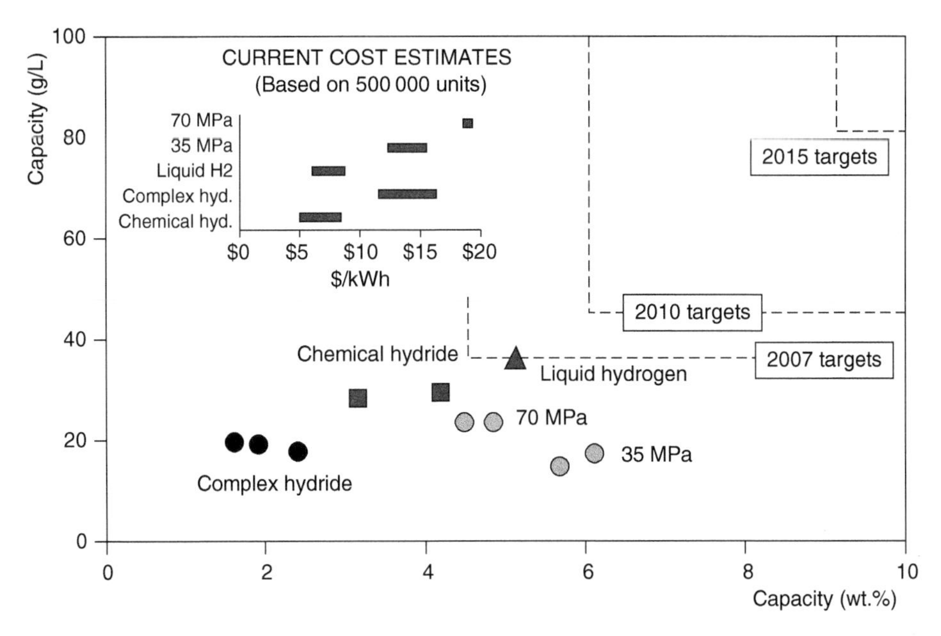

Figure 13.5 Hydrogen storage system capacities and cost. Note: Estimates are from developers, to be periodically updated by the DOE. Costs exclude regeneration processing. Complex hydride system data is projected. Data points include independent analysis results.

are designed, developed, and tested. Figure 13.5 showed the status of vehicular hydrogen storage systems in comparison to DOE 2010 and 2015 performance targets. Zhou and Liu provided a review of objectives, progress, and milestones of the research conducted on the topic of developing innovative metal-organic frameworks (MOFs) and porous organic polymers (POPs) for high-capacity and low-cost hydrogen-storage sorbents in automotive applications. The objectives of the research were to develop new materials as next-generation hydrogen storage sorbents that would meet or exceed the DOE's 2017 performance targets of gravimetric capacity of 0.055 kg H_2/kg_{system} and volumetric capacity of 0.040 kg H_2/L_{system} at a cost of $400/kg $H_{2stored}$. Different MOFs were development and studied: PCN–250, PCN–12, PCN–608, and PCN-609.

13.4.1 Reversible On-Board Approaches

13.4.1.1 Compressed Hydrogen Gas

Carbon fiber-reinforced composite tanks for 35 MPa and 70 MPa compressed hydrogen are under development and are already in use in prototype hydrogen-powered vehicles. The driving range of FC vehicles with compressed hydrogen tanks depends on the vehicle type, design, and amount of stored hydrogen. For example, the General Motors HydroGen 3 FC vehicle (Opel Zafira minivan with a target curb weight of 1590 kg) is specified for a 270 km driving range with 3.1 kg of hydrogen at 70 MPa. By increasing the amount of compressed hydrogen gas beyond 3 kg, a longer driving range can be achieved, but at more cost, higher weight, and reduced passenger and cargo space on the vehicle. Volumetric capacity, limits of high pressure, and cost are thus key challenges for compressed hydrogen tanks. Refueling or filling time, compression energy penalty (e.g. 15–20% of the lower heating value of hydrogen), and heat management requirements during refilling also need to be considered.

13.4.1.2 Liquid Hydrogen Tanks

Liquid hydrogen (LH_2) tanks can, in principle, store more hydrogen in a given volume than compressed gas tanks, since the volumetric capacity of liquid hydrogen is 0.070 kg/l (compared to 0.039 kg/l at 70 MPa). Key issues with LH_2 tanks are hydrogen boil-off, the energy required for hydrogen liquefaction, as well as tank cost. However, the driving range for vehicles using liquid hydrogen, excluding the effects of boil-off, can be longer than that for compressed hydrogen. For example, the General Motors HydroGen 3 Opel Zafira minivan is specified with a driving range of 400 km with 4.6 kg liquid hydrogen, versus 270 km described for the 70 MPa tank.

13.4.1.3 Metal Hydrides

Some metal hydrides have the potential for reversible on-board hydrogen storage and release the relatively low temperatures and pressures required for FC vehicles. Complex metal hydrides such as alanates have the potential for higher gravimetric hydrogen capacities in the operational window than conventional metal hydrides such as $LaNi_5H_6$. Sodium alanate can store and release hydrogen reversibly through chemical reactions conducted at modest temperature and pressure when catalyzed with titanium dopants, as discovered by Bogdanovic and co-workers, according to the following reactions:

$$NaAlH_4 \rightarrow 1/3\,Na_3AlH_6 + 2/3\,Al + H_2 \tag{13.15a}$$

$$Na_3AlH_6 \rightarrow 3NaH + Al + 3/2\,H_2 \tag{13.15b}$$

In practice, material (not system) gravimetric capacities are currently only 3–4%.

In metal hydride systems based on lithium amide, the following reversible reaction takes place at 325 °C and 0.1 MPa:

$$Li_2NH + H_2 \rightleftarrows LiNH_2 + LiH \tag{13.16}$$

In this reaction, 6.5 wt.% hydrogen can be reversibly stored. However, the temperature is outside of the vehicular operating window using the waste heat of a PEM FC.

Another example of a system that received significant attention since the recent work of Vajo et al. is “destabilized” lithium borohydride ($LiBH_4$) with over 9 wt.% material capacity demonstrated. Further improvements will be pursued by nanoengineering and catalyst development.

13.4.1.4 High-Surface-Area Sorbents and Carbon-Based Materials

While metal hydrides offer high volumetric capacities through dissociative absorption of hydrogen, high-surface-area sorbents offer the advantages of fast hydrogen kinetics and low hydrogen binding energies and, hence, potentially fewer thermal management issues during hydrogen charging and discharging. Single-walled carbon nanotubes (SWNTs), among several other high-surface-area sorbents (e.g. carbon nanofibers, graphite materials, metal–organic frameworks, aerogels, etc.) are being studied for hydrogen storage within the DOE program and by others. Transition metal atoms bound to fullerenes (Figure 16.2) have recently been proposed as potential adsorbents for high-density, room temperature, ambient pressure storage of hydrogen based on theoretical studies. It is indicated that stable scandium-based organometallic buckyball fullerenes might adsorb and desorb as many as 11 hydrogen atoms per scandium atom, leading to a theoretical maximum reversible hydrogen storage density of close to 9 wt.%. These materials have yet to be synthesized to confirm theoretical predictions. At a pressure of 2.5 GPa, the mole ratio between the stored hydrogen atoms and C atoms on the (20,0) nanotube (nanocontainer) is an impressive 1 : 1, corresponding to a weight ratio of 7.7%.

13.4.2 Chemical Hydrogen Storage: Regenerable Off-Board

Chemical hydrogen storage may offer options with high energy densities and potential ease of use, particularly if systems involve liquids that may be easily dispensed using infrastructure similar to today's gasoline refueling stations. Most of these reactions are irreversible, so the spent storage material would have to be regenerated off-board the vehicle because it cannot be reconstituted simply by applying an overpressure of hydrogen gas at modest temperature and pressure.

13.4.2.1 Hydrolysis Reactions

Hydrolysis reactions involve the reaction of chemical hydrides with water to produce hydrogen. The reaction of sodium borohydride solutions has been the most studied to date. This reaction is:

$$NaBH_4 + 2H_2O \rightarrow NaBO_2 + 4H_2 \tag{13.17}$$

Another hydrolysis reaction that is presently being investigated is the reaction of MgH_2 with water to form $Mg(OH)_2$ and H_2. In this case, particles of MgH_2 are contained in a non-aqueous slurry to inhibit premature water reaction when hydrogen generation is not required. Material-based capacities for the MgH_2 slurry reaction with water can be as high as 11 wt.%. However, water must also be carried on-board the vehicle in addition to the slurry, and the $Mg(OH)_2$ must be regenerated off-board.

13.4.2.2 Hydrogenation/Dehydrogenation Reactions

The hydrogenation and dehydrogenation of organic liquids offer a potential advantage by not requiring water on-board as a co-reactant. For simple organic compounds, dehydrogenation is endothermic, so external heat must be applied. One early example is the decalin-to-naphthalene reaction, which can release 7.3 wt.% hydrogen at 210 °C via this reaction:

$$C_{10}H_{18} \rightleftarrows C_{10}H_8 + 5H_2 \tag{13.18}$$

Recently, new organic liquid hydrogen storage media have been developed by Air Products that demonstrate the beneficial effect of heteroatom substitution on the thermodynamics of dehydrogenation. These liquids, including an example of N-ethyl carbazole, have shown 5–7 wt.% gravimetric hydrogen storage capacity and greater than 0.050 kg/l hydrogen volumetric capacity (material capacities only). Because this hydrogen-release reaction is endothermic, it can use waste heat from the FC (or internal combustion engine) and on-board heat rejection may not be an issue. Furthermore, the spent fuel regeneration (hydrogenation) reaction is endothermic, so it may be possible to couple the reaction efficiently at the regeneration plant to optimize energy recovery and reduce cost.

13.4.2.3 Ammonia Borane and Other Boron Hydrides

A number of boron hydride materials have high hydrogen content. Ammonia borane (AB, NH_3BH_3) is isoelectronic with ethane and has a high hydrogen storage capacity (up to 19.6 wt.% for release of three hydrogen molecules). Unlike ethane, hydrogen release from AB is exothermic. Products of dehydrogenation can include compounds such as cyclotriborazane (one hydrogen molecule released), borazine (two hydrogen molecules released), and polymeric analogues. The thermodynamics of these pathways is being determined both theoretically and experimentally.

Autrey and co-workers have shown that incorporating solid ammonia-borane into a mesoporous silica scaffold enhances hydrogen release through the formation of $(NH_2BH_2)_n$ and $(NHBH)_n$ compounds at relatively low temperatures (even at 80 °C, with a 6 wt.% material capacity, including the scaffold).

13.4.2.4 Ammonia

Ammonia (NH_3), which has a boiling point of −33.5 °C, has a high capacity for hydrogen storage, 17.6 wt.%. However, in order to release hydrogen from ammonia (an endothermic reaction), high fuel-processing temperatures, and therefore large reactor mass and volume, would be required. Other considerations include safety and toxicity issues, both actual and perceived, as well as the incompatibility of PEM FCs in the presence of trace levels of ammonia (>0.1 ppm).

13.4.2.5 Alane

Alane (AlH_3) is another metal hydride being investigated as a material for hydrogen storage. The chemical formula of alane contains a theoretical 10 wt.% of H_2 and a theoretical density of hydrogen in the compound (148 g H_2/l) that is more than double the density of liquid H_2. In a collaborative effort at Brookhaven National Laboratory, it was found that the addition of LiH reduces the desorption temperature. The onset of hydrogen desorption of alane has been lowered to below 125 °C, with hydrogen yields of 7–8 wt.% (based on material only) below 175 °C. However, in order to utilize AlH_3 as an on-board storage technology, significant issues need to be resolved. First, the desorption should be lowered further to make the release of H_2 compatible with the waste heat generated by the PEM FC system (~80 °C). Second, there is no practical, low-cost method to regenerate the spent Al powder back into AlH_3. Finally, the infrastructure implications of a solid-state hydride storage option that is not rechargeable on-board the vehicle have yet to be fully explored.

13.5 Gas Conversion Technologies/Natural Gas Upgrading

The evolution of known crude oil and natural gas reserves worldwide indicates a dramatic increase in the latter compared to a leveling off concerning crude oil. This trend is expected to continue, which will – in addition to price developments with respect to crude oil–based upgrading (UG) – most likely generate a gradual shift toward the application of natural gas as a feedstock for the production of fuels and petrochemicals. This situation has forced enhanced global interest in processes that can convert natural gas into liquids and higher value-added products without going via methanol as intermediate. This route is known as gas to liquids (GTL) technology, based on the Fischer–Tropsch route. Interest in manufacturing fuels and petrochemicals from natural gas is driven by a desire to apply this technology directly, for example at remote natural gas field sites, in order to minimize transportation costs and gas burning at recovery sites.

The following subsections deal with GTL technology based on FTS as well as methanol to hydrocarbon conversions, where zeolites and related microporous materials have been demonstrated to be superior catalysts.

13.5.1 GTL Conversion of Syngas to Fuel

Fuels production directly from syngas (in former times obtained from coal) was reported for the first time by Fischer and Tropsch in 1923, using an alkali-promoted iron catalyst (Section 12.2.10). Fuels manufactured via FTS have excellent quality since they consist mainly of linear paraffins and α-olefins and do not contain sulfur and aromatics. A Co-containing catalyst is applied for the production of heavy paraffins via FTS starting with natural gas: a technology developed by Shell and named Shell Middle Distillate Synthesis (SMDS). In addition, diesel fuel (or gasoline) is produced by hydrocracking the more or less sulfur- and nitrogen-free wax obtained through the SMDS process using zeolites containing noble metals. The more restrictive fuel specifications currently being introduced to reduce the environmental impact of hazardous emissions represent a driving force with respect to increased use of fuels prepared via FTS as a blending component of gasoline and diesel pools in the future.

Besides SMDS, an alternative has been presented by Sasol/Chevron: the Slurry Phase Distillate (Sasol SPD) process, again based on FTS, producing wax (using a Co-containing catalyst) followed by a hydrocracking step in order to get diesel or gasoline.

The methanol or gasoline (MTG) plant in New Zealand has been combined with a methane steam reforming unit for production of syngas and a methanol plant to produce gasoline from natural gas. The process economics can be improved considerably by a clever combination and close integration of the different steps. In the Topsoe Improved Gasoline Synthesis (TIGAS) process developed by Haldor Topsoe for the manufacture of gasoline in a pilot-scale plant, methanol synthesis and MTG reactions are integrated -without the separation of methanol as an intermediate product. A multifunctional catalyst has been developed; however, these process technologies do not usually apply catalysts based on zeolites or related microporous materials. Finally, ExxonMobil has introduced the Advanced Gas Conversion for the 21st Century (AGC-21) technology, again based on the FTS.

Further Reading

Abashar, M. (2004). Coupling of steam and dry reforming of methane in catalytic fluidized bed membrane reactors. *Int. J. Hydrog. Energy* 29: 799–808.

Aguiar, P., Adjiman, C.S., and Brandon, N.P. (2004). *J. Power Sources* 132: 113.

Ahmed, K. and Foger, K. (2000). *Catal. Today* 63: 479.
Amendola, S.C. et al. (2000). *Int. J. Hydrog. Energy* 25: 969.
Aupretre, F., Descrome, D., Duprez, D. et al. (2005). *J. Catal.* 233: 464.
Bajus, M. (1997). Current trends and the process oil and petrochemical technologies for the future. *Pet. Coal* 19 (2): 8–14.
Bajus, M. (1998). Natural gas conversion 5. In: *Studies in Surface Science Catalysis*, vol. 119 (eds. A. Parmaliana, D. Sanfilipo, F. Frusteri and F. Arena), 289–293. Elsevier.
Bajus, M. (2000). Ropa a alternatívne energetické zdroje. *Ropa, uhlie, plyn a petrochémia* 42 (2): 24–28.
Bajus, M. (2001). Reformulované a alternatívne palivá- súčasnosť a budúcnosť. *Ropa, uhlie, plyn a petrochémia* 43 (2): 20–24.
Bajus, M. (2002). Hydrocarbon technologies for the future, current trends in oil and petrochemical industry. *Pet. Coal* 44 (3-4): 112–119.
Bajus, M. (2002). Alternatívne palivá. *Energia* 4 (1): 42–47.
Bajus, M. (2002). Súčasnosť a budúcnosť alternatívnych palív. *Slovgas* 11 (5-6): 28–31.
Bajus, M. (2003). Hydrocarbon technologies. *Pet. Coal* 45 (1-2): 10–18.
Bajus, M. (2006). Proceedings of the International Symposium on Motor Fuels, Part 2, GTL Technologies. VÚRUP.
Bajus, M. and Back, M. (1995). *Appl. Catal. A Gen.* 128: 61.
Basile, A., Galucci, F., and Paturzo, L. (2005). *Catal. Today* 104 (2-4): 244.
Basile, A., Galucci, F., and Paturzo, L. (2005). *Catal. Today* 104 (2-4): 251.
Battle, P.D., Claridge, J.B., Coppiestone, F.A. et al. (1994). *Appl. Catal. A Gen.* 118: 217.
Bogdanovic, B. and Sandrock, G. (2002). *MRS Bulletin* (September): 712.
Bolt, P.H., Harbraken, F.H.P.M., and Geus, J.W. (1995). *J. Catal.* 151: 300.
Bowman, R.C. Jr. and Fultz, B. (2002). *MRS Bulletin* (September): 688.
Bradford, M.C.J. and Vannice, M.A. (1996). *Appl. Catal.* 142: 97.
Bravo, J., Karim, A., Conant, T. et al. (2004). *Chem. Eng. J.* 273: 125–132.
Brown, S. (1998). Fortune Magazine (30 March). http://www.pathfinder.com/fortune
de Bruin, F. (2005). *Green Chem.* 7: 132.
Cameron, D.S. (2003). *Platin. Met. Rev.* 47 (1): 28.
Chen, P., Xiong, Z., Luo, J. et al. (2002). *Nature* 420: 302.
Choudhary, V.R., Uphade, B.S., and Mamman, A.S. (1995a). *Catal. Lett.* 32: 387.
Choudhary, V.R., Rajput, A.M., and Prabhakar, B. (1995b). *Catal. Lett.* 32: 391.
Choudhary, V.R., Uphade, B.S., and Belhekar, A.A. (1996). *J. Catal.* 163: 312.
Cui, Y., Zhang, H., Xu, H., and Li, W. (2007). *Appl. Catal. A Gen.* 318: 79–88.
Darwish, N.A., Hilal, N., Wersteeg, G., and Heesing, B. (2004). *Fuel* 83: 409–417.
Dixon, D.A. and Gutowski, M. (2005). Thermodynamic properties of molecular borane amines and the [BH{sub}][NH {sub 4}{sup}] salt for chemical hydrogen storage systems from ab initio electronic structure theory. *J. Phys. Chem.* 109: 5129.
US Department of Energy (2005). Fuel cells and infrastructure technologies program multi-year research, development and demonstration plan. Office of Energy Efficiency and Renewable Energy. http://www.eere.energy.gov/hydrogenanandfuelcells/mypp.
Erdohelyi, A., Cserenyi, J., and Solymosi, F. (1993). *J. Catal.* 141: 287.
Erdöhelyi, A., Raskó, J., Kecskés, T. et al. (2006). *Catal. Today* 116: 367.
Fierro, W., Akdim, O., and Mirodatos, C. (2003). *Green Chem.* 5: 20.
Fischer, F. and Tropsch, H. (1923). *Brennstoff-Chemie* 4: 276.

FreedomCAR and Fuel Partnership (2009). Highlights of Technical Accomplishments.
Fukuhara, C., Ohkura, H., Kamat, Y. et al. (2004). *Appl. Catal. A Gen.* 273: 125–132.
Gao, L.Z., Dong, B.L., Li, W.S. et al. (1994). *J. Fuel. Chem. Tech. (China)* 22: 113.
Gao, L.Z., Li, J.T., Wan, H.L., and Tsai, K.R. (1997). *Sci. Technol. Rev. (China)* 3: 35.
Goula, M.A., Lemonidou, A.A., and Efstathiou, A.M. (1996). *J. Catal.* 161: 626.
Gross, K.J., Thomas, G.J., and Jensen, C.M. (2002). *J. Alloys Compd.* 330-332.
Gunardson, H. (1998). *Industrial Gases in Petrochemical Processing*. New-York-Basel, USA: Marcel Dekker, Inc.
Haga, F., Makajima, T., Miya, H., and Mishima, S. (1997). *Catal. Lett.* 48: 223.
Haga, F., Makajima, T., Miya, H., and Mishima, S. (1998). *React. Kinet. Catal. Lett.* 63: 253.
Haryanto, A., Fernardo, S., Murali, N., and Adhikari, S. (2005). *Energy Fuel* 19: 2098.
Hayakawa, T., Andersen, A.G., Shimizu, M. et al. (1993). *Catal. Lett.* 22: 307.
Hodoshima, S., Arai, H., and Saito, Y. (2003). *Int. J. Hydrog. Energy* 28: 197.
Holladay, J., Wainright, J., Jones, E., and Gano, S. (2004). *J. Power Sources* 130: 111.
Horný, C., Kiwi-Minsker, L., and Renken, A. (2004). *Chem. Eng. J.* 101: 3–9.
Itoh, N., Kaneko, Y., and Igarashi, A. (2002). *Ind. Eng. Chem. Res.* 41: 4702.
Jorgensen, S.L., Nielsen, P.E.H., and Lehrmann, P. (1995). *Catal. Today* 25: 303.
Kawamura, Y., Ogura, N., Yamamoto, T., and Igarashi, A. (2006). *Chem. Eng. Sci.* 61: 1088–1097.
Klein, J.-M., Bultel, Y., Georges, S., and Pons, M. (2007). *Chem. Eng. Sci.* 62: 1636.
Koukou, M.K., Papayannakos, N., Markatos, N.C. et al. (1998). *Trans. Ichem E* 76A: 911–920.
Kroll, V.C.H., Swaan, H.M., Lacombe, S., and Mirodates, C. (1997). *J. Catal.* 164: 387.
Kundu, A., Park, J.M., Ahn, J.E. et al. (2007). *Fuel* 86: 1331.
Laosiripojana, N., Sutthisripok, W., and Assabumrungrat, S. (2007). *Chem. Eng. J.* 127: 31.
Lee, S.-H., Cho, W., Ju, W.-S. et al. (2003). *Catal. Today* 87: 133–137.
Lim, M.S., Kim, M.R., Noh, J., and Woo, S.I. (2005). *J. Power Sources* 140: 66–71.
Lin, Y.-M. and Rei, M.-H. (2000). *Hydrog. Energy* 25: 211.
Lindermeir, A., Kah, S., Kavurucu, S., and Mühlner, M. (2007). *Appl. Catal. B Environ.* 70: 488–497.
Malwell, I.E. and Stork, W.H.J. (2001). Hydrocarbon processing with zeolites. In: *Introduction to Zeolite Science and Practise*, Studies Science and Catalysis, 2e, vol. 137 (eds. H. van Bekkum, E.M. Flamigen, P.A. Jacobs and J.C. Jansen), 747. Amsterdam: Elsevier.
Matsukata, M., Nishiyama, N., and Ueyama, K. (1994). *J. Chem. Soc. Chem. Commun.* 339.
Matter, P., Braden, D., and Ozkam, U. (2004). *J. Catal.* 223: 340.
McCarty, R.D., Hord, J., and Roder, H.M. (1981). Selected properties of hydrogen. Report no. 168. Center for Chemical Engineering, National Engineering Laboratory, National Bureau of Standards.
Men, Y., Gnaser, H., Zapfc, R. et al. (2004). *Catal. Commun.* 5: 671–675.
Men, Y., Gnaser, H., Zapfc, R. et al. (2004). *Appl. Catal. A Gen.* 277: 83–90.
Momirlan, M. and Vezirogh, T.N. (2005). *Int. J. Hydrog. Energy* 30: 795.
Montini, T., de Rogatis, L., Gombac, V. et al. (2007). *Appl. Catal: B Environ.* 71: 125.
Morgenstern, D.A. and Formango, J.P. (2005). *Energy Fuel* 19: 1708.
Neomagus, H.W.P. (1999). A catalytic membrane reactor for partial oxidation reactions. PhD thesis. Department of Chemical Engineering, Twente University of Technology.
Noble, R.D. and Falconer, J.L. (1995). *Catal. Today* 25: 209.
Pacheco, M., Sira, J., and Kopasz, J. (2003). *Appl. Catal. A Gen.* 250: 161.
Palo, D., Holladay, J., Rozmiarek, R. et al. (2002). *J. Power Sources* 108: 28.

Park, G.-G., Seo, D.J., Park, S.-H. et al. (2004). *Chem. Eng. J.* 101: 113–121.
Pfeifer, P., Schubert, K., Liauw, M.A. et al. (2003). *Chem. E.* 81 (Part A): 711–720.
Pfeifer, P., Schubert, K., Liauw, M.A., and Emig, G. (2004). *Appl. Catal. A Gen.* 270: 165 175.
Prakash, D., Vaidya, P.D., and Rodriques, A.E. (2006). *Chem. Eng. J.* 117: 39–49.
Prestig, H., Konle, J., Starkov, V. et al. (2004). *Mater. Sci. Eng.* B108: 162.
Prigent, M. (1997). *Revue De L'Institut Francais Du Petrole* 52 (3): 349.
Quantum Fuel Systems (2005). Annual progress report for the DOE hydrogen program. http://hydrogen.energy.gov/annual_progress05_storage.html.
Rostrup-Nielsen, J.R. and Bak Hansen, J.H. (1993). *J. Catal.* 144: 38.
Sahoo, D.R., Vajpai, S., Patel, S., and Pant, K.K. (2007). *Chem. Eng. J.* 125: 139.
Sandrok, G., Reilly, J., Graetz, J. et al. (2005). *Appl. Phys. A Mater. Sci. Process.* 80: 687.
Saracco, G. and Specchia, V. (1994). *Catal. Rev. Sci. Eng.* 36: 305.
Satyapal, S., Read, C., Ordaz, G., and Thomas, G. (2006). DOE hydrogen program annual merit review proceedings. http://www.hydrogen.energy.gov/annual_review06_plenary.html.
Satyapal, S., Petrovic, J., Read, C. et al. (2007). *Catal. Today* 120: 246–256.
Sie, S.T., Senden, M.M.G., and van Wechum, H.M.H. (1991). *Catal. Today* 8: 371.
Takahashi, K. (1998). Development of fuel cell electric vehicles. Paper presented at Fuel Cell Technology Conference, London, September.
Tanable, T., Kameoka, S., and Tsai, H.P. (2006). *Catal. Today* 111: 153.
Tsang, S.C., Claridge, J.B., and Green, M.L.H. (1995). *Catal. Today* 23: 3.
Uhlhorn, R.J.R., Keizer, K., and Burggraaf, A.J. (1992). *J. Membr. Sci.* 66: 271.
Vaidya, P. and Rodrigues, A.E. (2006). *Chem. Eng. J.* 117: 39.
Vajo, J.J. et al. (2004). *J. Phys. Chem. B* 108: 13977.
Vajo, J.J., Skeith, S., and Mertens, F. (2005). *J. Phys. Chem. B* 109: 3719.
Veldsink, W., Versterg, G.F., and Van Swaaij, W.P.M. (1995). *Ind. Eng. Chem. Res.* 34: 763–772.
Vernoux, P., Guillodo, M., Fouletier, J., and Hammou, A. (2000). *Solid State Ionics* 135: 425.
Vornoux, P., Guindet, J., and Kleitz, M. (1998). *J. Electrochem. Soc.* 145 (10): 3487.
Wang, H.Y. and Au, C.T. (1996). *Catal. Lett.* 38: 77.
Wang, H.Y. and Au, C.T. (1996). *Chin. Chem. Lett.* 7: 1047.
Wang, H.Y. and Au, C.T. (1997). *Appl. Catal.* 155: 239.
Wang, L., Murta, K., and Inaba, M. (2004). *Appl. Catal. A Gen.* 257: 443.
Wang, Z., Pan, Y., Dong, T. et al. (2007). *Appl. Catal. A Gen.* 320: 24.
Wieland, S., Melin, T., and Lamm, A. (2002). *Chem. Eng. Sci.* 57: 1571.
de Wild, P.J. and Verhaak, M.J.F.M. (2000). *Catal. Today* 60: 3–10.
Williams, M.C., Strakey, J.P., and Surdoval, W.A. (2005). *J. Power Sources* 143: 19.
Yan, Y., Davis, M.E., and Gavalas, G.R. (1995). *Ind. Eng. Chem. Res.* 34: 1652.
Yan, W.G., Gao, L.Z., and Yu, Z.L. (1997). *J. Nat. Gas Chem. (China)* 6: 93.
Ye, X., Gu, X., Gong, X.G. et al. (2007). *Carbon* 45: 315.
Yildirim, T. and Ciraci, S. (2005). *Phys. Rev. Lett.* 94: 175501.
Zalc, J.M. and Loffler, D.G. (2002). *J. Power Sources* 111: 58.
Zapf, R., Becker-Willinger, C., Berresheim, K. et al. (2003). *Trans IChemE* 81 (Part A): 721–729.
Zhao, Y., Kim, Y.-H., Dillon, A.C. et al. (2005). *Phys. Rev. Lett.* 94: 155504.
Zhou, H. and Liu, D.-J. (2017). Final project report for DOE/EERE high-capacity and low-cost hydrogen-storage sorbents for automotive applications. DE-EE0007049.
Zuttel, A. (2004). *Naturwissenschaften* 91: 157.

14

Biorefineries

CHAPTER MENU

14.1 Introduction

The second generation of biofuel processes should differ from the first in (i) utilizing the whole plant as a feedstock and (ii) use of non-food perennial crops (woody biomass and tall grasses) and lignocellulosic residues and wastes. Possible options for the conversion of these lignocellulosic plant materials include: thermal cracking, catalytic cracking, pyrolysis, carbonization, catalytic reforming, steam reforming, gasification, Fischer–Tropsch synthesis, hydro-dehydrogenation, hydrocracking, hydrorefining (see Sections 5.5.1, 7.5, and 12.2.1), and decarboxylation. The main goal of a biorefinery is to produce high-value, low-volume chemicals (levoglucosan) and low-value, high-volume fuels with a series of unit operations.

Thermal processes must be included among the attractive basic recycling technologies for polymers, for which thermal cracking and pyrolysis enable the conversion of polymer materials into fuels, monomers, and other valuable products. The subjects of the research in Section 14.5 and Chapter 15 were thermal and catalytic processes for the production of motor fuels from polymer materials from industrial materials or municipal trash sources, turning them into sulfur-free, nonaromatic, ecological fuel via chemical recycling to replace fossil fuels mainly from oil sources. The key is the liquefaction of polymer materials to oil/waxes that can be distilled to provide gasoline, diesel fuels, and heating oils that can be used directly or after hydrorefining.

We found a way to incorporate polymer waste into conventional liquid steam cracking (SC) feedstocks. Polyalkene oils and waxes decompose during copyrolysis (Chapter 15). The amount of desired alkenes (ethylene, propene) increases or is slightly less depending

Petrochemistry: Petrochemical Processing, Hydrocarbon Technology, and Green Engineering,
First Edition. Martin Bajus.

on the polymer type. A mixture of waxes in heavy naphtha (10–20% mass) exhibits a small tendency to coking. Feedstock and chemical recycling of polyalkene oils and waxes via copyrolysis is a very promising method for treating polymer waste.

We have developed the (Deep Scavenger steam cracking or Steam cracking activation process [DSSC/SCA]) for thermal cracking of used tires and rubber waste, which works in a flow reactor (Chapter 15). Basically, we received three fractions from the thermal cracking of used tires: gases, liquid oils (d,l-limonene), and solid residues (coke, carbon, steel).

14.2 Petrochemistry

The discovery that alkenes can be produced in high yields from both alkanes present in natural gas and from petroleum fractions has laid the foundation for what is now known as the *petrochemical industry*. The principal process used to convert the relatively unreactive alkanes into more reactive alkenes is *thermal cracking*, often referred to as *steam cracking*. The term *steam cracking* is slightly illogical; under the reaction conditions, steam is not cracked – it functions primarily as a diluent, allowing higher conversion. A more accurate description of the process is *pyrolysis*, which stems from Greek and means "bond breaking by heat." Pyrolysis is an endothermic cracking process that requires a large energy supply at a high temperature with a short residence time of the cracked products (Chapter 7). Equilibrium resulting in the formation of C and H_2 must be prevented by the short residence time (quenching). A hydrocarbon stream is thermally cracked in the presence of steam, yielding a complex product mixture. This process produces mainly ethene, but valuable coproducts such as propene, butadiene, and pyrolysis gasoline, with benzene as a major constituent, are also produced.

Since the late 1930s, when the petrochemical industry stared to take shape, ethene has almost completely replaced coal-derived acetylene and now is the largest-volume building block for the petrochemical industry (Chapters 7–9). Its main outlet is the production of polyethylene (>50%), followed by vinyl chloride, ethene oxide, and styrene.

14.3 Carbonization of Coal

Coal was important to the chemical industry in the nineteenth and early twentieth centuries. It provided calcium carbide and hence acetylene, synthesis gas and hence ammonia and methanol, petroleum-like fuels, and all the aromatic chemicals contained in coke oven distillate. When coal is heated in the absence of air to a temperature of about 1000 °C (pyrolysis; high-temperature carbonization of coal), coke forms, together with many liquid and gaseous products. This distillate, also called *coal tar*, provided aromatics and many other chemicals for the early chemical industry. Coke is almost pure carbon and is used in steel manufacturing. As the process changes and becomes more efficient, it requires less coke. Nonetheless, some coke will always be needed, and therefore the chemical industry will always have available the chemicals that volatilize from the coke ovens. A typical coking operation produces 80% coke by weight, 12% coke oven gas, 3% tar, and 1% light oil consisting of crude benzene, toluene, and xylenes.

14.4 Manufacturing of Activated Carbon

Activated carbon is the collective name for a group of porous carbons manufactured by the treatment of a char with gases or by carbonization of carbonaceous materials with simultaneous activation by chemical activation. All these carbons are prepared to exhibit a high degree of porosity and an extended internal surface area.

Almost any carbonaceous material may be used as a precursor for the preparation of activated carbon. However, in practice, wood, nut shells and fruit stones, peat, charcoal, brown coal, lignite, bituminous coal, petroleum coke, etc. are inexpensive materials that have a high carbon content and are low in inorganic components and consequently are adequate for the production of activated carbon, although the properties of the final products will differ depending on the nature of the raw material and activating agent used, and the conditions of the activation process.

The preferred raw materials for manufacture are lignocellulosic materials, which account for 47 wt. % of the total raw material used for the actual preparation of activated carbon; but there are significant differences between the various types that can be used. Low-density materials like wood, which also contain high volatile content, produce activated carbons with large pore volume but low bulk density; but if the process is modified to decrease the loss of carbon during carbonization, or the carbon is densified, the quality can be considerably increased. On the other hand, coconut shells, fruit stones (pits), and similar materials, which have high bulk density and volatile content, produce hard, granular carbons with large pore volume that are suitable for many applications.

Activated carbon from wood (waste biomass) may very well be a leading technology to catalyze the development of a bio-economy. Activated carbon derived from wood for industrial uses in filtration and purification applications presents a better opportunity for higher-value-added products in a market that already has an established infrastructure. By supplying inexpensive charcoal feedstock from the thermochemical process with effective activation techniques, wood can be turned into a-cost competitive renewable alternative to traditional activated carbon, thus creating a significant revenue stream for biorefineries.

There are two principal methods of manufacturing activated carbon: physical and chemical activation. The former implies the carbonization of the raw material in an inert atmosphere followed by partial gasification of the resulting char with steam, carbon dioxide, or a mixture of them. In chemical activation, the raw material is impregnated with a chemical such as phosphoric acid or zinc chloride, and the impregnated product is pyrolyzed and then washed to remove the activating agent.

The furnaces used for the production of activated carbon can be of different types: rotary kiln, multiple hearth, vertical shaft, or fluidized bed. Most companies use internally heated rotary kilns, since they are suitable for production of activated carbon of a large range of particle sizes. These are furnaces of large dimensions. The residence time of the carbon in rotary kilns is usually much longer than in other types of furnaces. Multiple-hearth furnaces with rotary arms and stationary floors on each stage have the advantage over rotary kilns of better temperature control (400–900 °C); the residence time in this type of furnace is much shorter than in rotary kilns.

Pyrolytic carbons, e.g. those deposited slowly from methane in helium, are a well-ordered form of carbon, approaching the quality of single-crystal graphite. They were developed together with nuclear-grade graphite as a cladding material for spherical fuel elements. Their applications then waned until the advent of carbon fiber/carbon composites, when they became one matrix of carbon materials within the weaves of carbon fiber systems. Such carbons are finding significant application in the carbon composites of modern aircraft disc brakes.

Ever-optimistic carbon scientists have their attention firmly attached to two new areas of development, i.e. diamond and diamond-like film and graphitic film, and fullerenes and nanotubes and rods. The case for fullerenes and nanotubes is still being developed (Section 16.5.3). Efforts to make fullerenes into viable commercial materials are strenuous; they came to us from outer space, and this generated many new possibilities.

14.5 Chemicals and Fuels from Biomass

14.5.1 Degasification

Pyrolysis can by defined as the direct thermal decomposition of matter in the absence of oxygen. When applied to biomass, an array of useful products can be produced, including liquid and solid derivatives and fuel gases. By the 1950s, more than 200 chemical compounds had been identified from the pyrolysis of wood. Before the onset of the petrochemical era at the beginning of the twentieth century, pyrolysis processes were utilized for the commercial production of a wide range of fuels, solvents, chemicals, and other products from biomass feedstocks. Recently, global problems associated with the intensive use of fossil fuels (global warming, depletion of natural resources, security of supply of energy and materials) have led to a renewed interest in (modern varieties of) these processes. The fact that different biomass constituents react differently at different temperatures to yield different spectra of products can be exploited to extract value-added chemicals from biomass as renewable-route products that can be regarded as petrochemical substitutes. *Staged degasification* is a low-temperature thermochemical conversion route to generate value-added chemicals from lignocellulosic biomass. Figure 14.1 presents a schematic overview of staged degasification and its place in a thermochemical biorefinery.

The main biomass consists of hemicellulose, cellulose, and lignin and can be selectively devolatilized into value-added chemicals. This thermal breakdown (BD) is guided by the order of thermochemical stability of the biomass constituents, which ranges from hemicellulose (fast degassing/decomposition from 200 to 300 °C) as the least stable polymer to the more stable cellulose (fast degassing/decomposition from 300 to 400 °C). Lignin exhibits intermediate thermal degradation behavior (gradual degassing/decomposition from 250 to 500 °C). Results obtained for beech wood in principle acknowledge the view that chemical wood components are decomposed in the order of hemicellulose-cellulose-lignin, with a restricted decomposition of the lignin at relatively low temperatures. In the further course of heating, recondensation of the lignin takes place, whereby thermally largely stable macromolecules develop. Whereas both hemicellulose and cellulose exhibit a relatively high devolatilization rate over a relatively narrow temperature range, thermal degradation of lignin is a slow-rate process that commences at a lower temperature when compared to cellulose.

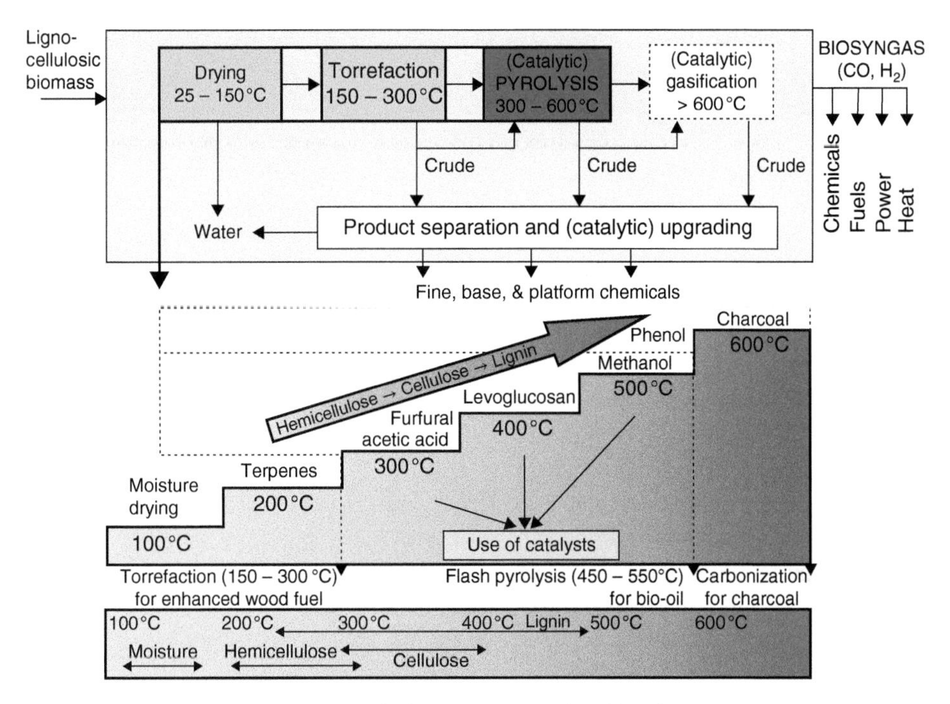

Figure 14.1 Staged degasification within a thermochemical biorefinery.

Since the thermal stabilities of the main biomass constituents partially overlap and the thermal treatment is not specific, a careful selection of temperatures, heating rates, and gas and solid residence times is required to make discrete degasification possible when applying a step-wise increase in temperature. To enhance the selectivity toward wanted products, catalysts can be applied as impregnates of the biomass or as external aids, e.g. in the form of the fluidization material in a fluidized bed reactor or as a catalytically active fluidization gas (steam, hydrogen, oxygen, CO_2). Downstream (DS) treatment of the primary product vapors in a fixed bed of catalyst is another possibility.

Depending on these process conditions and parameters like biomass composition, and the presence of catalytically active materials, the product mixture is expected to contain more or less degradation fragments from hemicellulose, cellulose, or lignin. The staged degasification approach stands in contrast with fast pyrolysis technology, in which the biomass is rapidly heated to temperatures around 500 °C, causing an almost instantaneous release of a myriad of thermal degradation products that are quickly quenched to a so-called *pyrolysis oil*. It is obvious that the extraction of value-added chemicals from this complex mixture of thermal degradation products is a challenge. In a critical review of pyrolysis-oil, Ravichandran et al. presented an extensive overview of pyrolysis oil and related issues.

Whereas fast pyrolysis of biomass has primarily been developed to maximize liquid product yield, staged degasification aims at the gentle devolatilization of thermal degradation products from the biomass. Hereby it is assumed that the type, yield, and selectivity of the liberated products can be influenced by matching the process conditions of the degasification process with the thermal stability of the main biomass constituents. Due to the relatively mild conditions, the overall product spectrum may be less complex, more stable, and less prone to unwanted secondary reactions when compared to the harsher pyrolysis process where all three biomass components are degraded simultaneously and at the same temperature.

14.5.2 Oxygenation

Recently, a limited number of value-added chemicals from biomass have been identified in an extensive study by the National Renewable Energy Laboratory / Pacific Northwest National Laboratory (NREL/PNNL). For the carbohydrate fraction of the biomass (hemicellulose and cellulose), furfural and levoglucosan are interesting value-added chemicals that can possibly be produced by direct thermochemical conversion (Figure 14.2).

Although several (dry) thermochemical processes for furfural production have been explored in recent decades, modern commercial processes to produce furfural involve mostly aqueous phase hydrolysis/dehydration processes operating at relatively low temperatures (around 200 °C) and often using catalysts like sulfuric acid. This leaves the anhydrosugar levoglucosan (dehydrated glucose) as the most interesting candidate that could be directly produced from the carbohydrate fraction of biomass by (staged) degasification or pyrolysis. Alternatively, staged degasification could be targeted at the production of groups of chemicals that can be upgraded (UG) using existing (petro) chemical technology like selective hydrogenation (Sections 11.3 and 11.3.1). Examples of these groups are carboxylic acids (formic, acetic, propionic), furans (furfural, furfuryl alcohol, furanone, hydroxymethylfurfural), C_2, C_3, and C_4 oxygenates (hydroxyacetaldehyde glyoxal, acetol), anhydrosugars (predominantly levoglucosan), and hydroxylated aromatics and aromatic aldehydes, which constitute potential thermochemical degradation products from lignin. It is obvious that the separation and subsequent upgrading of these groups of chemicals is easier and cheaper than the isolation of a single chemical from the complex mixture of thermal degradation products.

Product gas samples were analyzed offline with standard gas chromatography, flame ionization detection, and mass spectrometric detection (GC/FID/MS) for most organic compounds, ion-chromatography (IC) for formic acid, and Karl-Fisher titration for water. This standardized method has been developed for the following set of species that are representative for typical thermal degradation products from lignocellulosic biomass: methanol, carboxylic acids (acetic acid, formic acid), other C_2–C_4 oxygenates (acetaldehyde, methylformate, methylacetate, propanal, acetone, 2-butenal, hydroxy-acetone (acetol), 1-hydroxy-2-butanone), furans (alpha-angelica lactone, 5-methyl-2(3H)-furanone, furfural, 5-methyl-2-furaldehyde, furfuryl alcohol, 2(5H)-furanone, hydroxymethylfurfural), levoglucosal, phenols (2-methoxyphenol [guaiacol,], 4-methylguaiacol, phenol, eugenol, 3-ethylphenol, 2,6-dimethoxyphenol, iso-eugenol, pyrocatechol, syringaldehyde, hydroquinone), and other aromatics (3,4,5-trimethoxytoluene, 1,2,4-trimethoxybenzene). Typically, 40–60% of the total GC peak area is attributed to unidentified components.

De Wild et al. describe experimental proof of principle activities for staged degasification, a simple and elegant thermochemical conversion option to valorize lignocellulosic biomass. Due to the overlapping thermal stabilities of the main biomass constituents,

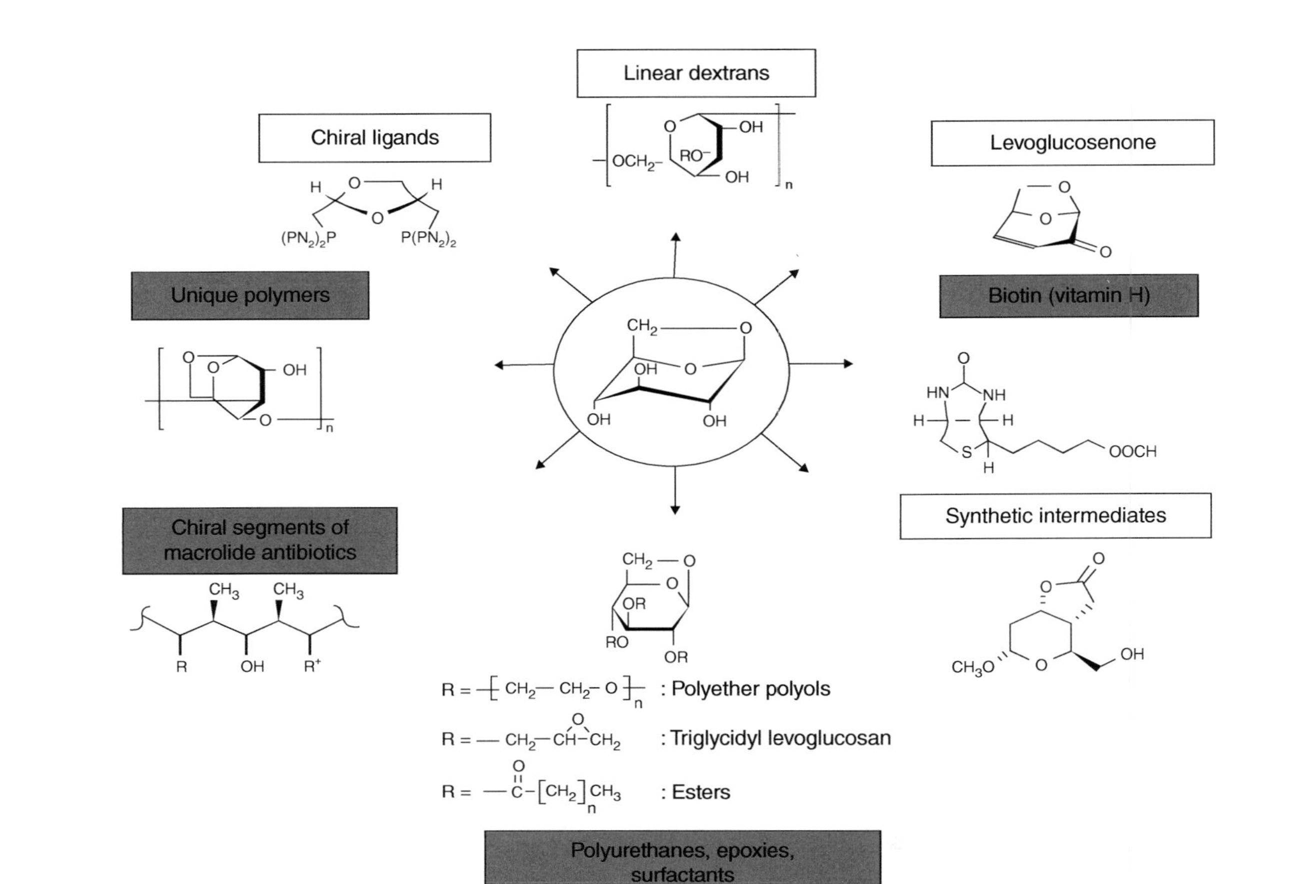

Figure 14.2 Conversion of levoglucosan.

degasification of the feedstock during a discontinuous step-wise temperature ramp in an auger reactor (a single screw moving-bed type of reactor) leads to complex mixtures of degradation products, with each staged degasification mixture consisting of small amounts of degradation products that originate from all three main biomass constituents. Except for acetic acid, yields of individual chemicals are generally below 1 wt.% (based on the dry feedstock weight). However, certain groups of thermal degradation products like C_2–C_4 oxygenates and phenols are formed in higher yields up to 3 wt.%. These results are roughly similar for the four selected biomass types: beech, poplar, spruce, and wheat straw. The only major difference is the higher yield of methanol and phenols for the deciduous beech and poplar woods when compared to spruce and straw. Slow pyrolysis of beech in a bubbling fluidized bed typically yields more water, less permanent gases and char, less methanol and phenols, and more levoglucosan when compared to conventional pyrolysis in a screw reactor. The main reasons for these differences are the longer solid residence time and the large gradients in the screw reactor when compared to the fluidized bed. Depending on the added value of the product, a limited yield is not necessarily a drawback for a cost-effective process, provided that product selectivity is sufficient for effective separation and upgrading. To conclude, staged degasification is an elegant thermochemical conversion option to valorize lignocellulosic biomass; but to increase product yields and/or selectivity, more research and development (R&D) efforts are needed, especially toward optimization of reactor conditions, application of catalysts, and/or specific biomass pretreatments. Indeed, results of a hybrid degasification approach involving a specific hydrothermal pretreatment and subsequent solid state ^{13}C nuclear magnetic resonance (NMR) characterization of the solid products have indicated that significantly higher yields of value-added condensable organic compounds can be achieved.

14.5.3 Levoglucosan

Levoglucosan (1,6-anhydro-ß-D-glucopyranose) is a potential chemical obtained from pyrolysis of cellulosic biomass. The utilization in synthesis of various chemicals (especially chiral chemicals) and materials is very promising. Its unique chemical nature, i.e. 1,6-acetal ring and $^1C_4$ conformation, makes it an attractive chiral raw material. Notably, the rigid conformation and sterically hindered ß-D-face of the molecule are advantageous for regioselective protection of the OH groups and stereoselective functionalization. Preparation of various biologically active compounds such as antibiotic macrolides has been approached starting from levoglucosan. Sacchardication through rapid or vacuum pyrolysis of cellulosic biomass has also been proposed as a pretreatment method for ethanol fermentation.

Since Pictet's report in 1918, polymerization of levoglucosan has been studied extensively. Many kinds of polysaccharides have been prepared from polymerization of levoglucosan and its derivatives.

Yield of levoglucosan from cellulosic biomass is known to be influenced by various factors. The pyrolysis temperature and inorganic impurities are the most important factors. Effective production of levoglucosan requires >300 °C. Many papers have reported that very small amounts of the inorganic impurities reduced the levoglucosan yield greatly, although the mechanism has not been clarified yet. As a factor relating to cellulose itself, the levoglucosan yield increases with increasing crystallinity of cellulose. Some papers also indicate the influences of other constituent polymers in cellulosic biomass; scientists have reported that most of the identified volatile products from cellulose and lignin increased in their yields in pyrolysis of cellulose-lignin mixtures. Figure 14.3 shows a schematic of the production of bio oil from biomass and biowaste.

Mass- and heat-transfer efficiencies have also been discussed with levoglucosan yield. The levoglucosan yield increases under vacuum conditions, since levoglucosan effectively vaporizes before suffering from secondary reactions. Fast pyrolysis conditions, which are characteristic in rapid heating and quick recovery of the product vapor, are also preferable in production of levoglucosan.

Accordingly, the relationships between pyrolysis conditions and levoglucosan yield are well documented. However, it has not been discussed how coexisting products from cellulose and other cellulosic biomass affect the recovering process of levoglucosan. Levoglucosan is reported to be stabilized up to 350 °C in some aromatic substances with high π-electron densities, probably through complexation with the CH/π interaction. Such alternation of reactivity may be included in pyrolytic production of levoglucosan; understanding secondary decomposition behavior in the presence of other pyrolysis products will be useful to improve that pyrolytic production.

For pyrolysis of biomass, a widely variety of reactor configurations have been studied. Fluid beds offer robust and scalable reactors, but the problem of heat transfer at large scale is not yet proven. Circulating fluid beds and transported beds may overcome the heat-transfer problem, but scaling is not yet proven, and there is an added problem of char attrition.

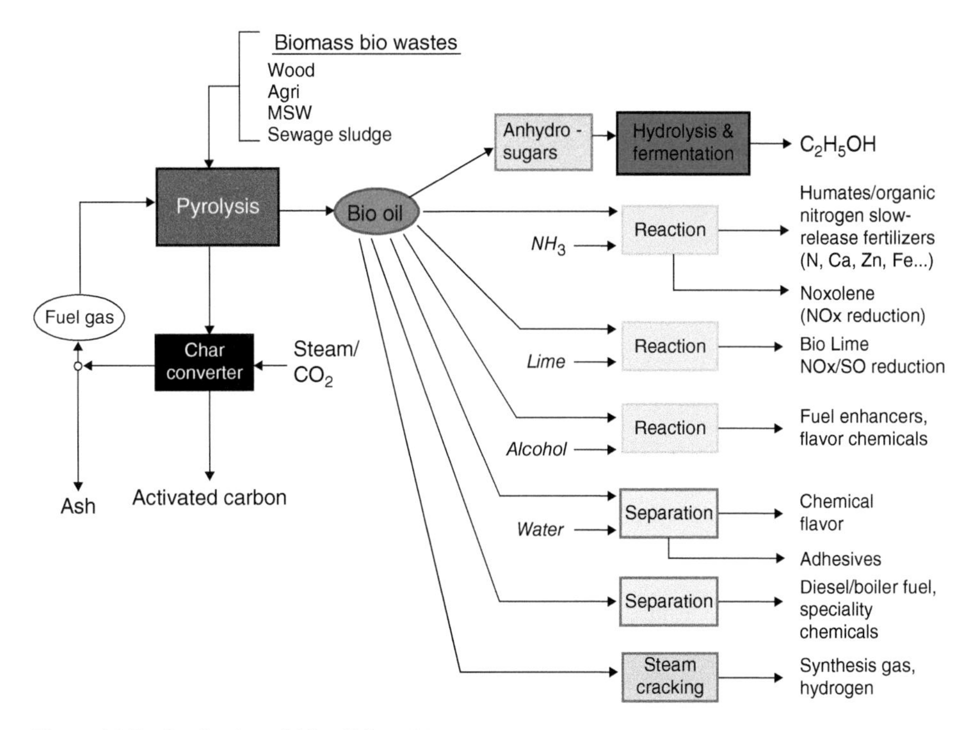

Figure 14.3 Production of bio-oil from biomass.

Mechanical devices such as ablative, rotating cone, and screw reactors offer advantages of compactness and absence of fluidizing gas, but may suffer from scaling problems as well as issues associated with moving parts at high temperatures. Liquid bio-oil products have the considerable advantage of being storable and transportable as well as the potential to supply a number of valuable chemicals.

There are specific challenges facing pyrolysis products that relate to the technology, products, and applications, including:

- Cost of bio-oil (10–100% more than fossil fuel).
- Availability (there are limited supplies for testing and development of applications).
- Lack of standards for use and distribution of bio-oil; and inconsistent quality, which inhibits wider usage.
- Incompatibility with conventional fuels.
- Lack of user familiarity with bio-oils, and the need for dedicated fuel-handling systems.

Further Reading

Bajus, M. (2008). *Pet. Coal* 50 (3): 27–48.

Bajus, M. (2010). Pyrolysis of woody material. *Petroleum & Coal* 52 (3): 207–214.

Bowen, C.P. and Jones, D.F. (2008). Mega-olefin plant design: reality now. *Hydrocarbon Processing*: 71.

Kawamoto, H., Morisaki, H., and Saka, S. (2009). *J. Anal. Appl. Pyrolysis* 85: 247–251.

Marsh, H., Hintz, E., and Rodriguez-Reinoso, F. (1997). *Introduction to Carbon Technologies*. Spain: University of Alicante.

Meier, D. (2008). Applied pyrolysis of biomass for energy and chemicals. Paper presented at the 18th International Symposium on Analytical and Applied Pyrolysis, Lanzarote, Canary Islands.

Nakamura, D.N. (2008). *Oil and Gas Journal* July 28: 46.

Pictet, A. (1918). Sur la transformation de la levoglucosane en dextrine. *Helv. Chim. Acta* 1 (1).

Ravichandran, P., Sugumaran, P., Seshadri, S. et al. (2018). Optimizing the route for production of activated carbon from Casuarina equisetifolia fruit waste. *Royal Society Open Science* 5 (7).

de Wild, P.J., den Uil, H., Reith, J.H. et al. (2009). *J. Anal. Appl.Pyrolysis.* 85: 124–133.

15

Recycling Technologies

CHAPTER MENU

15.1 Feedstock Recycling of Plastic Wastes

The severe limitations on the mechanical recycling of plastic wastes highlight the interest in and potential for feedstock recycling, also called *chemical* or *tertiary* recycling. It is based on the decomposition of polymers by means of heat, chemical agents, and catalysts to yield a variety of products ranging from starting monomers to mixtures of compounds, mainly hydrocarbons, with possible applications as a source of chemical or fuels. The products derived from plastic decomposition exhibit properties and quality similar to those of their counterparts prepared by conventional methods.

A wide variety of procedures and treatments have been investigated for feedstock recycling of plastics and rubber wastes (Figure 15.1). For the purposes of this book, these methods have been classified into the following categories:

- Chemical depolymerization by reaction with certain agents to yield the starting monomers.
- Gasification with oxygen and/or steam to produce synthesis gas.
- Thermal decomposition of polymers by heating in an inert atmosphere.
- Catalytic cracking and reforming. The polymer chains are broken down by the effect of a catalyst, which promotes cleavage reactions.
- Hydrogenation. The polymer is degraded by the combined actions of heat, hydrogen, and, in many cases, catalysts.

The progress and current status of these alternative methods of feedstock recycling are described in this chapter. At present, feedstock recycling is limited by the process economy rather than for technical reasons. Three main factors determine the profitability of these alternatives: the degree of separation required in raw wastes, the value of the products obtained, and capital investment in processing facilities. In most of the methods

Petrochemistry: Petrochemical Processing, Hydrocarbon Technology, and Green Engineering, First Edition. Martin Bajus.

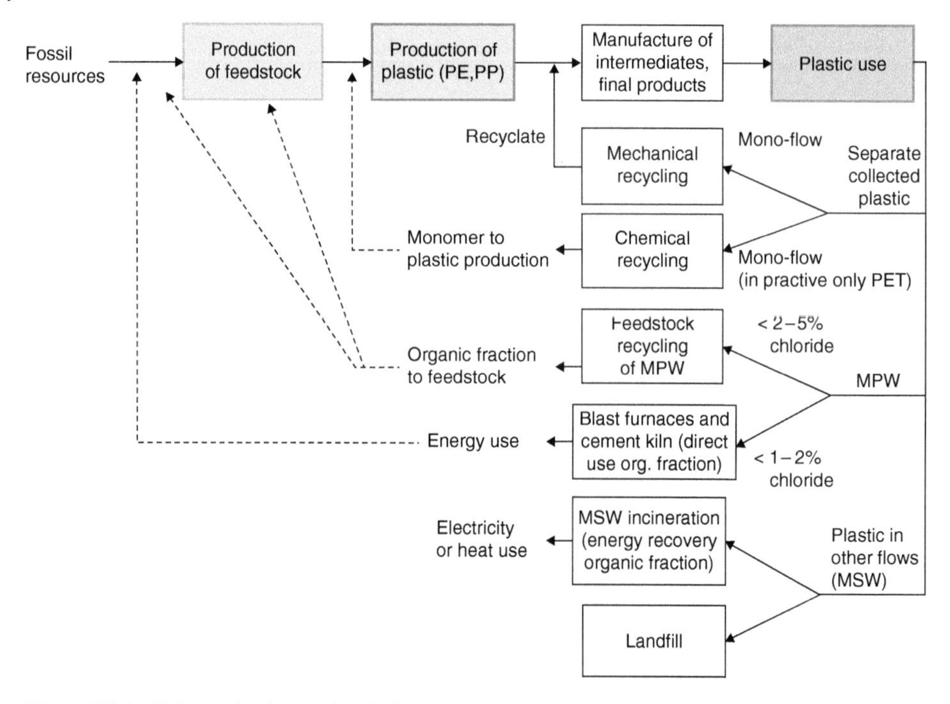

Figure 15.1 Schematic of material chains related to plastics, from production to waste-disposal routes.

listed, some pretreatment and separation operations must be carried out on the plastic wastes prior to feedstock recycling, which results in an increase in recycling costs. According to the separation steps required, the different methods of feedstock recycling can be ordered as follows: gasification < thermal treatments ≅ hydrogenation < catalytic cracking < chemical depolymerization.

Many projects concerned with chemical recycling of waste plastics have failed in the past due to the relatively low price of the derived products. In recent years, there has been a trend toward the production of value-added compounds such as olefin gases, paraffins, active carbons, etc. In general terms, the commercial value of the products obtained in the different treatments can be ordered as follows: thermal oils ≅ synthesis gas < hydrogenation oils ≅ catalytic oils < monomers.

It is interesting to note that the required pretreatments and product value are in almost reverse orders However, many other factors should be included for an adequate comparison of these alternatives: the possibility of carrying out the treatment in existing or new facilities, minimum size of industrial plants needed to be profitable, required investment, plant location, etc. Hofmann and Gebauer have identified the main problems involved in several of the feedstock recycling methods:

- High capital expenditure, especially in hydrogenation plants.
- Lack of a regional plastic waste volume to support the continuous operation of large-scale plants. For gasification, a minimum capacity of about 0.4–0.5 MTA is necessary.
- Pyrolysis and hydrogenation lead to a wide range of end products, which then have to be further upgraded and processed, mainly in refinery units.

Several petrochemical companies have considered possible feedstock recycling of plastic wastes in existing refinery facilities, which would avoid the need to invest in and build new processing plants. This alternative is based on the similarity of elementary composition between plastics and petroleum fractions. Moreover, taking into account the differences in production between plastics and the total of petroleum-derived products, plastic wastes could be incorporated into the refinery stream in relatively low amounts. The main problem associated with this approach is the possible presence of undesired elements and compounds (Cl, N, metals, etc.) in the plastic wastes, which would be introduced into the refinery. In such cases, even a small drop in yield or efficiency, when multiplied by hundreds of tons, would have dramatic effects on the refinery economy. Accordingly, plastic wastes should be intensively pretreated and conditioned prior to being added to petroleum fractions. Main refinery units may be used in the processing of plastic wastes according to Hofmann and Gebauer (visbreaking, hydrogenation, and gasification). Processing of plastic wastes in the cookers and catalytic cracking units of refineries has also been proposed. The economic viability of this alternative greatly depends on the proximity of the plastic waste generation points and the refineries, due to the high cost of plastic transportation.

15.2 Fuels and Chemicals from Polymer Waste

The goal of our research was to study chemical and feedstock recycling of polymer waste. We have studied:

- Whether oils/waxes obtained from thermal cracking of individual and mixed polymers are able to produce valuable fuels (automotive gasoline and diesel fuel)
- Whether it is possible to obtain low-molecular alkenes (ethene and propene) of solutions (oil/wax in heavy naphtha) during copyrolysis

Individual and mixed polymers are suitable for thermal cracking in an inert nitrogen atmosphere at 450 °C. Gaseous, liquids (oils/waxes), and solid products are produced (Figure 15.2 and Figure 1.5).

The prevailing gaseous hydrocarbons from thermal cracking of individual and mixed polymers are C_3 (propane, propene), C_2 (ethane, ethene), C_4 (butane, 1-butene and methylpropene), and C_5 (pentane, 1-pentene). Formation of aromatics present in gas is low, or no aromatics are present. In the case of thermal cracking of simulated plastic waste, production of aromatics is also low.

Polyethylene oils/waxes are composed predominantly of linear alkanes and 1-alkenes, but polypropylene oils/waxes are also composed of branched alkanes and alkenes. The addition of polypropylene to the individual polyethylene (LDPE, LLDPE, HDPE) influences the product distribution in oils/waxes and gases. In addition to linear alkanes and alkenes, branched alkanes and alkenes are also present in mixtures.

Oils/waxes are suitable for distillation. The first fraction obtained from each feedstock was distilled at atmospheric pressure up to 180 °C, corresponding to gasoline. The second fraction underwent a vacuum distillation, which was stopped at the temperature of 190 °C that corresponds to 330 °C at atmospheric pressure. This fraction corresponds to diesel fuel.

Unsaturation of oils/waxes is represented by the value of the bromine number. The values of bromine numbers for distillates (gasoline and diesel fuel) obtained by distillation of

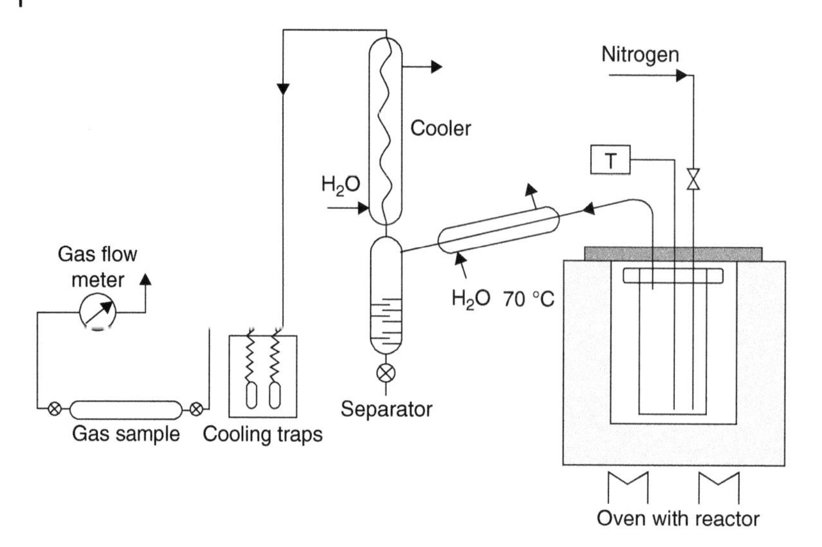

Figure 15.2 Schematic diagram of the thermal decomposition reactor.

oils/waxes are higher than for original oils/waxes. They contain higher amounts of unsaturated hydrocarbons. It is interesting that the bromine numbers of distillates and oils/waxes of LDPE/PP (1 : 1) are similar to individual polypropylene. The presence of aromatics in gasolines and diesel fuels is low, or they are not present. Another advantage of obtained fractions is that they do not contain sulfur.

The octane number is a measure of gasoline quality. The value of the research octane number (RON) for gasoline is high despite a low amount of aromatics. The RON values for original gasolines are higher than for hydrogenated gasolines. Polypropylene is an exception. RONs for original gasoline of polyethylenes are about 81 to 82 units and for hydrogenated naphtha fractions are from 67 to 71 units. The RON for polypropylene is about 93 before hydrogenation and after hydrogenation.

The value of the diesel index increases after hydrogenation of diesel fuel. The Diesel index for original diesel fuels (from 74 to 76) and hydrogenated diesel fuels (from 98 to 105) presents excellent combustion properties. The amount of aromatics is low.

Oils/waxes obtained by thermal cracking of polymers are dissolve better in the liquid petroleum fraction of heavy naphtha because they are in the form of smaller molecules. Thanks to their composition (alkanes, alkenes, a small amount of [or no] aromatics), they are suitable as feedstock for pyrolysis/copyrolysis. The presence of PET, PVC, and PS does not worsen the quality of oils/waxes.

The addition of oils/waxes influences yields of ethene and propene during copyrolysis in the presence of steam. The yields of ethene and propene are higher or comparable with the yields of virgin heavy naphtha. The finding proving that the solution of LLDPE/LDPE/PP/HDPE/PVC/PET/PS is able to provide high yields of propene and similar or higher yields of ethene in the comparison with virgin heavy naphtha, at both temperature levels, is rather positive. The yields of propene are similar to yields of a solution of HDPE with naphtha. We have confirmed that even if the solution contains PET, PVC, and PS, it is suitable for copyrolysis.

The addition of oils/waxes influences the formation of coke as does the amount and composition of pyrolysis gas and liquid during copyrolysis of a solution (oil/wax in naphtha) in severe reaction conditions (without steam). Heavy naphtha was used as a baseline. The amount of produced coke obtained from pyrolysis of heavy naphtha was compared to the amount of produced coke obtained from copyrolysis of solutions.

The shapes of kinetic curves of formed coke have the same trend for all feedstocks. Kinetics of coke deposition confirms that the coking rate is the highest stage in the initial stage of the experiment. The catalytic influence of the metal reactor surface causes high initial coking rates that evidence the coking carried out on the clean reactor surface. The high initial coking rate also depends on the properties of the feedstock. The initial contribution of coke deposition for different feedstocks is minimal. The differences in the formed amount of coke are evident in other stages. The formation of carbon oxides is higher under copyrolysis conditions.

Oils/waxes obtained in thermal cracking of polymers (individual, mixed polymers, polymers simulating the composition of plastic waste) in 10 and 15 mass % amounts in heavy naphtha are suitable as feedstocks for copyrolysis at 820 °C, which is evident from the formed depositions of coke.

The obtained results suggest the possibility of polymer recycling using copyrolysis. The amount of formed coke obtained in copyrolysis for 10 or 15 mass % solutions doesn't vary a lot in comparison to the amount of formed coke obtained from pyrolysis of individual heavy naphtha. Adding 10 mass % (and more) of waste polymers to virgin heavy naphtha can spare a substantial amount of this valuable liquid raw material.

15.3 Fuels and Chemicals from Used Tires

Used tires are problematic waste materials on one hand and valuable potential secondary raw materials on the other hand. The European Union intends to limit the amount of organic matter being landfilled in the near future, so the only disposal alternatives left for rubber wastes will be incineration and recycling. While incineration utilizes the energetic content of the wastes, recycling may partially conserve their chemical structure and therefore possibly represent a better ecological approach.

In recent years, about 40% of tire and rubber waste was incinerated, 17% was reused or exported to countries outside the European Union, 9% was used as recyclate, and 20% was used in other ways. Due to the irreversible chemical structure of the rubber, primary or secondary recycling is only possible to a limited extent. Research in the field of chemical recycling of rubbers by pyrolysis has focused on scrap tires as the most important rubber waste. Apart from scrap tires, pyrolysis was selected to thermally decompose rubbers for non-tire applications (technical rubbers represent an important disposal problem).

Whole tires were pyrolyzed successfully on a laboratory and technical scale in a flow reactor using the Deep Scavenger steam cracking process or steam cracking activation process (DSSC/SCA), which was developed by the Slovak University of Technology in Bratislava and successfully commercialized by DRON Industries, Mliečany, Slovakia. The laboratory batch reactor has been used for decomposition of used tires (Figure 15.2).

At a temperature from 400 to 700 °C, fuel gas (18% mass), oil (35% mass) with a high content of BTX aromatics and d,l-limonene (Figure 15.3), coke (40% mass), and iron-steel (7% mass) were yielded. The design of the pilot plant reactor throughput of 200 kg/h of used tires is shown schematically in Figure 15.4. The separator for producing gases with a throughput 15 000 t/year is shown in Figurc 15.5.

The pyrolysis oil shows a high heating value of about 42.5 MJ/kg. Pyrolysis gases are composed of methane, C_2–C_5-hydrocarbons, hydrogen, carbon monoxide, carbon dioxide, and hydrogen sulfide. These gases are comparable with natural gas and are therefore a good source for heating plants. Energy-producing fuel is being utilized for the tire-processing plant or as an energy input for other purposes.

The oil fraction is composed of aromatic hydrocarbons, polyaromatic species, limonenes, aldehydes, ketones, and carboxylic acids. This oil fraction has variable applications and in particular is used as an energy-producing fuel.

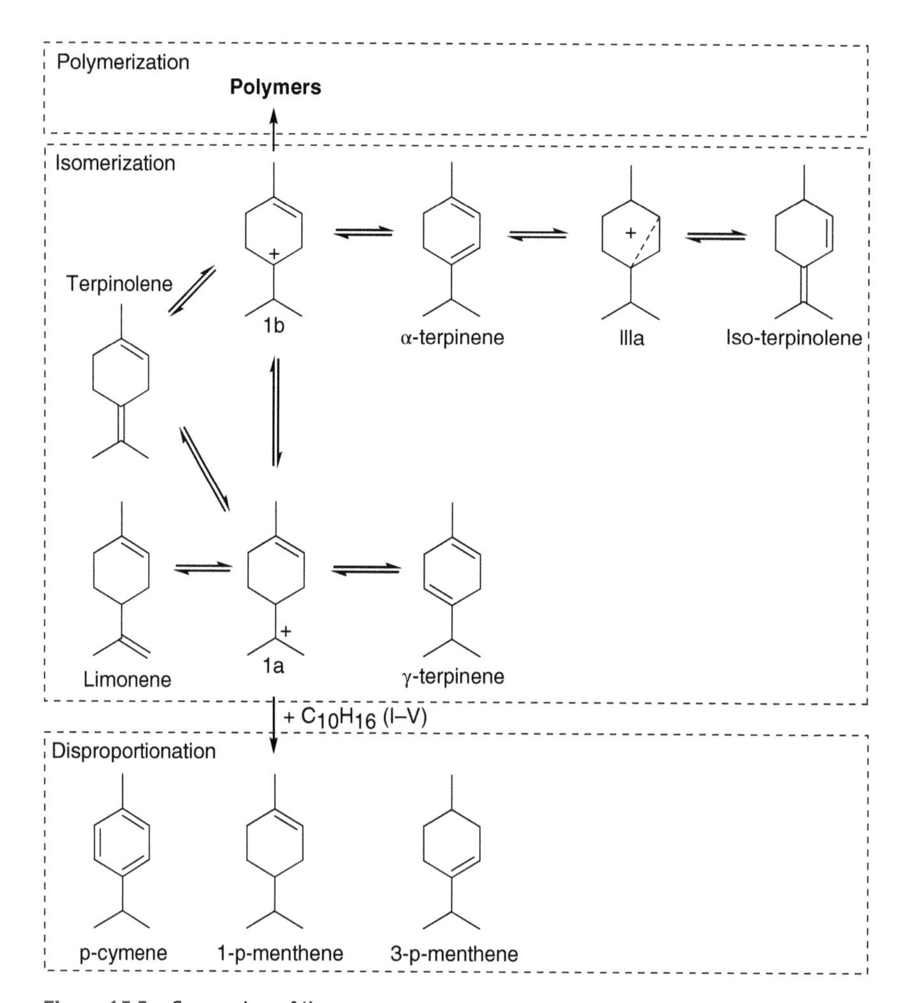

Figure 15.3 Conversion of limonene.

Figure 15.4 The design of the pilot plant reactor.

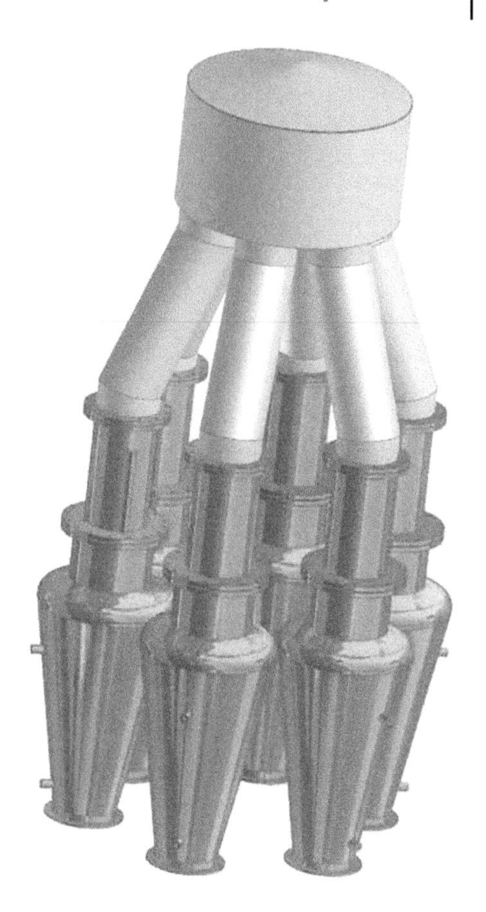

Figure 15.5 The view of the technological unit in Mliečany.

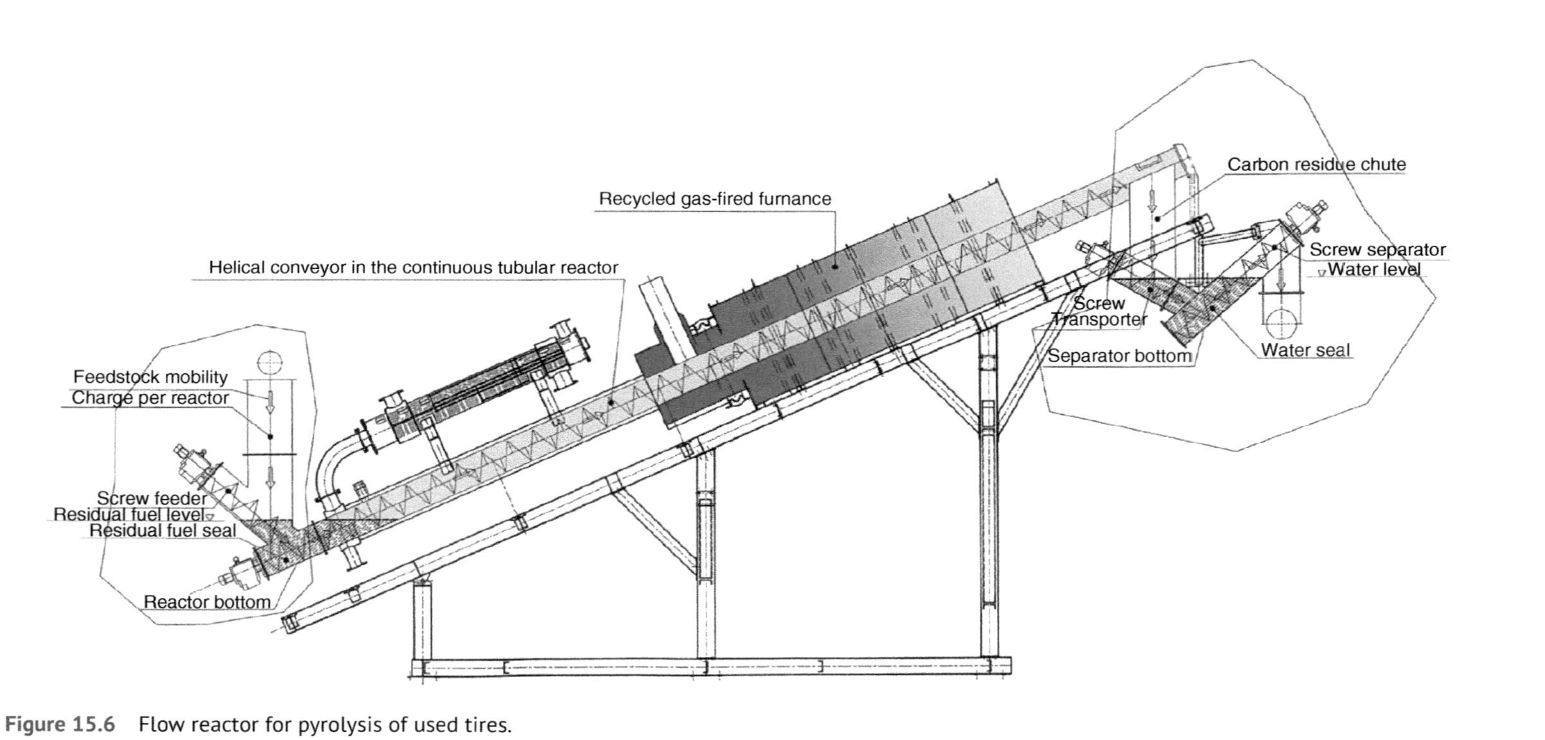

Figure 15.6 Flow reactor for pyrolysis of used tires.

We received coke at a temperature from 400–700 °C with area 70–80 m^2/g. Surface areas are comparable with raw carbon black, which is used in tire manufacture. Carbon black, as the most valuable fraction, can be reused after activation with steam as a filler material in technical rubber production. Pyrolysis char contains a higher proportion of ash and has a large particle size, which make this product an alternative to carbon black. Coke has a heating value of 28–34 MJ/kg. Iron material is a high-quality feedstock waste for steel production.

Advantages of DSSC/SCA process are as follows:

- New, effective technology for thermal cracking of used tires with steam activation in the flow reactor (Figure 15.6).
- Full exploitation and processing of rubber waste and used tires.
- Total material conversion of rubber waste and used tires.
- We have used produced gases to heat the reactor.
- High potential in the near future for the recycling of plastic material.

Further Reading

Aranzo, J. (2008). Pyrolysis technology for wastes treatment and energy production. Paper presented at the 18th International Symposium on Analytical and Applied Pyrolysis, Lanzarote, Canary Islands.

Farkas, L. (2016). Method of thermal decomposition of organic material and device for implementing this method. Slovak Republic patent 288 338.

Hájeková, E. and Bajus, M. (2005). *J. Anal. Appl. Pyrolysis* 73: 177–188.

Hájeková, E. and Bajus, M. (2005). *J. Anal. Appl. Pyrolysis* 74: 270–281.

Hájeková, E., Mlynková, B., Bajus, M., and Špodová, L. (2007). *J. Anal. Appl. Pyrolysis* 79: 196–204.

Hájeková, E., Špodová, L., Bajus, M., and Mlynková, B. (2007). *Chem. Pap.* 61 (4): 262–270.

Hájeková, E., Mlynková, B., Fáberová, S., and Bajus, M. (2008). *Pet. Coal* 50 (1): 52–55.

Hofman, U. and Gebauer, M. (2002). Chemical recycling of used plastic materials. In: Plastics Waste-Feedstock Recycling, Chemical Recycling and Incineration, report 148, vol. 13, number 4. Rapra Review Reports.

Mlynková, B., Hájeková, E., and Bajus, M. (2008). *Fuel Process. Technol.* 89: 1047–1055.

Mlynková, B., Bajus, M., Hájeková, E. et al. (2010). *Chem. Pap.*, ISSN 0366-6352 64 (1): 15–24.

Soják, L., Kubinec, R., Jurdáková, H. et al. (2007). *J. Anal. Appl. Pyrolysis* 78: 387–399.

Tukker, A. (2002). *Plastic Waste: Feedstock Recycling, Chemical Recycling and Incineration*, Rapra Review Reports, vol. 13. Rapra Technology Ltd.

16

Microchannel Technologies and Nanotechnology

CHAPTER MENU

16.1 Introduction

The miniaturization of devices achieves small process volume and hold-up, large specific transport rates, and high surface-to-volume ratios. The combination of these features results in extremely short response times, considerably simplifying process control, and eases the handling of large heat and mass fluxes, thus making the synthesis of dangerous compounds safer while reducing hazardous waste. For chemical technologies, the applications of miniaturized reaction systems, or *microreactors*, is centered on gas- and liquid-phase reactions covering simple microscale mixing of different fluids in volumetric titrations, heterogeneous and homogeneous catalysis, catalytic oxidation, pyrolysis, heterocyclic synthesis, photochemical reactions, and micro-fuel cell applications (Chapter 13). Due to higher yields and selectivities, the application of more environmentally favorable reaction routes, and the possibility of replacing large plants with small plants for distributed production according to actual demand, it is to be expected that microtechnologies will eventually contribute to microreactors' sustainable development.

In the last two decades, microtechnologies and microequipment have been objects of great interest in chemical engineering. A small series of high-performance micro–heat expanders, reactors, mixers, extractors, pumps, and valves has been developed and produced. Such equipment has a range of cross-section dimensions between 10 μ and 1–3 mm.

Petrochemistry: Petrochemical Processing, Hydrocarbon Technology, and Green Engineering,
First Edition. Martin Bajus.

Microfluids – micro or microchannel technologies, as they are sometimes called – essentially accomplish the miniaturization of conventional process devices and enable the integration of reaction and unit operation elements directly with sensors and actuators. Research in the field of microscale devices has accelerated during the past 10–15 years. Researchers from various fields are intensively analyzing the possibilities offered by miniaturization of equipment to achieve a fundamental change in design philosophy for chemical plants. The smaller characteristic dimensions of microscale devices lead to an extremely high surface-to-volume ratio (S/V), which in turn provides several advantages over conventional-size reactors, mixers, and heat transfer equipment:

(a) Higher transport (e.g. heat and mass transfer) rates.
(b) Safe environment for hazardous or toxic chemicals (due to the low amount of chemicals used in the process).
(c) Simplified process control for effective process/materials screening (due to the extremely short response time).
(d) On-demand or on-site synthesis of critical chemicals such as H_2O_2, ethylene oxide.
(e) No strict limit on size reduction or expansion of plant components, since any production capacity is achievable by means of parallel operation.
(f) Development of integrated chemical analytical platforms, as in micro total analysis systems.

For chemical technologies, the application of miniaturized reaction systems, or microreactors, is centered on gas- and liquid-phase reactions covering simple microscale mixing of different fluids to volumetric titrations, heterogeneous and homogeneous catalysis, catalytic oxidation, heterocyclic synthesis, photochemical reactions (Chapter 1, Figure 1.1, Section 1.2.7) and micro–fuel cell applications (Chapter 13). Due to higher yields and selectivities, the applications of more environmentally favorable reaction routes, and the possibility of replacing large plants for distributed production according to actual demand, it is to be expected that microtechnologies will eventually contribute to their sustainable development.

Revolutionary advances offered by miniaturized devices for real-world applications are demanding a paradigm shift in approach. In many cases, it is interesting to find that reality has overtaken imagination in microtechnology research. The interests of established researchers from multidisciplinary backgrounds have led to a broad spectrum of research in the field of microchannel technologies and how these technologies are opening new pathways and reinventing approaches in chemical technologies, microreaction engineering, and medical sciences.

Microscale devices will influence the future of the process industry. Therefore, understanding and proper evaluation of the flow and mixing behavior in microscale devices such as microreactors and mixers is critical for their effective design and optimization. The accurate, optimized design of microreactors and mixers relies heavily on flow characteristics (laminar, turbulent, and laminar-to-turbulent flow regimes), and pressure-drop correlations are still discussed controversially in the literature; therefore, a careful analysis of experimental data reported by various researchers is highly desirable.

Research articles and patents published between 2000 and 2017 on microreactors and mixers have been analyzed in terms of their field of application. Many publications do not include actual information about microreactors and mixers; they are only mentioned in the introduction or the main text. Many research articles and patents are assigned to the field of chemistry (physical, analytical, organic, and inorganic), biochemistry, nanoscience and nanotechnology, instruments and instrumentations, etc. A number of chemical companies are trying to exploit the advantages of microreactors. It is observed that microreactors and mixer technologies are prominent in fine chemicals, specialty chemicals, pharmaceuticals, and consumer products applications. However, microreactors still present challenges for industrial production, including the fundamental, basic understanding of hydrodynamics and mixing processes in microscale devices. For industrially designed and optimized microscale devices, it is required to understand the geometrical, flow, and mixing parameters.

16.2 Fluid Flow in Microchannels

Over the course of the past two decades, many publications have given conflicting results on the validity of classical macroscale equations for microchannel fluid flow and heat transfer. Among this research are investigations of the validity of the macroscale equations number and the Nusselt number on the microscale. Some authors claim that new phenomena occur in microchannels, while many others report that several effects exist, which are usually neglected in deriving correlations and lead to larger discrepancies with observed experimental results in microchannels. Usually, several assumptions are made in literature studies when flow and heat transfer are modeled: (i) steady-state, fully developed flow; (ii) thermophysical properties of the fluid do not vary with temperature; (iii) simplified boundary conditions (constant wall temperature or constant heat flux); and (iv) the fluid heating due to viscous dissipation can be neglected (Section 1.2).

An accurate estimate of the pressure-drop and flow behavior in microscale devices requires precise measurement techniques. In spite of the existence of numerous experimental investigations for flow resistance and transition from laminar-to-turbulent flow in microchannels, there are discrepancies in experimental data with the conventional theories, which results in different arguments. The database comprises experimental papers that deal directly with pressure-drop measurements and the transition from laminar-to-turbulent flow in microchannel for various cross-sectional geometries (circular, rectangular, triangular, trapezoidal, hexagonal, etc.), flow rates (in terms of Reynolds number [Re]), and fluids (gases or liquids). In the case of gas (both compressible and incompressible) flow in microchannels, air, nitrogen, helium, argon, etc. are reported; while in the case of liquid flows, water, isomers of alcohols silicon oils, R134a, etc. are analyzed. The hydraulic diameter of microchannels investigated by different researchers varies between 1 and 4010 μm. The experimental friction factor data is compared with conventional theory, and it is observed that there is no unequivocal answer, although microchannel dimensions, process fluids, and conditions considered were similar in different studies. The discrepancy, which has been reported by many authors, was the transition from laminar-to-turbulent flow, even though the surface roughness is ≤1%. In some case, even though the microchannel cross-section was not circular, the experimental Poiseuille number was compared with the conventional results for micropipes (Re = 64).

As most microreactors operate at elevated temperatures, it is often the case that temperature differences in microchannels can create a non-uniform flow distribution due to differences in pressure drop, which is temperature dependent. This situation can be observed even when a proper flow distributor is applied. Therefore the problem of flow distribution is closely related to the temperature distribution in a downstream microreactor or micro–heat exchanger. On the other hand, in most heat-transfer and pressure-drop calculations for microstructure reactors, it is presumed that the inlet flow and temperature distribution across the reactor cross-section are uniform.

16.3 Intensified Superheated Processing

By reconsideration of a novel chemistry concept introduced in 2005, called *novel process windows* (NPWs), six major process-intensification pathways were identified:

- New chemical transformation
- Routes at greatly elevated temperatures
- Routes at greatly elevated pressures
- Routes at greatly increased concentrations, or solvent-free
- Process integration and simplification
- Routes in the explosive thermal runaway regime

Here, effects due to operation at temperatures above the atmospheric boiling point are reported. There are two kind of processing at greatly elevated temperature. The first involves bringing strongly cooled (cryogenic) chemical processes to ambient temperature or slightly above/below. Here, the intrinsic kinetic potential of these reactions is released, which cannot be exploited by conventional technology since mixing masking is inherent (owing to the overly long mixing times – in the seconds range). The second and more NPW relevant issue is to perform organic reactions under much higher temperatures of >150 °C, most favorably in the range from 180 to 250 °C, which is significantly above the solvent's boiling point and uses overpressure to maintain a single-phase operation: superheated processing. Here, the reaction rates are accelerated according to the temperature dependence given in the Arrhenius equation. There is considerable enlargement of the operational regime with regard to temperature and pressure as compared to batch operation and even to the encased harsh processing in microwave chemistry, which nicely demonstrates the potential of and justifies the naming of NPWs.

Diverse gas-phase reactions have been investigated in microreactors, among them (partial) oxidation, hydrogenation, dehydrogenation, dehydration, and reforming processes. Particular attention has been drawn to achieving excellent temperature control and preventing hotspots. For many reactions, increases in selectivity were found: in particular, many examples of partial oxidations were described, including processes of utmost industrial importance such as ethylene oxide synthesis. With consecutive processes, as e.g. given for multiple hydrogenations, high selectivity was achieved for species that are thermodynamically not the most stable molecules of all species serially generated, such as monoenes yielded by hydrogenation of polyenes. Increases in conversion were also achieved, e.g. by processing at higher pressure and high temperature, often in an explosive regime. As a consequence, high space–time yields were reported as well. In many cases, reactor performance was achieved that was better than fixed-bed technology. Process safety was found to be high when using microreaction devices. With respect to process optimization, fast serial screening of process parameter variation was conducted, at low sample consumption.

16.3.1 Oxidative Dehydrogenation of Hydrocarbons

Microchannel technology seems to be very useful for performing fast, strongly exothermic, heterogeneously catalyzed gas-phase reactions. The oxidative dehydrogenation of propane was found to be a well-suited, sensitive test reaction, which shows many typical features of this class of reactions (Section 8.1). Due to their superior heat-transfer properties, microstructure reactors are well suited for performing strongly exothermic heterogeneously catalyzed gas-phase reactions. In order to utilize the full potential of this reaction technology, a new low-cost manufacturing concept was developed, using a Ni-Ag-Sn solder system for bonding the individual structured steel pallets. Three different methods for depositing a VO_x/ γ-Al_2O_3 material on the microchannels were investigated with respect to morphology, mechanical stability, and catalytic behavior of the obtained coatings. To evaluate the performance of the coatings, oxidative dehydrogenation of propane served as a sensitive test reaction. The modules and catalytic coatings withstood the applied reaction conditions (400–600 °C at ambient pressure), which makes them safe, flexible tools for research activities and small-scale production processes.

Microchannel chemical reactors are a new type of structured reactor. The common feature of such reactors is small single channel or many parallel channels with submillimeter cross-section. The small cross-section of microchannels permits one to obtain a high surface-to-volume ratio and high mass-and-heat transfer rates, which can be 1–2 orders of magnitude higher than those observed for systems with a fixed catalyst bed. As a result, microchannel reactors can provide nearly isothermal conditions even for extremely exothermic or endothermic reactions. In the microchannel plates for the heterogeneously catalyzed reactions, the catalyst is generally deposited onto the walls. For the catalytic partial oxidation of methane, reactors are traditionally made of the FeCrAl alloy (Section 8.1), the surface of which is covered with a thin catalyst layer using sol–gel methods. Microchannel catalytic reactors have repeatedly proved their high efficiency in the process of partial oxidation of methane (at 650–780 °C) as compared to traditional fixed-bed catalytic reactors.

16.3.2 Steam Reforming of Ethanol

Steam reforming of ethanol was studied over $Rh/CeO_2/Al_2O_3$ catalysts in a microchannel reactor. First the catalyst support was deposited on the metallic substrate by wash coating, and then CeO_2 and active metal were sequentially impregnated. The effect of the support composition as well as the active metal composition on steam reforming of ethanol was studied at atmospheric pressure, with an ethanol to water molar ratio 1 : 6, over a temperature range of 400–600 °C. The microchannel reactor's performance was compared with a packed-bed reactor using a 2% Rh/20% CeO_2/Al_2O_3 catalyst at identical operating conditions. Activity was similar in both reactors, but selectivity to desired products was higher in the microchannel reactor. The hydrogen yield obtained in the microchannel reactor was higher: ~65 l/g h, as compared to 60 l hydrogen/g h in the packed reactor.

16.3.3 Fischer-Tropsch Synthesis and GTL

FTS is a well-known catalytic process for the conversion of synthesis gas into liquid fuels (Section 12.2.10). The main product of the process is a mixture of hydrocarbons of variable molecular weight. Different metallic supports (aluminum foams, honeycomb monolith, and micro-monolith, respectively) have been loaded with a 20% Co – 0.5% Re/γ-Al_2O_3 catalyst by the wash-coating method. Layers of different thicknesses have been deposited onto the metallic supports. The catalytic coatings were characterized, measuring their textural properties, adhesion, and morphology. These structured catalysts have been tested in FTS and compared with a microchannel block presenting perpendicular channels for reaction and cooling. Selectivity depends on the type of support used and mainly on the thickness of the layer deposited. The C_5^+ selectivity of the microchannel reactor is higher than that of the structured supports and the powder catalyst (Figure 16.1).

The most efficient path to syngas is the conversion of natural gas via steam methane reforming (Section 12.2.1). Both steam methane reforming and FTS benefit from process intensification offered by microchannel technology, resulting in smaller, less costly processing hardware and thus enabling cost-effective production of synthetic fuels from smaller facilities appropriate for stranded and associated gas resources, both on and offshore. The products from FTS processes can be upgraded into diesel or synthetic paraffinic kerosene, or simply blended with crude oil for transport to the word market. Compared to conventional tube-style reactors, the reaction passages in microchannel FTS have characteristic dimensions that are orders of magnitude smaller, which greatly improves heat and mass transfer (Section 3.4.4, Figures 3.6 and 3.7).

This allows optimal temperature control across the catalyst bed, which minimizes catalyst activity and life, and leads to far higher reactor productivity. Microchannel steam methane reforming offers similar advantages, reducing reactor size by up to 90%. Due to improved volumetric and catalytic productivity, microchannel steam methane reforming and FTS enable lower capital and operating costs compared to conventional gas-to-liquids (GTL) processes. Because conventional GTL technologies are not economically viable at small scale, the current focus is on large, land-based natural fields, such as those in Qatar. Microchannel technology permits economic production at this smaller scale (Section 13.5.1). In addition to steam methane reforming and FTS, microchannel processing technology is being applied to the final step of a synthetic fuel process: hydrocracking.

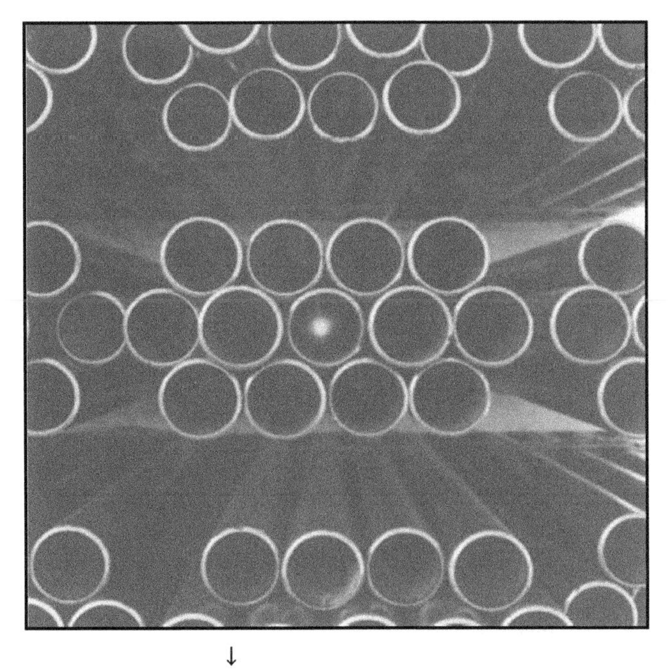

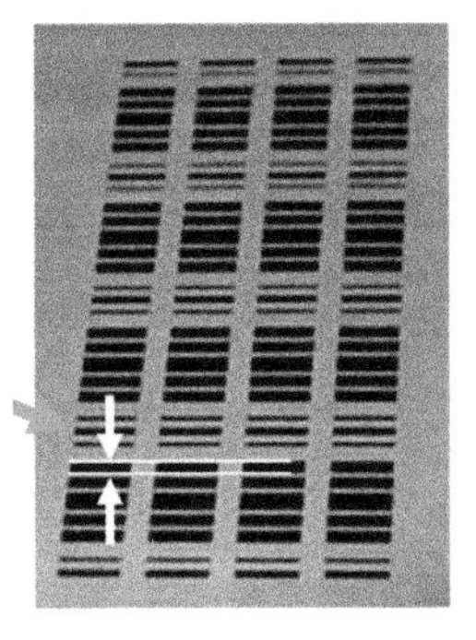

Figure 16.1 Conventional vs. microchannel reactors.

16.4 Steam Cracking of Hydrocarbons

The advantages of microchannel technology are very useful for performing fast, strongly endothermic homogeneously and heterogeneously catalyzed gas-phase reactions. The pyrolysis of hydrocarbons (from crude oil and natural gas) was found to be a well-suited, sensitive test reaction, which shows many typical features of this class of reaction. In addition, the pyrolysis of hydrocarbons produces low-molecular olefins as opposed to conventional production processes (Chapter 7). Although ethylene is a very important base chemical, it is primary obtained as a main product from steam cracking and other refinery processes (e.g. fluid catalytic cracking and deep catalytic cracking; Section 5.5.1). However, all of these processes are thermodynamically limited, are extremely energy-intensive, and commonly suffer from catalyst deactivation by heavy coking. Therefore, a route that avoids most of these problems is attractive.

The main route to ethylene production is by the thermal cracking of hydrocarbons. This reaction is carried out in tubular coils located in the radiant zone of fired heaters. Steam is added to reduce the partial pressure of the hydrocarbons in the radiant coils. The reactions that result in the transformation of mostly saturated hydrocarbons to olefins are highly endothermic and require a reactor temperature in the range of 750–900 °C depending on the feedstock and design of the reactor coils.

The on-stream availability of thermal cracking reactors is determined by fouling of either the cracking coils or the cracked-effluent transfer line exchanger (TLE). Coke is produced as a side product of thermal cracking, and it deposits on the radiant coil walls and inside the tubes of TLEs. This limits heat transfer and decreases the pressure drop, thus reducing olefin selectivity. The run length is normally determined by the tube metal temperature increase of the radiant coil, the outlet temperature of the TLE, or the increased pressure drop (Section 7.3.3.1).

Coke is believed to be formed by two mechanisms: catalytic and condensation. The metal surface of the cracker coil catalyzes the growth of a filamentary type of coke that contains metal granules. The second type of coke is formed by condensation, polymerization, and/or agglomeration of heavies in the gaseous phase. The precise modeling of coking in commercial cracking coils is a highly complex process, and the mechanism is not fully understood.

The radiant coils are centrifugally cast from 25 Cr/35 Ni or 35 Cr/45 Ni materials for their carburization and creep resistance. These materials have a maximum service temperature of up to 1150 °C (Section 7.2.2.5). The short residence time (SRT) heater design is characterized by a configuration that maximizes the coil surface-to-volume ratio at the coil inlet, where coking tendencies are low. This is done by using small-diameter parallel tubes in the inlet pass or passes. At the outlet, where coking tendencies are high, large-diameter tubes are used. Typical radiant coil dimensions for high-selectivity furnaces are in the range of 40–120 mm ID (S/V, surface-to-volume ratio = 1.0–0.33 cm^{-1}).

The main question connected with the use of the pyrolysis tubular reactors deals with deviations from ideality. In the laboratory, it is very difficult to realize ideal plug flow, characterized by a Reynolds number greater than 10^4 and a length/diameter ratio greater than 100. Generally, the Reynolds number is less than 100, the internal diameter of the reactor tube is around 5 mm S/V (surface-to-volume ratio = 4 cm^{-1}), and the gas flow is laminar, with a parabolic profile. In the use of a laminar-flow tubular reactor, however, it is customary to assume that (i) entering gas molecules warm up instantaneously to reaction temperature, (ii) there is a single residence time, (iii) the reactant mixture is quenched instantaneously at the outlet of the reactor, and (iv) there is no pressure drop along the tube. Normally, the pressure drop is negligible in laminar-flow experiments, and measurements of pressures permit ready confirmation of this fact. But it is more difficult to correct the results of deviations from plug flow due to axial and radial diffusion, and of no constant temperature due to imperfect radial heat transfer.

With smaller-diameter tubes, surface reactions in pyrolysis units are relatively more important and have hence received greater attention from designers. Recent evidence obtained in both laboratory and industrial units indicates that these surface reactions are more important than was thought. Surface reactions produce the following results:

1. Pyrolysis coils using smaller-diameter tubes have resulted in higher selectivity toward production of olefins and diolefins (1,3-butadiene) because of shorter residence time.
2. At least some coke is produced.
3. Part of the coke is gasified. Apparently all of the CO and CO_2 produced occur as a result of surface reactions involving steam. In laboratory experiments in Vycor glass tubes, no CO or CO_2 was produced.
4. Carburization reactions weaken stainless steel and are a factor in eventual tube failure.
5. Surface roughening and corrosion occur as a result of repeated oxidizing, reducing, and sulfiding-desulfiding steps (Section 7.4.1–7.4.2). Coke production at metal surfaces also destroys the integrity of the surface.

Extrapolating these laboratory results to industrial units involved the following problems, among others. Surface-to-volume ratios of the laboratory coil were 4–20 times greater than those of industrial coils in the 2.5–13 cm range. The ratio of surface to mass of reacting gases is inversely proportional to the total pressure of the reactant. Because commercial units normally operate at 0.2–0.4 MPa in the reaction zone (radiant section), the relative importance of surface reactions would be two to four times less because of the pressure effects as compared to laboratory units operated at atmospheric pressure. At least seven surface reactions occur during most industrial pyrolysis. In a complex and as yet not completely understood manner, they all contribute directly or indirectly to coke formation (Section 7.4.1–7.4.2).

Elemental sulfur, hydrogen sulfide, light mercaptans such as methyl and ethyl mercaptan, thiophene, dimethylsulfide, and diethylsulfide are suitable sulfur-containing compounds that react readily with stainless-steel surfaces in ethylene furnaces to produce metal sulfides on the surface (Section 1.2.8.3). Reactions with these sulfur-containing gases are complicated and have not been completely clarified. Certainly the gases react with either oxidized or somewhat reduced metal surfaces to produce metal sulfides. Oxidized metal surfaces and H_2S react to produce metal sulfides on the surface, steam, sulfur dioxide, hydrogen, and sulfur; at least some H_2S decomposes to form sulfur and hydrogen. Sulfur also reacts with the metal surfaces to produce metal sulfides. No information was found in the literature as to the relative rates of formation of the several metal sulfides or as to specific sulfides produced. Sulfiding to a limited extent of the inner surface of a coil that has just been decoked and is hence oxidized is often beneficial for subsequent pyrolysis of hydrocarbons. As a result, production of coke, carbon oxides, and hydrogen is significantly reduced while ethylene yields are improved (Section 7.4.1–7.4.2).

However, among a number of challenges, high production costs for microstructure devices have prevented a breakthrough of this innovative technology in the past. In order to utilize microreaction technology for possible industrial applications, a new flexible and scalable manufacturing concept for microstructure reactors was developed and optimized. Special attention was paid to the possibilities of affecting radical transformation of hydrocarbons by homogeneous additives and the heterogeneous surface of the reactor, with the intention of spreading the findings dealing with chemical reactions of free radicals and critical evaluation of the possibilities of their practical application for industrial pyrolysis processes.

Many studies have been carried out investigating of reactivity of some organic and inorganic substances that have, in their molecule sulfur, nitrogen, oxygen, phosphorus, and chlorine with orientation to sulfur and its compounds. We studied the initiated pyrolysis from the point of acceleration of radical decomposition of hydrocarbons, selectivity improvement for desired alkenes, and a decrease in the course of secondary reactions, the consequence of which is the formation of coke. The reactor wall modifies in many directions the course of radical decomposition of hydrocarbons with the application of heterogeneous reaction stages. The surface treatment (coating) of the reactor with sulfur substances and oxidizing agents (history of reactor = new, clean, coking, decoking, oxidation with $O_2 + H_2O$, carburization, after the reactor is used and old), which precedes pyrolysis, is an important factor in the pyrolysis of hydrocarbon feedstocks, when heterogeneous stages affect the rate of decomposition and coke formation. For the same reason, we studied unconventional materials with the aim of contributing to the knowledge and intensification of industrial processes.

Many studies have been carried out investigating different means of catalyst preparation/deposition on microchannels, and relevant results have been recently reviewed. However, only a few reports combine catalyst deposition with reactor fabrication in a single study, which is a major prerequisite for a coherent manufacturing concept. In particular, methods suitable for mass production of microstructure reactors such as soldering and spray-coating technologies are rarely described. Our research activities were not exclusively focused on the thermal decomposition reaction of hydrocarbons. All methods and processes were developed with respect to the general applicability of microstructure reactors to heterogeneously catalyzed gas phase reactions. The results of this study will be published.

16.5 Nanotechnology

16.5.1 Definition

Nanotechnology is the manipulation of matter on an atomic, molecular, and supramolecular scale. The earliest, widespread description of nanotechnology referred to the particular technological goal of precisely manipulating atoms and molecules for the fabrication of macroscale products, also now referred to as *molecular nanotechnology*. A more generalized description of nanotechnology was subsequently established by the US National Nanotechnology Initiative, which defines nanotechnology as the manipulation of matter with at least one dimension sized from 1–100 nm. This definition reflects the fact that quantum mechanical effects are important at this quantum-realm scale, and so the definition shifted from a particular technological goal to a research category inclusive of all types of research and technologies that deal with the special properties of matter that occur below the given size threshold. It is therefore common to see the plural form *nanotechnologies* as well as *nanoscale technologies* to refer to the broad range of research and applications whose common trait is size. Because of the variety of potential applications (including industrial and military), governments have invested billions of dollars in nanotechnology research. From

FY 2001–2014, the US federal government invested approximately $20 billion in nanoscale science, engineering, and technology through the US National Nanotechnology Initiative.

Nanotechnology as defined by size is naturally very broad, including fields of science as diverse as surface science, organic chemistry, molecular biology, semiconductor physics, microfabrication, molecular engineering, etc. The associated research and applications are equally diverse, ranging from extensions of conventional device physics to completely new approaches based upon molecular self-assembly, from developing new materials with dimensions on the nanoscale to direct control of matter on the atomic scale.

Scientists are currently debating the future implications of nanotechnology. Nanotechnology may be able to create many new materials and devices with a vast range of applications, such as in nanomedicine, nanoelectronics, biomaterials, energy production, and consumer products. On the other hand, nanotechnology raises many of the same issues as any new technology, including concerns about the toxicity and environmental impact of nanomaterials, and their potential effects on global economics, as well as speculation about various doomsday scenarios. These concerns have led to a debate among advocacy groups and governments on whether special regulation of nanotechnology is warranted.

K.E. Drexler used the term *nanotechnology* in his 1986 book *Engines of Creation: The Coming Era of Nanotechnology,*" which proposed the idea of a nanoscale "assembler." The emergence of nanotechnology as a field occurred through the convergence of Drexler's theoretical and public work, which developed and popularized a conceptual framework for nanotechnology, and high-visibility experimental advances that drew additional widescale attention to the prospects of atomic control of matter. In the 1980s, two major breakthroughs sparked the growth of nanotechnology in the modern era. First, the invention of the scanning tunneling microscope in 1981 provided unprecedented visualization of individual atoms and bonds, and was successfully used to manipulate individual atoms in 1989. The microscope's developers, Gerd Binnig and Heinrich Rohrer at IBM Zurich Research Laboratory, received a Nobel Prize in Physics in 1986. Binnig, Quate, and Gerber also invented the analogous atomic force microscope that year.

Second, *fullerenes* were discovered in 1985 by Harold Kroto, Richard Smalley, and Robert Curl, who together won the 1996 Nobel Prize in Chemistry. Members of the fullerene family are a major subject of research falling under the nanotechnology umbrella (Figure 16.2). Buckminsterfullerene C_{60}, also known as a *buckyball*, is a representative fullerene. It was not initially described as nanotechnology; the term was used regarding subsequent work with related graphene tubes (called *carbon nanotubes* or sometimes *buckytubes*), which suggested potential applications for nanoscale electronics and devices.

In the early 2000s, the field garnered increased scientific, political, and commercial attention that led to both controversy and progress. Controversies emerged regarding the definitions and potential implications of nanotechnologies, exemplified by the Royal Society's report on nanotechnology. Challenges were raised regarding the feasibility of applications envisioned by advocates of molecular nanotechnology, which culminated in a public debate between Drexler and Smalley in 2001 and 2003.

Meanwhile, commercialization of products based on advancements in nanoscale technologies began emerging. These products are limited to bulk applications of nanomaterials and do not involve atomic control of matter. Some examples include the Silver Nano platform for using silver nanoparticles as an antibacterial agent, nanoparticle-based transparent

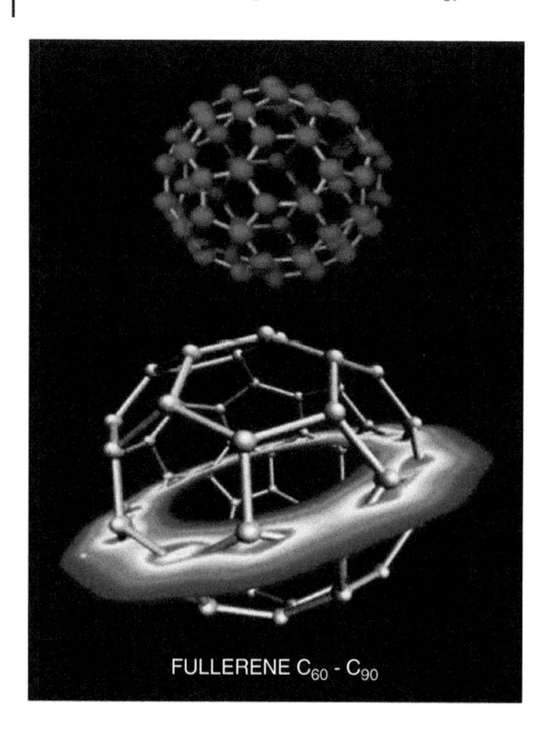

Figure 16.2 Fullerene $C_{60}-C_{90}$.

sunscreens, carbon-fiber strengthening using silica nanoparticles, and carbon nanotubes for stain-resistant textiles. Governments moved to promote and fund research into nanotechnology, such as in the United States with the National Nanotechnology Initiative, which formalized a size-based definition of nanotechnology and established funding for research on the nanoscale; and in Europe via the European Framework Programmers for Research and Technological Development.

By the mid-2000s, new and serious scientific attention began to flourish. Projects emerged to produce nanotechnology roadmaps, which center on atomically precise manipulation of matter and discuss existing and projected capabilities, goals, and applications.

16.5.2 Fundamental Concepts

Nanotechnology involves the engineering of functional systems at the molecular scale. This covers both current work and concepts that are more advanced. In its original sense, nanotechnology refers to the projected ability to construct items from the bottom up, using techniques and tools being developed today to make complete, high-performance products.

One nanometer (nm) is one billionth, or 10^{-9}, of a meter. By comparison, typical carbon–carbon bond lengths, or the spacing between these atoms in a molecule, are in the range 0.12–0.15 nm, and a deoxynucleic acid (DNA) double helix has a diameter around 2 nm. On the other hand, the smallest cellular life forms, bacteria of the genus *Mycoplasma*, are around 200 nm in length. As mentioned earlier, by convention, nanotechnology is taken as the scale range 1–100 nm, following the definition used by the National Nanotechnology Initiative. The lower limit is set by the size of atoms (hydrogen has the smallest atoms, which are approximately a quarter of an nm in diameter; Chapter 13) since nanotechnology must build its devices from atoms and molecules. The upper limit is more or less arbitrary but is around the size below which phenomena not observed in larger structures start to become apparent and can be made use of in nano devices. These new phenomena make nanotechnology distinct from devices that are merely miniaturized versions of an equivalent macroscopic device; such devices are on a larger scale and come under the description of microtechnology.

To put that scale in another context, the comparative size of a nanometer to a meter is the same as that of a marble to the size of the earth. In another way of putting it, a nanometer is the amount an average man's beard grows in the time it takes him to raise a razor to his face.

Two main approaches are used in nanotechnology. In the *bottom-up* approach, materials and devices are built from molecular components, which assemble themselves chemically by principles of molecular recognition. In the *top-down* approach, nano-objects are constructed from larger entities without atomic-level control. Several phenomena become pronounced as the size of the system decreases. These include statistical mechanical effects, as well as quantum mechanical effects: for example, the *quantum size effect*, where the electronic properties of solids are altered with great reductions in particle size. This effect does not come into play by going from macro to micro dimensions. However, quantum effects can become significant when the nanometer size range is reached, typically at distances of 100 nm or less – the so-called *quantum realm*. Additionally, a number of physical (mechanical, electrical, optical, etc.) properties change when compared to macroscopic systems. One example is the increase in surface area to volume ratio, altering mechanical, thermal, and catalytic properties of materials.

Areas of physics such as nanoelectronics, nanomechanics, nanophotonics, and nanoionics have evolved during the last few decades to provide a basic scientific foundation for nanotechnology. Diffusion and reactions at nanoscale, nanostructures materials, and nanodevices with fast ion transport are generally referred to *nanoionics*. Mechanical properties of nanosystems are of interest in nanomechanics research. The catalytic activity of nanomaterials also opens potential risks in their interaction with biomaterials.

Materials reduced to nanoscale can show different properties compared to what they exhibit on a macroscale, enabling unique applications. For instance, opaque substances can become transparent (copper); stable materials can turn combustible (aluminum); and insoluble materials may become soluble (gold). A material such as gold, which is chemically inert at normal scales, can serve as a potent chemical catalyst at nanoscale. Much of the fascination with nanotechnology stems from these quantum and surface phenomena that matter exhibits at nanoscale.

16.5.3 Nanomaterials

The nanomaterials field includes subfields, which develop or study materials having unique properties arising from their nanoscale dimensions.

16.5.4 Applications

An interesting application is the use of nanoparticles that are added to diesel fuel for vehicles. The nanoparticles help the fuel burn better in the engine, and the result is more kilometers per liter; and because the fuel burns more efficiently, less effluent comes out of the exhaust pipe. In addition, the engine does not need servicing as often because the engine stays cleaner. Nanomaterials are also increasingly being used to make cars lighter and stronger, so they use less fuel and fewer metals.

Scientists are now turning to nanotechnology in an attempt to develop diesel engines with cleaner exhaust fumes. Platinum is currently used as the diesel engine catalyst in these engines. The catalyst is what cleans the exhaust fume particles. First a reduction catalyst is employed to take nitrogen atoms from NOx molecules in order to free oxygen. Next the oxidation catalyst oxidizes the hydrocarbons and CO to form CO_2 and water. Platinum is used in both the reduction and the oxidation catalysts. Using platinum is inefficient, though, in that it is expensive and unsustainable.

The company Innovation Fund Denmark invested DKK 15 million in a search for new catalyst substitutes using nanotechnology. The goal of the project, launched in the autumn of 2014, is to maximize surface area and minimize the amount of material required. Objects tend to minimize their surface energy; two drops of water, for example, will join to form one drop and decrease surface area. If the catalyst's surface area that is exposed to the exhaust fumes is maximized, efficiency of the catalyst is maximized. The team working on this project aims to create nanoparticles that will not merge. Every time the surface is optimized, material is saved. Thus, creating these nanoparticles will increase the effectiveness of the resulting diesel engine catalyst – in turn leading to cleaner exhaust fumes – and will decrease cost. If successful, the team hopes to reduce platinum use by 25%.

Further Reading

Almeida, L.C., Echave, F.J., Sanz, O. et al. (2011). *Chem. Eng. J.* 157: 536–544.
Avila, P., Montes, M., and Miro, E.E. (2005). *Chem. Eng. J.* 109: 11–36.
Bajus, M. and Veselý, V. (1976). *Patent Czechoslovak Republic* 175 812.
Chaudhuri, A., Guha, C., and Dutta, T.K. (2007). *Chem. Eng. Technol.* 30: 425–430.
Hessel, V. (2009). *Chem. Eng. Technol.* 32 (11): 1655–1681.
Hessel, V., Cortese, B., and de Croon, M.H.J.M. (2011). *Chem. Eng. Sci.* 66: 1426–1448.
Hessel, V., Lõb, P., and Lőwe, H. (2005). *Curr. Org. Chem.* 9 (8): 765–787.
Illg, T., Hessel, V., and Lõb, P. (2010). *Bioorg. Med. Chem.* 18 (11): 627–4154.
Kolb, G. and Hessel, V. (2004). Review: micro-structured reactors for gas phase reactions. *Chem. Eng. J.* 98: 1–38.
Kumar, V., Parshivoiu, M., and Nigam, K.D.P. (2011). *Chem. Eng. Sci.* 66: 1329–1373.
Lerou, J.J. (2010). Plenary lecture 5. 9th Novel Gas Conversion Symposium.
Makarshin, L.L., Andreev, D., Gribovskyi, A.G., and Parmon, V.N. (2011). *Chem. Eng. J.* 178: 276–281.
Meilla, V. (2006). *Appl. Catal. A. Gen.* 315: 1–17.
Peela, N.R., Mubayi, A., and Kunzru, D. (2001). *Chem. Eng. J.* 167: 578–587.
Rebrov, E.V., Schouten, J.C., and Croon, M.H.J.M. (2011). *Chem. Eng. Sci.* 66: 1374–1393.
Schwarz, O., Duong, P.-Q., Schäfer, G., and Scomäcker, R. (2009). *Chem. Eng. J.* 145: 420–428.

Index

Petrochemistry: Petrochemical Processing, Hydrocarbon Technology, and Green Engineering, First Edition. Martin Bajus.